CYBER SECURITY FOR

INDUSTRIAL CONTROL SYSTEMS

from the viewpoint of close-loop

CYBER SECURITY FOR
INDUSTRIAL CONTROL SYSTEMS

from the viewpoint of close-loop

Edited by
Peng Cheng
Heng Zhang
Jiming Chen

CRC Press
Taylor & Francis Group
Boca Raton London New York

CRC Press is an imprint of the
Taylor & Francis Group, an **informa** business

CRC Press
Taylor & Francis Group
6000 Broken Sound Parkway NW, Suite 300
Boca Raton, FL 33487-2742

© 2016 by Taylor & Francis Group, LLC
CRC Press is an imprint of Taylor & Francis Group, an Informa business

No claim to original U.S. Government works

Printed on acid-free paper
Version Date: 20151116

International Standard Book Number-13: 978-1-4987-3473-8 (Hardback)

Library of Congress Cataloging-in-Publication Data

Names: Cheng, Peng, editor.
Title: Cyber security for industrial control systems : from the viewpoint of close-loop / edited by Peng Cheng, Heng Zhang, and Jiming Chen.
Description: Boca Raton : Taylor & Francis Group, 2016. | Includes bibliographical references and index.
Identifiers: LCCN 2015042965 | ISBN 9781498734738
Subjects: LCSH: Automation--Security measures. | Feedback control systems--Security measures. | Computer security. | Process control--Security measures. | Industries--Security measures.
Classification: LCC T59.5 .C93 2016 | DDC 629.8/9558--dc23
LC record available at http://lccn.loc.gov/2015042965

Visit the Taylor & Francis Web site at
http://www.taylorandfrancis.com

and the CRC Press Web site at
http://www.crcpress.com

Contents

SECTION III: SECURITY ISSUES IN APPLICATION FIELDS 251

Preface

By exploiting sensing, networking, and computation capabilities, the new-generation industrial control systems are able to better connect cyber space and the physical process in close-loop than ever before. However, such connections have also provided rich opportunities for adversaries to perform potential malicious attacks.

There has been extensive research on security issues from the viewpoint of networks and communication, and the secure defending approaches mainly consider how to guarantee network performance. Considerable efforts have been devoted to the design of secure defending approaches against malicious attacks with communication technologies. For example, channel hopping is often used to maintain network performance, for example, throughput, in a jamming attack environment.

Such network performance rarely considers the operation of physical plants without the features of automatic control or feedback. However, industrial control systems are characterized by feedback control and aim to optimize the system control performances, such as reducing state estimation errors, improving the stability of unstable plants, and enhancing the robustness against uncertainties and noise. Thus, it is equivalent or even more important to protect the system control performance while studying the cyber-security issues in industrial control systems. For example, when the communication of system entities is under a jamming attack, different from the existing design, such as the channel hopping algorithm, the secure state estimation and control algorithms may be better configured by exploiting the feedback information as well as the dynamics of the physical plants. As a result, it is of great research interest to develop novel theories and technologies from the viewpoint of close-loop in order to protect the industrial control system performance under various cyber and physical attacks.

Cyber Security for Industrial Control Systems: From the Viewpoint of Close-Loop is the first comprehensive and updated book on cyber security from the viewpoint of close-loop. This book provides a comprehensive technical guide on up-to-date secure defending theories and technologies, novel design, and systematic understanding of secure architecture and some practical applications. Specifically, it consists of 10 chapters, which are divided into three parts. The first part, consisting of Chapters 1 through 3, extensively introduces secure state estimation technologies, providing a systematic presentation on the latest progress in security issues regarding state estimation. The second part, composed of five chapters, focuses on the design of secure feedback control technologies in industrial control systems, showing its extraordinary difference from that of traditional secure defending approaches from the viewpoint of network and communication. The third part, with two chapters, elaborates on the systematic secure control architecture and algorithms for various concrete application scenarios.

This book has the following salient features:

1. Provides an extensive introduction to state-of-the-art cyber security theories and technologies from the viewpoint of close-loop

2. Identifies the quantitative characteristics of typical cyber attacks and analyzes the attack decision mechanisms in closed-loop industrial systems in depth

3. Proposes novel intrusion detection mechanisms against cyber attacks in industrial control systems

4. Presents a systematic understanding of the secure architectural design for industrial control systems

5. Addresses secure control approaches against cyber attacks for the representative applications in industrial control systems

This book provides detailed descriptions on attack model and strategy analysis, intrusion detection, secure state estimation and control, game theory in closed-loop systems, and various cyber-security applications. We expect the book to be favorable to those who are interested in secure theories and technologies for industrial control systems.

We would like to thank all the contributors of each chapter for their expertise and cooperation, and efforts invested, without which we would not have such an excellent book. Specially, we highly appreciate the support, patience, and professionalism of Ruijun He and Kathryn Everett from the very beginning to the final publication of the book. Last but not least, we are grateful for our families and friends for their constant encouragement and understanding throughout this project.

Peng Cheng
Heng Zhang
Jiming Chen

Contributors

Haiyong Bao
School of Electrical and Electronic Engineering
Nanyang Technological University
Singapore

Jiming Chen
College of Control Science and Technology
Zhejiang University
Zhejiang, China

Peng Cheng
College of Control Science and Technology
Zhejiang University
Hangzhou, China

Mo-Yuen Chow
Department of Electrical and Computer Engineering
North Carolina State University
Raleigh, North Carolina, USA

Kun Ji
Corporate Technology
Siemens Corporation
Princeton, New Jersey, USA

Xiaohua Jia
Department of Computer Science
City University of Hong Kong
Kowloon Tong, Hong Kong, China

Beibei Li
School of Electrical and Electronic Engineering
Nanyang Technological University
Singapore

Yuzhe Li
Department of Electronic and Computer Engineering
Hong Kong University of Science and Technology
Kowloon, Hong Kong, China

Rongxing Lu
School of Electrical and Electronic Engineering
Nanyang Technological University
Singapore

Sonia Martínez
Department of Mechanical and Aerospace Engineering
University of California
San Diego, California, USA

Yilin Mo
School of Electrical and Electronic
    Engineering
Nanyang Technological University
Singapore

Arash Mohammadi
Department of Electrical and
    Computer Engineering
University of Toronto
Toronto, Ontario, Canada

Konstantinos N. Plataniotis
Department of Electrical and
    Computer Engineering
University of Toronto
Toronto, Ontario, Canada

Daniel E. Quevedo
Department of Electrical
    Engineering
University of Paderborn
North Rhine-Westphalia, Germany

Xuemin (Sherman) Shen
Department of Electrical and
    Computer Engineering
University of Waterloo
Waterloo, Ontario, Canada

Ling Shi
Department of Electronic and
    Computer Engineering
Hong Kong University of Science
    and Technology
Kowloon, Hong Kong, China

Dong Wei
Corporate Technology
Siemens Corporation
Princeton, New Jersey, USA

Kan Yang
Department of Electrical and
    Computer Engineering
University of Waterloo
Waterloo, Ontario, Canada

Wente Zeng
Department of Electrical and
    Computer Engineering
North Carolina State University
Raleigh, North Carolina, USA

Heng Zhang
College of Control Science and
    Technology
Zhejiang University
Hangzhou, China

Jun Zhao
CyLab and Department of Electrical
    and Computer Engineering
Carnegie Mellon University
Pittsburgh, Pennsylvania, USA

Minghui Zhu
Department of Electrical
    Engineering
Pennsylvania State University
University Park, Pennsylvania, USA

Quanyan Zhu
Department of Electrical and
    Computer Engineering
New York University
New York, New York, USA

SECURE STATE ESTIMATION

Chapter 1

A Game-Theoretic Approach to Jamming Attacks on Remote State Estimation in Cyber-Physical Systems

Yuzhe Li

Department of Electronic and Computer Engineering,
Hong Kong University of Science and Technology

Ling Shi

Department of Electronic and Computer Engineering,
Hong Kong University of Science and Technology

Peng Cheng

College of Control Science and Technology, Zhejiang University

Jiming Chen

College of Control Science and Technology, Zhejiang University

Daniel E. Quevedo

Department of Electrical Engineering, University of Paderborn

CONTENTS

We consider security issues in remote state estimation of cyber-physical systems in this chapter. We first investigate the single-sensor case. A sensor node communicates with a remote estimator through a wireless channel that may be jammed by an external attacker. The interactive decision-making process of when to send and when to attack is studied when the sensor and the attacker are both subject to energy constraints. We formulate a game-theoretic framework and prove that the optimal strategies for both sides constitute a Nash equilibrium of this zero-sum game. To tackle the computational complexity issues, we present a constraint-relaxed problem and provide corresponding solutions using Markov chain theory. Under a constraint-relaxed formulation, taking the multisensor data fusion into consideration, the problem for the multiple sensors case is also studied.

1.1   Introduction

Cyber-physical systems (CPS) are systems that integrate sensing, control, communication, computation, and physical process. Typical CPS usually consist of a group of networked agents, including sensors, actuators, control processing units, and communication devices [27] (Figure 1.1), which have a wide spectrum of applications in areas such as aerospace, smart grids, civil infrastructure, and transportation. Significant advances in terms of efficiency, reliability, adaptability, and autonomy of engineered systems have been brought by the rapid development of CPS in recent years.

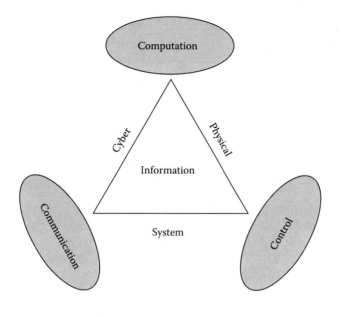

Figure 1.1: Architecture of cyber-physical systems.

Due to the increasing connection of CPS to many safety-critical applications, high risks of cyber attacks by adversaries around the globe have arisen. For example, the future electric power grids, that is, the smart grids, will be the largest and most complex CPS. Since grid operation and communication are mainly through a shared network, such systems are quite vulnerable to cyber-security threats. Any severe attack on the national power grids may have a significant impact on the environment, national economy, or national security or even the loss of human life [25]. Therefore, designing CPS taking into account security issues is of fundamental importance to ensure their safe operation, which has been a hot research area in recent years. Two possible types of attacks on CPS have been studied by Cardenas et al. [10]: denial-of-service (DoS) attacks and deception attacks, which correspond to the traditional security goals of *availability* and *integrity*, respectively. The DoS attack blocks the exchange of information, including sensor measurement data or control inputs, between each part of the CPS, while the integrity attack focuses on the integrity of the data by modifying the data packets. Other works in the literature that have dealt with specific types of attacks include DoS attacks [3, 11, 38, 39], false data injection attacks [23, 24], integrity attacks [27], and replay attacks [26]. Liu et al. [23] studied a new class of attacks on state estimation schemes in electric power grids, namely, false data injection attacks. Given the knowledge of the configuration and parameters of the power system, the attacker can launch such attacks to inject arbitrary errors into certain state variables without triggering the existing bad measurement detection alarm. Mo and Sinopoli [27] described the CPS model as a discrete linear time-invariant system running a Kalman filter, an LQG controller, and a χ^2 failure detector, which is under integrity attacks. The authors presented a quantitative index of the system resilience by investigating the set of nondetectable adversary's attack strategies and the corresponding state estimation error under certain attacks.

Though some fundamental frameworks have been proposed in existing literature, such as [3, 11, 39], they have focused only on one side, that is, either the attacker or the defender. If attackers have knowledge of system parameters, however, both parties (defender and attacker) will be involved in an interactive decision-making process. Each side chooses the optimal action based on all the information it has, including the understanding and prediction of the actions its opponent may take. To study such a situation, one requires a more comprehensive description of CPS security, which goes beyond a static one-sided analysis. In this chapter, we adopt a game-theoretic approach that provides an alternative way to handle these interactive decision issues.

Game theory is the study of mathematical models of conflict and cooperation between intelligent rational decision makers, that is, interactive decision theory [13]. While game theory was originally used for studying economic systems, it has since developed into a wide range of applications [2, 6, 8]. The work [19] studied the zero-sum game on multiple-input multiple-output (MIMO) Gaussian Rayleigh fading channels where both the jammer and the encoder are subject to power constraints. Gupta et al. [14] considered a dynamic game between a controller for a discrete-time linear time-invariant (LTI) plant and an attacker who could jam the communication between the controller and the plant. The equilibrium control and jamming strategies for both players were provided. The work [30] investigated existing results for enhancing network security under the game-theoretic framework and provided a classification of recent results based on the types of corresponding games. Agah et al. [1] formulated a cooperative game between sensor nodes in mobile wireless sensor networks and showed that through cooperation between two nodes the data communication between them will be more reliable.

In the work described in [21], the communication channels for transmitting system state information from a sensor to a remote controller can be partly jammed by a jammer. Multiple channels can be chosen to avoid the attack. The payoff function of the stochastic game investigated in [21] is in quadratic form, consisting of the weighted sum of the norms of the system state and action vector with discount factors. This objective (also used in [14]) and the assumption of noiseless sensor measurements, however, are not suitable for remote estimation scenarios, where the overall estimation quality is critical with trade-off to the energy constraints. A preliminary version of parts of this chapter (see [22]) investigated the jamming game in CPS, which studied the case where the data packet from the sensor always arrives at the remote estimator successfully without attack and drops under attack. To better describe the practical communication process, here we consider a more practical communication model that embeds [22] as a special case. Furthermore, as the computational complexity issue for solving the optimal solution is significant in [22], we propose a constraint-relaxed problem formulation and provide corresponding closed-form expressions that significantly reduce the calculation. Note that [14, 21, 22] only investigated the single-sensor case, which is not typical in general CPS, where many applications are dealing with sensor networks. We also consider the jamming game between the attacker and multiple sensors, which is a more interesting and practical challenge in CPS.

The remainder of this chapter is organized as follows. Section 1.2 presents the system model and states the main problem of interest. Section 2.3 presents some game theory preliminaries and studies the optimal strategies for both sides. Section 2.4 provides the dynamic updating algorithm. Section 2.5 provides a constraint-relaxed problem formulation that reduces the computational complexity. The multisensor case is considered in Section 1.6. Section 1.7 draws the conclusions.

Notations: $\mathbb{Z}$ denotes the set of all integers and $\mathbb{N}$ the positive integers. $\mathbb{R}$ is the set of real numbers. $\mathbb{R}^n$ is the n-dimensional Euclidean space. $\mathbb{S}_+^n$ (and $\mathbb{S}_{++}^n$) is the set of $n \times n$ positive semi-definite matrices (and positive definite matrices). When $X \in \mathbb{S}_+^n$ (and $\mathbb{S}_{++}^n$), we write $X \geqslant 0$ (and $X > 0$). $X \geqslant Y$ if $X - Y \in \mathbb{S}_+^n$. $\text{Tr}(\cdot)$ is the trace of a matrix. The superscript $'$ stands for transposition. For functions f, f_1, f_2 with appropriate domains, $f_1 f_2(x)$ stands for the function composition $f_1\big(f_2(x)\big)$, and $f^n(x) \triangleq f\big(f^{n-1}(x)\big)$, where $n \in \mathbb{N}$ and $f^0(x) \triangleq x$. δ_{ij} is the discrete-time Dirac delta function; that is, δ_{ij} equals 1 when $i = j$ and 0 otherwise. The notation $\mathbb{P}[\cdot]$ refers to probability and $\mathbb{E}[\cdot]$ to expectation.

1.2   Problem Setup

Consider a general discrete linear time-invariant (LTI) process of the form

$$x_{k+1} = Ax_k + w_k,$$

$$y_k = Cx_k + v_k,$$

where $k \in \mathbb{N}$, $x_k \in \mathbb{R}^{n_x}$ is the process state vector at time k, $y_k \in \mathbb{R}^{n_y}$ is the measurement taken by the sensor, and $w_k \in \mathbb{R}^{n_x}$ and $v_k \in \mathbb{R}^{n_y}$ are zero-mean independent and identically distributed (i.i.d.) Gaussian noises with $\mathbb{E}[w_k w_j'] = \delta_{kj} Q$ $(Q \geqslant 0)$, $\mathbb{E}[v_k v_j'] = \delta_{kj} R$ $(R > 0)$, and $\mathbb{E}[w_k v_j'] = 0$ $\forall j, k \in \mathbb{N}$. The initial state x_0 is a zero-mean Gaussian random vector with covariance $\Pi_0 \geqslant 0$ and is uncorrelated with w_k and v_k. The pair (A, C) is assumed to be observable and $(A, Q^{1/2})$ is controllable.

1.2.1 Local State Estimation

Our interest lies in security of remote state estimation as depicted in Figure 1.2. In CPS, sensors are typically equipped with onboard processors [17]. These capabilities can be used to improve system performance significantly. At each time k, the sensor first locally estimates the state x_k based on all the measurements it collects up to time k and then transmits its local estimate to the remote estimator. Denote $\hat{x}_k^s$ and P_k^s as the sensor's local minimum mean-squared error (MMSE) estimate of the state x_k and the corresponding error covariance, which are given by

$$\hat{x}_k^s = \mathbb{E}[x_k | y_1, y_2, ..., y_k],$$

$$\hat{P}_k^s = \mathbb{E}[(x_k - \hat{x}_k^s)(x_k - \hat{x}_k^s)' | y_1, y_2, ..., y_k]$$

and can be calculated by a Kalman filter as follows:

$$\hat{x}_{k|k-1}^s = A\hat{x}_{k-1}^s,$$

$$9K_k^s = P_{k|k-1}^s C'[CP_{k|k-1}^s C' + R]^{-1},$$

$$\hat{x}_k^s = A\hat{x}_{k-1}^s + K_k^s(y_k - C\hat{x}_{k|k-1}^s),$$

$$P_k^s = (I - K_k^s C)P_{k|k-1}^s,$$

where the recursion starts from $\hat{x}_0^s = 0$ and $P_0^s = \Pi_0 \geqslant 0$.

For notational ease, we introduce the functions $h, \tilde{g} : \mathbb{S}_+^n \to \mathbb{S}_+^n$ as

$$h(X) \triangleq AXA' + Q,$$

$$\tilde{g}(X) \triangleq X - XC'[CXC' + R]^{-1}CX,$$

$$h^k(X) \triangleq \underbrace{h \circ h \circ \cdots \circ h \circ h}_{k \text{ times}}(X).$$

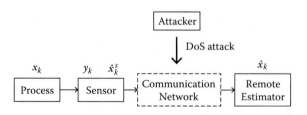

Figure 1.2: The communication network is jammed by a malicious attacker. This affects remote estimation performance.

It is well known that under suitable conditions the estimation error covariance of the Kalman filter converges to a unique value from any initial condition (proved in [36]); thus, the local estimation error covariance P_k^s will converge to a steady state. Without loss of generality (similar assumptions can be found in works such as [22, 32]), we assume that the Kalman filter at the sensor side has entered the steady state and simplify our subsequent discussion by setting

$$P_k^s = \overline{P}, \ k \geqslant 1, \tag{1.1}$$

where $\overline{P}$ is the steady-state error covariance given in [36], which is the unique positive semidefinite solution of $\tilde{g} \circ h(X) = X$. For convenience, we also assume that the remote estimator's error covariance is $\overline{P}$ before the process starts, that is,

$$P_0 = \overline{P}. \tag{1.2}$$

The error covariance $\overline{P}$ has the following property:

Lemma 1.1
(see [34].) For $0 \leqslant t_1 \leqslant t_2$, the following inequality holds:

$$h^{t_1}(\overline{P}) \leqslant h^{t_2}(\overline{P}).$$

In addition, if $t_1 < t_2$, then

$$\text{Tr}\left(h^{t_1}(\overline{P})\right) < \text{Tr}\left(h^{t_2}(\overline{P})\right).$$

1.2.2 Communication Channel

Denial-of-service (DoS) attacks are the most reachable attack pattern for the attacker [38] since the communication between sensors and remote estimators in CPS is mainly through a wired or wireless network. Typical DoS attacks can jam the communication between components in CPS and degrade the overall system performance [3, 11, 39]. In this chapter, we assume the attacker to be capable of conducting a DoS attack on the server to jam the communication channel between the sensor and the remote estimator, therefore worsening the system performance (see Figure 1.2).

Energy constraint is a natural concern for both sensors and attackers in practice, which affects the remote estimation performance and attacking policies [5, 7, 20, 29, 35]. To encompass energy limitations, we will assume that within a time horizon T, the sensor can send data packets at most $M \leqslant T$ times to the remote estimator, while the attacker can launch jamming attacks at most $N \leqslant T$ times.

Denote

$$\theta_S \triangleq \{\gamma_1, \gamma_2, ..., \gamma_T\} \tag{1.3}$$

as the sensor's data-sending strategy, where $\gamma_k = 1$ means that the sensor sends data packets at time k; otherwise, $\gamma_k = 0$. Consequently, we have

$$\sum_{k=1}^{T} \gamma_k \leqslant M. \tag{1.4}$$

Similarly, we denote

$$\theta_A \triangleq \{\lambda_1, \lambda_2, ..., \lambda_T\} \tag{1.5}$$

as the attacker's attack strategy, where $\lambda_k = 1$ means that the attacker launches a jamming attack at time k; otherwise, $\lambda_k = 0$. Then we have:

$$\sum_{k=1}^{T} \lambda_k \leqslant N. \tag{1.6}$$

Packet dropout may occur due to different reasons, including signal degradation, channel fading, and channel congestion in the practical communication process. The dropout probability of data packets from the sensor arriving at the remote estimator is assumed to be β_1 in the absence of attack, and $\beta_2(>\beta_1)$ under the DoS attack. The strategies of the sensor and the attacker at time k are assumed to be γ_k and λ_k, respectively. Thus, the conditional probability of the remote estimator receiving the data packets from the sensor, which is denoted as p_k, is given by

$$p_k = \gamma_k \lambda_k (1 - \beta_2) + \gamma_k (1 - \lambda_k)(1 - \beta_1). \tag{1.7}$$

Remark 1.1 In [22], we studied the case where the data packet from the sensor will always arrive at the remote estimator successfully without attack, and drop under attack. This is embedded in our current model as a special case when $\beta_1 = 0$ and $\beta_2 = 1$, and consequently, $p_k = \gamma_k (1 - \lambda_k)$.

1.2.3 Estimation Process

In recent literature on networked estimation [35, 36], state estimation with dropouts has been well studied. Define I_k as the total information and data packets collected by the remote estimator from time 1 to time $k \leqslant T$, that is,

$$I_k = \{\gamma_1(1 - \lambda_1), \gamma_2(1 - \lambda_2), ..., \gamma_k(1 - \lambda_k)\}$$

$$\bigcup \{\gamma_1(1 - \lambda_1)\hat{x}_1^s, \gamma_2(1 - \lambda_2)\hat{x}_2^s, ..., \gamma_k(1 - \lambda_k)\hat{x}_k^s\}.$$

Denote $\hat{x}_k$ and P_k as the remote estimator's MMSE state estimate (see Figure 1.2) and the corresponding error covariance based on I_k:

$$\hat{x}_k = \mathbb{E}[x_k|I_k],$$

$$P_k = \mathbb{E}[(x_k - \hat{x}_k)(x_k - \hat{x}_k)'|I_k].$$

Then $\hat{x}_k$ can be calculated following a procedure similar to that in [32]: once the sensor's local estimate packet arrives, the estimator synchronizes its own estimate with the sensor; otherwise, the estimator just predicts x_k based on its previous optimal estimate, that is,

$$\hat{x}_k = \begin{cases} \hat{x}_k^s & \text{if } \hat{x}_k^s \text{ arrives,} \\ A\hat{x}_{k-1} & \text{otherwise.} \end{cases} \tag{1.8}$$

As a result, the state estimation error covariance P_k satisfies

$$P_k = \begin{cases} \overline{P} & \text{if } \hat{x}_k^s \text{ arrives,} \\ h(P_{k-1}) & \text{otherwise.} \end{cases} \tag{1.9}$$

We assume that the remote estimator knows the system parameters A, C, Q, R. Thus, based on (1.7) and (1.9), the expected state estimation error covariance can be easily written in a recursive way as

$$\mathbb{E}[P_k] = p_k \overline{P} + (1 - p_k) h (\mathbb{E}[P_{k-1}]). \tag{1.10}$$

Remark 1.2 The Caltech multivehicle wireless testbed (MVWT-II) (see [18]) is a typical real-time networked control system experimental platform. Each vehicle (hovercraft) has an onboard processing unit that can implement certain local controllers. All the vehicles run in a laboratory environment with localization achieved using an overhead camera. The image of the camera is then converted into the positions of each vehicle. Next, the position information is sent to each vehicle for controlling the hovercraft to accomplish some team tasks, for example, the RoboFlag competition (see [12]).

1.2.4 Main Problem

To quantify estimation quality over a finite time horizon $T \in \mathbb{N}$, we introduce the cost function $J_\alpha(T)$ as

$$J_\alpha(T) \triangleq \alpha \frac{1}{T} \sum_{k=1}^{T} \text{Tr}(\mathbb{E}[P_k]) + (1 - \alpha) \text{Tr}(\mathbb{E}[P_T]),$$

where $\alpha = 1$ or 0, corresponding to the overall performance and terminal performance, respectively.

It is easy to see that when $\alpha = 1$, $J_1(T)$ is the trace of the average expected estimation error covariance at the estimator's side, thus focusing on the overall performance [29, 31, 33]. On the other hand, with $\alpha = 0$, the cost function becomes the trace of the final expected estimation error covariance at the estimator's side; thus, $J_0(T)$ focuses on the performance at the end of the process, [33, 35]. Note that choosing α as a real number between 0 and 1 is also feasible. When α is a real number, it provides a trade-off between the average cost and the terminal cost. Whether α is a real number or binary variable will not change the subsequent analysis.

The goal of the decision maker at the sensor's side is to minimize $J_\alpha(T)$, while the attacker tries to maximize the cost. Since the objective of the sensor is opposite that of the attacker, for convenience, we define the objective functions of the attacker and the sensor as

$$J_A(\theta_A) \triangleq J_\alpha(T) \tag{1.11}$$

and

$$J_S(\theta_S) \triangleq -J_A(\theta_A) = -J_\alpha(T), \tag{1.12}$$

where θ_A and θ_S are defined in (1.5) and (1.3), respectively. The goal of both sides is to maximize their objective functions. We are interested in finding the optimal strategies for each side, subject to the energy constraints (1.4) and (1.6). To be more specific, we consider the following optimization problem:

Problem 1.1
For the sensor,

$$\max_{\theta_S} \; J_S(\theta_S),$$

$$\text{s.t.} \; \sum_{k=1}^{T} \gamma_k \leqslant M,$$

where $\theta_S \triangleq \{\gamma_1, \gamma_2, ..., \gamma_T\}$.
For the attacker,

$$\max_{\theta_A} \; J_A(\theta_A),$$

$$\text{s.t.} \; \sum_{k=1}^{T} \lambda_k \leqslant N,$$

where $\theta_A \triangleq \{\lambda_1, \lambda_2, ..., \lambda_T\}$.

Since more energy is always beneficial for improving the performance, it is not difficult to show that the optimal strategies for each side remain the same if the constraint of Problem 1.1 is changed to $\sum_{k=1}^{T} \gamma_k = M$ and $\sum_{k=1}^{T} \lambda_k = N$; see [31]. Thus, we will focus on the following problem:

Problem 1.2

For the sensor,

$$\max_{\theta_S} \quad J_S(\theta_S),$$

$$\text{s.t.} \quad \sum_{k=1}^{T} \gamma_k = M.$$

For the attacker,

$$\max_{\theta_A} \quad J_A(\theta_A),$$

$$\text{s.t.} \quad \sum_{k=1}^{T} \lambda_k = N.$$

For future reference, we denote the optimal strategies for each side as $\theta_S^\star$ and $\theta_A^\star$, respectively.

Remark 1.3   In previous works [27, 38], the sensor is assumed to have sufficient energy to send the data packet during the entire time horizon, that is, $M = T$. In this special case, one only needs to investigate the optimal attack strategy for the attacker, that is, $\theta_A^\star$, which is given by [38].

We will next consider the more interesting case, where both the attacker and the sensor have energy constraints, and try to maximize their respective objectives. To study this situation, in this chapter, the tools provided in [38] are not adequate, and we will adopt a game-theoretic framework, as revised Section 1.3.

1.3   Game-Theoretic Framework

In this section, we model the decision-making process of the sensor and the attacker in a game-theoretic framework. Based on the objective functions of both sides, that is, (1.11) and (1.12), the sensor and the attacker are regarded as the two players of one zero-sum game. The details of this game are discussed as follows.

1.3.1   Nash Equilibrium

To develop our results, we need the following basic assumptions and definitions from game theory [13]:

Assumption 1.1   Both the sensor and the attacker are intelligent rational decision makers: given information available, they pursue the maximization of their objective functions (1.12) and (1.11), respectively.

Remark 1.4 This "rational maximizing behavior" is a basic assumption in game theory. It does not necessarily mean that people always make "100% perfect decisions," since people may be limited by the amount of information they have.

Definition 1.1 Given one event e, if all the players from G know e, they all know that they know e, they all know that they all know that they know e, and so on, then we say there is common knowledge of e in group G.

Assumption 1.2 The decision makers at each side act simultaneously without knowing the current actions of the other. However, they know the objective function and the possible action set of each other, which is common knowledge among them.

Definition 1.2 A game consists of a set of players, a set of strategies available to those players, and a specification of payoffs for each player on the condition of each combination of strategies.

Definition 1.3 A player's *strategy* refers to one of all the options he can choose in the game, which will determine all the actions to take at any stage of the game.

A *pure strategy* provides a complete definition of how a player will play a game.

A player's *strategy set* is the set of all the pure strategies available to that player.

A *mixed strategy* is an assignment of probability to each pure strategy in the strategy set, which allows the player to randomly select a pure strategy.

A *strategy profile* (strategy combination) is a set of strategies for each player that fully specifies all actions in a game.

Remark 1.5 We can regard the pure strategy as a degenerate case of the mixed strategy, where the particular pure strategy is selected with probability 1 and every other strategy with probability 0.

If, in a game, each player has chosen a strategy and no player can benefit by changing his own strategy while the other players keep theirs unchanged, then the current strategy profile, that is, the current set of strategy choices, constitutes a Nash equilibrium (defined in [13]). Nash defined a mixed strategy Nash equilibrium for any game with a finite set of strategies and proved that at least one mixed strategy Nash equilibrium must exist in such a game in [28].

Theorem 1.1 [28]
For any game with a finite set of strategies, there exists at least one mixed strategy Nash equilibrium in the game.

In this chapter, for the case with energy constraint for both sides, that is, $M < T$ and $N < T$, the tools provided in the existing literature that focus on only one side with energy constraint cannot be used directly. Since both sides have many different strategies and have to take the opponent's strategy into consideration, we can investigate the problem from a game-theoretic point of view:

- *Player*: Two players: the sensor and the attacker

- *Action*: θ_S and θ_A for the sensor and the attacker, respectively

- *Payoff*: $J_S(\theta_S)$ and $J_A(\theta_A)$ for the sensor and the attacker, respectively

1.3.2 Existence of the Nash Equilibrium

To analyze the situation where both the sensor and the attacker have limited energy, we first note that the number of all pure strategies for the sensor is

$$C_T^M = \binom{T}{M} \triangleq K.$$

For future reference, we denote those pure strategies as $\theta_S^{\text{pure}}(1)$, $\theta_S^{\text{pure}}(2)$, ..., $\theta_S^{\text{pure}}(K)$. Though the number of pure strategies is finite, there are infinitely many mixed strategies for each side. Mixed strategies for the sensor can be written as

$$\theta_S^{\text{mixed}}(\pi_1, \pi_2, ..., \pi_K) = \{\theta_S^{\text{pure}}(k) \text{ with probability } \pi_k\},$$

$$k = 1, 2, ..., K,$$

where

$$\sum_{k=1}^{K} \pi_k = 1, \quad \pi_k \in [0, 1].$$

Note that different combinations of $\{\pi_k\}$ constitute different mixed strategies.

For the attacker, we use a similar notation:

$$\theta_A^{\text{mixed}}(\mu_1, \mu_2, ..., \mu_L) = \{\theta_A^{\text{pure}}(k) \text{ with probability } \mu_k\},$$

$$k = 1, 2, ..., L,$$

where

$$C_T^N = \binom{T}{N} \triangleq L$$

and

$$\sum_{k=1}^{L} \mu_k = 1, \mu_k \in [0, 1].$$

The optimal solutions $\theta_S^\star$ and $\theta_A^\star$ for the case examined in [38] are the so-called pure strategy and are indeed a special form of the general mixed strategy investigated next.

We now introduce the following result, which together with Theorem 1.1 shows that a Nash equilibrium exists for the considered two-player zero-sum game between the sensor and the attacker.

Proposition 1.1   The optimal strategies for the sensor and the attacker constitute a Nash equilibrium of this two-player game.

Proof 1.1   The optimal strategies for each side are denoted as $\theta_S^\star$ and $\theta_A^\star$, respectively. Given the optimal strategy $\theta_A^\star$ chosen by the attacker, the optimal strategy $\theta_S^\star$ for the sensor is the one that maximizes $J_S(\theta)$, that is, $J_S(\theta_S^\star|\text{given } \theta_A^\star) \geqslant J_S(\theta_S|\text{given } \theta_A^\star)$, $\forall \theta_S$.

For the attacker, we have a similar conclusion. Since the objective functions $J_S(\theta)$ and $J_A(\theta)$ can be regarded as each side's payoff function, respectively, from the definition in [13], $\theta_S^\star$ and $\theta_A^\star$ constitute a Nash equilibrium.   □

Remark 1.6   Note that $\theta_S^\star$ and $\theta_A^\star$ are mixed strategies, which are only an assignment of probability to each pure strategy. Thus, although there is a common knowledge of $\theta_S^\star$ and $\theta_A^\star$, both sides still do not know the exact following action taken by their opponent. This is indeed the reason why both sides can reach the Nash equilibrium.

1.3.3   Finding the Nash Equilibrium

We devote this subsection to find the Nash equilibrium of the game. From Theorem 1.1, there exists at least one Nash equilibrium. Suppose that $\theta_A^\star \triangleq \{\mu_1^\star, \mu_2^\star, ..., \mu_L^\star\}$ is one equilibrium mixed strategy of the attacker.

Given $\theta_A^\star$, we can easily calculate the objective function $J_S\left(\theta_S^{\text{pure}}(k)\right) \triangleq M_k$ for each $\theta_S^{\text{pure}}(k)$ based on the recursion introduced in (1.9) and (1.10). Thus, we can write the objective function of the mixed strategy θ_S^{mixed} of the sensor as

$$J_S(\theta_S^{\text{mixed}}) = \sum_{k=1}^{K} \pi_k M_k, \quad \sum_{k=1}^{K} \pi_k = 1.$$

Based on the definition of Nash equilibrium, given $\theta_A^\star$, the equilibrium strategy of the sensor $\theta_S^\star = \{\pi_1^\star, \pi_2^\star, ..., \pi_K^\star\}$ is the one that maximizes $J_S(\theta_S^{\text{mixed}})$ under the constraint (1.3.2), which can be calculated easily using the Lagrange multiplier method [4, 15].

The above provides $\theta_S^\star = \{\pi_1^\star, \pi_2^\star, ..., \pi_K^\star\}$, where $\pi_k^\star$ is a function of $\theta_A^\star$, that is, $\mu_1^\star, \mu_2^\star, ..., \mu_L^\star$. Then for the attacker, we can run the same procedure given $\theta_S^\star$, and solve $\theta_A^\star = \{\mu_1^\star, \mu_2^\star, ..., \mu_L^\star\}$, where $\mu_k^\star$ is a function of $\theta_S^\star$, that is, $\pi_1^\star, \pi_2^\star, ..., \pi_K^\star$.

By combining the two solutions to obtain both $\mu_k^\star$ and $\pi_k^\star$, one can then derive the optimal solutions for both sides, $\theta_S^\star$ and $\theta_A^\star$.

1.4 Dynamic Update Based on Online Information

In the game investigated in the Section 1.3, though randomly chosen, the decisions of both sides are made before the process begins and thus can be regarded as an off line schedule. In some practical situations, both sides may be able to monitor the status of the network environment. For example, after the attacker launches a jamming attack, the network server may be able to record the abnormality and inform the sensor of the attack [16]. The attacker can also detect whether the data packet is sent based on the network status. Thus, though each side is not sure what strategy will be taken by their opponent, they can still detect (in a causal manner) the opponent's action and therefore narrow the scope of their opponent's action sets. In the present section, we will study such scenarios.

As both sides can detect their opponent's past actions, at each time step after each observation, a new game with new constraints will arise. For our subsequent analysis, we will denote via $\theta_S^\star(T, M, N, \Phi_0)$ and $\theta_A^\star(T, M, N, \Phi_0)$ the optimal mixed strategies for the sensor and the attacker, respectively, in the game with parameters T, M, N, Φ_0, where T is the time horizon, M, N are the energy constraints, and $\Phi_0 = \bar{P}$ is the expected initial state estimate error covariance. Note that $\theta_S^\star(T, M, N, \Phi_0)$ and $\theta_A^\star(T, M, N, \Phi_0)$ provide the action sequences for the whole time horizon, but both sides only employ the actions for the first time step of the new game. After both sides move to the next step, the information, for example, time horizon and energy constraints, is updated, and thus a new game with new constraints will arise.

To update the whole decision-making process, a recursive algorithm (Algorithm 1.1) can be used. Following this algorithm, both sides are involved in a series of games with varying time horizon, energy constraints, and initial states.

Example 1.1　Consider a time horizon $T = 3$ and the constraint for the sensor $M = 1$. After time $k = 1$, the attacker is assumed to observe that the sensor sent the data packet ($\gamma_1 = 1$). Since the mixed strategy for the sensor is chosen from $\{1, 0, 0\}$, $\{0, 1, 0\}$, and $\{0, 0, 1\}$, the attacker can deduce that the sensor's sending scheme is $\{1, 0, 0\}$. Then the attacker can make adjustments to his strategy and thus distribute the remaining energy more efficiently. A similar mechanism also applies to the sensor.

1.5 Relaxation: Average Energy Constraints

In this section, we modify the constraints in Problem 1.2 and propose a new problem formulation that can reduce the computation complexity significantly.

Algorithm 1.1 Updating Algorithm for Both Sides

1: Process begins;

2: $H_{\text{time}} = T$;

3: $\Phi_0 = \mathbb{E}[P_0]$;

4: **for** $k = 1 : H_{\text{time}}$ **do**

5: Solve for $\theta_S^\star(T, M, N, \Phi_0)$ and $\theta_A^\star(T, M, N, \Phi_0)$;

6: Employ the actions of $\theta_S^\star(T, M, N, \Phi_0)$ and $\theta_A^\star(T, M, N, \Phi_0)$ designed for the first time step for the new game as the action of the current time step k;

7: Observe the actions taken by both sides at time k;

8: **if** $\gamma_k = 1$ **then**

9: $M = M - 1$;

10: **else**

11: $M = M$;

12: **end if**

13: **if** $\lambda_k = 1$ **then**

14: $N = N - 1$;

15: **else**

16: $N = N$;

17: **end if**

18: $T = T - 1$;

19: $\Phi_0 = \mathbb{E}[P_k]$;

20: **end for**

1.5.1 Constraint-Relaxed Problem Formulation

When using total energy constraints as considered in Problem 1.2, during the calculation of the optimal mixed strategy for each side, typically we need three steps and consider $C_T^N \times C_T^M$ different combinations of pure strategies from both sides to obtain closed-form expressions of the objective functions, that is, the order of computational complexity is $O\big((T!)^2\big)$, where T is the time horizon.

For example, when $T = 10$ and $M = N = 5$, the total number of pure strategies is $C_{10}^5 \times C_{10}^5 = 252 \times 252 = 63{,}504$; that is, we need to calculate 63,504 different combinations of J_S and J_A. Therefore, complexity issues must be taken into account when dealing with the general case in practice.

In the sequel, we will consider expected (average) energy constraints. Thus, the constraints in Problem 1.2 are relaxed to

$$\sum_{k=1}^{T} \mathbb{E}[\gamma_k] = M$$

and

$$\sum_{k=1}^{T} \mathbb{E}[\lambda_k] = N,$$

leading to the following constraint-relaxed problem.

Problem 1.3
For the sensor,

$$\max_{\theta_S} \quad J_S(\theta_S),$$

$$\text{s.t.} \quad \sum_{k=1}^{T} \mathbb{E}[\gamma_k] = M.$$

For the attacker,

$$\max_{\theta_A} \quad J_A(\theta_A),$$

$$\text{s.t.} \quad \sum_{k=1}^{T} \mathbb{E}[\lambda_k] = N.$$

Remark 1.7   This type of constraint relaxation is attractive in some practical applications, where instead of only one time game, the estimation jamming process will repeat many times under a total energy constraint. In view of a long time horizon, the expected (average) energy constraint on each short period is equivalent to the total energy constraint, and thus is more flexible and practical.

Under the relaxed constraints, the number of feasible pure strategies to be considered is changed from either C_T^M or C_T^N to 2^T. Though the amount of pure strategies is reduced from $C_T^M \times C_T^N$ to $2^T \times 2^T$, and thereby the order of computational complexity is changed from $O\big((T!)^2\big)$ to $O(4^T)$, we can simplify the computational complexity even more. When considering the optimal mixed strategy, all 2^T possible pure strategies are now feasible to choose. Thus, we can express the mixed strategy of both sides in a more explicit way as follows.

The sensor's data-sending mixed strategy now can be defined as

$$\tilde{\theta}_S^{\text{mixed}} \triangleq \{\tilde{\gamma}_1, \tilde{\gamma}_2, ..., \tilde{\gamma}_T\},$$

where $\tilde{\gamma}_k \in [0, 1]$ is denoted as the probability of the sensor sending data packet at time k, that is,

$$\tilde{\gamma}_k = \mathbb{P}[\gamma_k = 1].$$

Similarly, the attacker's strategy can be expressed as

$$\tilde{\theta}_A^{\text{mixed}} \triangleq \{\tilde{\lambda}_1, \tilde{\lambda}_2, ..., \tilde{\lambda}_T\},$$

where

$$\tilde{\lambda}_k = \mathbb{P}[\lambda_k = 1] \in [0, 1].$$

Then at each time instant k, similar to (1.7), the probability of the sensor data packet arriving successfully at the remote estimator, denoted as $\tilde{p}_k$, can be written as

$$\tilde{p}_k = \tilde{\gamma}_k \tilde{\lambda}_k (1 - \beta_2) + \tilde{\gamma}_k (1 - \tilde{\lambda}_k)(1 - \beta_1). \tag{1.13}$$

Therefore, we can write Problem 1.3 in an equivalent, but more convenient, form as

Problem 1.4
For the sensor,

$$\max_{\tilde{\theta}_S^{\text{mixed}}} \quad J_S(\tilde{\theta}_S^{\text{mixed}}),$$

$$\text{s.t.} \quad \sum_{k=1}^{T} \tilde{\gamma}_k = M, \ \tilde{\gamma}_k \in [0, 1].$$

For the attacker,

$$\max_{\tilde{\theta}_A^{\text{mixed}}} \quad J_A(\tilde{\theta}_A^{\text{mixed}}),$$

$$\text{s.t.} \quad \sum_{k=1}^{T} \tilde{\lambda}_k = N, \ \tilde{\lambda}_k \in [0, 1].$$

Interestingly, with average energy constraints, a closed-form expression for the objective functions $J_S(\tilde{\theta}_S^{\text{mixed}})$ and $J_A(\tilde{\theta}_A^{\text{mixed}})$ can be derived, by studying an underlying Markov chain, as discussed next.

1.5.2 Markov Chain Model

Based on the updating procedure of the error covariance P_k in (1.9) and the steady-state assumption (1.13), it is easy to see that at any time instant $k_2 \geqslant k_1$, the error covariance at the remote estimator side can be written as $P_{k_2} = h^{k_2-k_1}(\overline{P})$, where k_1 is the latest time when the sensor data packet arrives successfully.

Definition 1.4 During the time horizon T, if at time k, the state error covariance at the remote estimator $P_k = h^{i-1}(\overline{P})$, for some $i = 1, 2, ..., T+1$, then we denote the state of the remote estimator $S_k \triangleq Z_{i,k}$.

Thus, the state space for the remote estimator becomes

$$\mathbf{S} \triangleq \{Z_{i,k}\}, \quad 1 \leqslant i \leqslant T+1, \ 1 \leqslant k \leqslant T,$$

which includes all the possible states during the entire time horizon T. Denote the state sets for time k as

$$\mathbb{Z}_k = \{S_k | S_k = Z_{i,k}, 1 \leqslant i \leqslant T+1\}, \quad k = 1, 2, ..., T.$$

For convenience, as we already assumed that the remote estimator's error covariance is $\overline{P}$ before the process starts, that is, $P_0 = \overline{P}$, we denote $\mathbb{Z}_0 = \{Z_{1,0}\}$ as the initialization state set before the process begins.

Due to the updating equation in (1.8), at any time $k+1$, the state S_{k+1} is only related to the previous state S_k. Thus, the stochastic process $\{S_k\}, k = 1, 2, ..., T$, constitutes a Markov chain [9] (as shown in Figure 1.3).

Denote $\mathbb{T}_k$ as the transition matrix from state set $\mathbb{Z}_{k-1}$ to $\mathbb{Z}_k$; then each entry of $\mathbb{T}_k$ can be expressed as

$$\mathbb{T}_k(i_1, i_2) = \mathbb{P}[Z_{i_2,k} | Z_{i_1,k-1}]. \tag{1.14}$$

Thus, the process is described by T transition matrices $\{\mathbb{T}_k\}_{k=1,2,...,T}$, and each element can be easily computed as follows.

If the sensor data packet arrives at the remote estimator, we have $P_k = \overline{P}$. Based on (1.14), this gives

$$\mathbb{T}_k(i_1, 1) = \mathbb{P}[Z_{1,k} | Z_{i_1,k-1}]$$

$$= \mathbb{P}[P_k = \overline{P} | Z_{i_1,k-1}]$$

$$= \tilde{p}_k, \quad \forall \ 1 \leqslant i_1 \leqslant T+1,$$

where $\tilde{p}_k$ is defined in (1.13).

On the other hand, if the packet is dropped (even if the channel is not attacked), then $P_k = h(P_{k-1})$, and we have

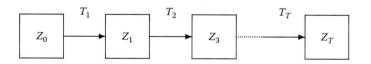

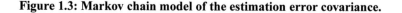

Figure 1.3: Markov chain model of the estimation error covariance.

$$\mathbb{T}_k(i_1, i_1 + 1) = \mathbb{P}[Z_{i_1+1,k}|Z_{i_1,k-1}]$$

$$= \mathbb{P}[P_k = h(P_{k-1})|Z_{i_1,k-1}]$$

$$= \tilde{q}_k, \quad \forall \, 1 \leqslant i_1 \leqslant T,$$

where $\tilde{q}_k = 1 - \tilde{p}_k$ is the corresponding packet jamming probability.

Other entries of $\mathbb{T}_k$ are 0, since the corresponding state transitions are not possible. This gives

$$
\mathbb{T}_k = \begin{bmatrix}
\tilde{p}_k & \tilde{q}_k & & \\
\tilde{p}_k & & \tilde{q}_k & \\
\vdots & & & \ddots \\
\tilde{p}_k & & & & \tilde{q}_k
\end{bmatrix}_{(T+1)\times(T+1)}, \tag{1.15}
$$

where the missing entries are 0.

Denote $\pi_{i,k}$ as the probability of state $Z_{i,k}$ occurring at time k, that is,

$$\pi_{i,k} = \mathbb{P}[S_k = Z_{i,k}]; \tag{1.16}$$

then we can construct the probability matrix

$$\Pi = [\pi_{i,k}]_{(T+1)\times T}.$$

From the definition of $\pi_{i,k}$, it is straightforward to show that

$$\sum_{i=1}^{T+1} \pi_{i,k} = 1, \quad k = 1, 2, ..., T. \tag{1.17}$$

As we assumed $P_0 = \overline{P}$, at time $k = 1$, we readily obtain

$$\pi_{1,1} = \mathbb{P}[P_1 = \overline{P}] = \tilde{p}_1,$$

$$\pi_{2,1} = \mathbb{P}[P_1 = h(\overline{P})] = \mathbb{P}[P_1 = h(P_0)] = \tilde{q}_1,$$

and

$$\pi_{i,1} = 0; \quad \forall i = 3, 4, ..., T+1.$$

To calculate an explicit form of Π, we first recall the definition of the transition matrix $\mathbb{T}_k$ in (1.14), from where the following relationship between $\mathbb{T}_k$ and Π can be established:

$$\Pi[k]^{\mathrm{T}} = \Pi[k-1]^{\mathrm{T}}\mathbb{T}_k, \tag{1.18}$$

where $\Pi[k]$, $k = 1, \ldots, T$, denotes the kth column of Π.

As we have already shown $\Pi[1] = [\tilde{p}_1, \tilde{q}_1, 0, 0, \ldots, 0]^T$, following the recursion in (1.18), through some basic calculation, we can obtain the exact form of Π as a $T + 1 \times T$ matrix:

$$
\Pi = \begin{bmatrix}
\tilde{p}_1 & \tilde{p}_2 & \tilde{p}_3 & \cdots & & \tilde{p}_T \\
\tilde{q}_1 & \tilde{p}_1\tilde{q}_2 & \tilde{p}_2\tilde{q}_3 & \cdots & & \tilde{p}_{T-1}\tilde{q}_T \\
0 & \tilde{q}_1\tilde{q}_2 & \tilde{p}_1\tilde{q}_2\tilde{q}_3 & \cdots & & \tilde{p}_{T-2}\tilde{q}_{T-1}\tilde{q}_T \\
0 & 0 & \tilde{q}_1\tilde{q}_2\tilde{q}_3 & \cdots & & \tilde{p}_{T-3}\tilde{q}_{T-2}\tilde{q}_{T-1}\tilde{q}_T \\
\vdots & \vdots & \vdots & \ddots & & \vdots \\
0 & 0 & 0 & 0 & & \tilde{q}_1\tilde{q}_2\tilde{q}_3\ldots\tilde{q}_T
\end{bmatrix}.
\tag{1.19}
$$

The above result significantly alleviates the computational issues. Once we have the probability matrix Π, we can easily obtain the closed-form expected error covariance for each time slot within the entire time horizon.

From the definition of $\pi_{i,k}$ (see [1.16]), we obtain

$$
\mathbb{E}[P_k] = \sum_{i=1}^{T+1} \pi_{i,k} h^{i-1}(\overline{P}).
\tag{1.20}
$$

Thus, we have

$$
J_\alpha(T) = \alpha \frac{1}{T} \sum_{k=1}^{T} \mathrm{Tr}\left(\mathbb{E}[P_k]\right) + (1-\alpha)\mathrm{Tr}\left(\mathbb{E}[P_T]\right)
$$

$$
= \alpha \frac{1}{T} \sum_{k=1}^{T} \mathrm{Tr}\left(\sum_{i=1}^{T+1} \pi_{i,k} h^{i-1}(\overline{P})\right) + (1-\alpha)\mathrm{Tr}\left(\sum_{i=1}^{T+1} \pi_{i,T} h^{i-1}(\overline{P})\right)
\tag{1.21}
$$

$$
= \alpha \frac{1}{T} \sum_{k=1}^{T} \left(\sum_{i=1}^{T+1} \pi_{i,k} \mathrm{Tr}\left(h^{i-1}(\overline{P})\right)\right) + (1-\alpha) \sum_{i=1}^{T+1} \pi_{i,T} \mathrm{Tr}\left(h^{i-1}(\overline{P})\right).
$$

Consequently, we can readily write $J_S(\tilde{\theta}_S^{\mathrm{mixed}})$ and $J_A(\tilde{\theta}_A^{\mathrm{mixed}})$ in a closed form. These objective functions depend on $\{\pi_{i,k}\}$, which are functions of parameters $\{\tilde{\gamma}_k\}$ and $\{\tilde{\lambda}_k\}$. Thus, following a procedure similar to that in Section 1.3.3, by using Lagrange multipliers, one can obtain the optimal solutions $\tilde{\theta}_S^\star$ and $\tilde{\theta}_A^\star$.

1.5.3 Comparison and Analysis

We will compare the constraint-relaxed problem formulation in Problem 1.4 with the original one (Problem 1.2) using examples.

As mentioned before, when $T = 10$ and $M = N = 5$, one needs to consider $C_{10}^5 \times C_{10}^5 = 252 \times 252 = 63{,}504$ different combinations of pure strategies from both sides to obtain the closed-form expression of the objective functions. When the time horizon T increases, the combinations will increase exponentially, which becomes impractical. Under the current constraint-relaxed problem formulation (Problem 1.4), however, computational complexity can be significantly reduced. This is due to the fact that the probability matrix can be obtained in a quite systematical way, as shown in (1.19). Thus, one can easily obtain the closed-form expression of the objective function based on (1.20), as shown in (1.21).

Once we have the closed-form expression of the objective function consisting of only $2T$ variables, $\{\tilde{\lambda}_i, \tilde{\gamma}_i\}_{i=1}^T$, we can apply the Lagrange multiplier method to find the optimal strategies.

Now we use a simple example to illustrate the constraint-relaxed problem formulation. Suppose $T = 2$, $M = 1$, $N = 1$, $\alpha = 1$. The system parameters A, C, Q, R and the steady-state error covariance $\bar{P}$ are all scalars. Then one has the optimal solution to Problem 1.4:

$$\begin{cases} \tilde{\gamma}_1^\star = \tilde{\gamma}_2^\star = \frac{1}{2}, \\ \tilde{\lambda}_1^\star = \tilde{\lambda}_2^\star = \frac{1}{2}; \end{cases} \tag{1.22}$$

that is, the optimal strategy for the sensor under the relaxed constraints is to send data packets with probability 0.5 at both times $k = 1$ and $k = 2$. For the attacker, we have the same conclusion.

One can interpret the optimal strategy under the relaxed constraints as a mixed strategy that randomly chooses the pure strategies $\{0,0\}$, $\{0,1\}$, $\{1,0\}$, $\{1,1\}$, with the same probability 0.25.

As a comparison, the optimal solutions for the original Problem 1.2 are

$$\begin{cases} \mu_1^\star = \mu_2^\star = \frac{1}{2}, \\ \pi_1^\star = \pi_2^\star = \frac{1}{2}; \end{cases} \tag{1.23}$$

that is, the optimal strategies for both sides for the original problem are to randomly choose pure strategies $\{0,1\}$ and $\{1,0\}$, each with probability 0.5.

When $T > 2$, the optimal strategy is different under different parameters and a closed-form solution only exists when $T = 2$. For example, consider a one-dimensional process with parameters $A = 1.2$, $C = 0.7$, $Q = R = 0.8$, $\beta_1 = 0$, $\beta_2 = 1$, $T = 3$, $M = N = 1$. The optimal solution turns out to be

$$\begin{cases} \mu_1^\star = 0.2748, \\ \mu_2^\star = 0.4504, \\ \mu_3^\star = 0.2748, \end{cases}$$

and

$$
\begin{cases}
\pi_1^\star = 0.3626, \\
\pi_2^\star = 0.2748, \\
\pi_3^\star = 0.3626.
\end{cases}
$$

As we can see, though the solutions for the constraint-relaxed problem and the original problem, that is, (1.22) and (1.23), look similar, their meanings are different.

1.6 Multisensor Scenario

Most previous literature related to jamming attack on CPS investigated the single-sensor case [21, 22, 38]. However, many typical applications in CPS are dealing with sensors networks, where multiple sensors measure the states of the same process and then send their local estimates to the remote estimator for information fusion. This motivates us to consider the jamming game between the attacker and multiple sensors.

1.6.1 Multiple Sensor Formulation

In the multisensor scenario, instead of (1.13), we assume that there are H sensors measuring linear combinations of a common underlying process as described in (1.24):

$$
y_k^i = C_i x_k + v_k^i, \quad i = 1, 2, ..., H, \tag{1.24}
$$

where $y_k^i \in \mathbb{R}^{n_y^i}$ is the measurement taken by sensor i, and $v_k^i \in \mathbb{R}^{n_y^i}$ are zero-mean i.i.d. Gaussian noises with $\mathbb{E}[v_k^i v_j^{i'}] = \delta_{kj} R_i \, (R > 0)$ and $\mathbb{E}[w_k v_j^{i'}] = 0 \, \forall j, k \in \mathbb{N}, i = 1, 2, ..., H$. Furthermore, we also assume that w_k and $v_k^i, i = 1, 2, ..., H$, are uncorrelated to each other. In addition, $C_i \in \mathbb{R}^{n_y^i \times n_x}$ is the measurement matrix for sensor i.

Every time step k, each sensor first calculates the local estimate $\hat{x}_k^{si}$ based on its own measurements, and then sends $\hat{x}_k^{si}$ to the remote estimator for information fusion. We assume that the remote estimator has sufficient computational ability for data fusion.

In practice, the number of sensors in CPS can be quite large [25, 37], which makes the tracking of the packet arrival sequences and updating of corresponding estimates following the procedure in (1.20) for each sensor neither realistic nor effective. Thus, at every time step, as in [37], the remote estimator will simply integrate all the data packets that arrive successfully. At time step k, suppose that there are $H_k (\leqslant H)$ data packets (to simplify the notation, denoted as sensors 1 to H_k) that arrive from all the H sensors; based on [37], the optimal information fusion (that is, in the linear minimum variance sense) estimator with matrix weights and its corresponding variance are given as

$$\hat{x}_k^o = \sum_{i=1}^{H_k} \bar{A}_i^k \hat{x}_k^{si}$$

and

$$P_k^o = (e^T \Sigma_k^{-1} e)^{-1}, \tag{1.25}$$

where Σ_k and e are matrices defined in Theorem 1 in [37], and the optimal matrix weights A_i^k are the blocks of the following matrix $\bar{A}$:

$$[\bar{A}_1^k, \bar{A}_2^k, ..., \bar{A}_H^k]^T \triangleq \bar{A}_k = \Sigma^{-1} e (e^T \Sigma^{-1} e)^{-1}.$$

1.6.2 Constraint-Relaxed Game Formulation

Following a description similar to that used in Section 1.5 for the single-sensor case, under the relaxed constraints, we assume that the data-sending mixed strategy for sensor i with energy constraint M_i is

$$\tilde{\theta}_{S_i}^{\text{mixed}} \triangleq \{\tilde{\gamma}_1^i, \tilde{\gamma}_2^i, ..., \tilde{\gamma}_T^i\}, \quad i = 1, 2, ..., H, \tag{1.26}$$

where $\tilde{\gamma}_k^i \in [0, 1]$ is denoted as the probability of the sensor i sending data packet at time k and

$$\sum_{k=1}^{T} \tilde{\gamma}_k^i = M_i.$$

Thus, the overall strategy for the sensor side can be expressed with the strategy matrix Θ_S as

$$\Theta_S = [\tilde{\gamma}_k^i]_{k \times i},$$

where $\tilde{\gamma}_k^i$ is defined in (1.26).

Similarly, we can define the strategy matrix Θ_A for the attacker:

$$\Theta_A = [\tilde{\lambda}_k^i]_{k \times i},$$

where $\tilde{\lambda}_k^i$ is the probability of the attacker launching a DoS attack on the communication channel between sensor i and the remote estimator at time step k, and the energy constraint for the attacker is

$$\sum_{k=1}^{T} \sum_{i=1}^{H} \tilde{\gamma}_k^i = N.$$

Based on (1.24), the arrival probability matrix, denoted as Ω, can be written as

$$\Omega = [\tilde{p}_k^i]_{T \times H} \triangleq (1 - \beta_1)\Theta_S - (\beta_2 - \beta_1)\Theta_S \circ \Theta_A,$$

where $\circ$ is the Hadamard product (the entrywise product) and the element $\tilde{p}_k^i$ is the probability of the data packet of sensor i arriving successfully at the remote estimator at time step k (assumed here as the Bernoulli process).

As the arrival probability of each $\hat{x}_k^{si}$ and P_k^o, as in (1.25), can be expressed in terms of the elements of Ω, the procedure for obtaining the optimal solution will then be similar to the one-sensor case investigated in Section 1.5.

1.7 Conclusion

We have studied a CPS scenario where a malicious agent carries out jamming attacks on the communication channel between a sensor and a remote estimator. For fixed horizon cost functions and total energy constraints, we investigated the interactive decision-making process between the sensor node and the attacker. We first considered a situation where the sensor and attacker fix their strategies (e.g., the transmission and attack probabilities to be used) a priori. Using a game-theoretical framework, we proved that the optimal strategies for both sides constitute a Nash equilibrium. For the case where the sensor and the attacker have online information about the previous transmission outcomes, we provided an algorithm that performs the game dynamically. We also introduced an alternative problem formulation, which considers average energy constraints. The associated optimization problems require significantly less computations. Then under constraint-relaxed formulation, taking the multisensor data fusion into consideration, the problem for the multisensor case was also studied.

References

1. A. Agah, S.K. Das, and K. Basu. A game theory based approach for security in wireless sensor networks. In *2004 IEEE International Conference on Performance, Computing, and Communications*, pages 259–263. IEEE, 2004.

2. T. Alpcan and T. Başar *Network Security: A Decision and Game-Theoretic Approach*. Cambridge University Press, Cambridge, 2010.

3. S. Amin, A. Cardenas, and S. Sastry. Safe and secure networked control systems under denial-of-service attacks. *Hybrid Systems: Computation and Control*, pages 31–45, 2009.

4. T.M. Apostol. *Mathematical Analysis, 2nd edition*. Addison-Wesley, Boston, 1974.

5. R. Bansal and T. Başar. Communication games with partially soft power constraints. *Journal of Optimization Theory and Applications*, 61(3):329–346, 1989.

6. T. Başar and G.J. Olsder. *Dynamic Noncooperative Game Theory*, volume 200. SIAM, 1995.

7. T. Başar and Y.W. Wu. Solutions to a class of minimax decision problems arising in communication systems. *Journal of Optimization Theory and Applications*, 51(3):375–404, 1986.

8. S. Bhattacharya and T. Başar. Game-theoretic analysis of an aerial jamming attack on a UAV communication network. In *American Control Conference (ACC), 2010*, pages 818–823. IEEE, 2010.

9. P. Brémaud. *Markov Chains*. Springer, New York, 1999.

10. A.A. Cardenas, S. Amin, and S. Sastry. Secure control: Towards survivable cyber-physical systems. In *28th International Conference on Distributed Computing Systems Workshops*, pages 495–500, 2008.

11. G. Carl, G. Kesidis, R.R. Brooks, and S. Rai. Denial-of-service attack-detection techniques. *IEEE Internet Computing*, 10(1):82–89, 2006.

12. R.D'Andrea and R.M. Murray. The roboflag competition. In *Proceedings of the 2003 American Control Conference*, volume 1, pages 650–655. IEEE, 2003.

13. R. Gibbons. *A Primer in Game Theory*. Harvester Wheatsheaf, Hemel Hempstead, 1992.

14. A. Gupta, C. Langbort, and T. Başar. Optimal control in the presence of an intelligent jammer with limited actions. In *Proceedings of IEEE Conference on Decision and Control (CDC)*, pages 1096–1101, 2010.

15. M. Hazewinkel. *Encyclopaedia of Mathematics: Supplement*, volume 3. Springer, New York, 2002.

16. L.T. Heberlein, G.V. Dias, K.N. Levitt, B. Mukherjee, J. Wood, and D.Wolber. A network security monitor. In *1990 IEEE Computer Society Symposium on Research in Security and Privacy*, pages 296–304. IEEE, 1990.

17. P. Hovareshti, V. Gupta, and J.S. Baras. Sensor scheduling using smart sensors. In *2007 46th IEEE Conference on Decision and Control*, pages 494–499. IEEE, 2007.

18. Z. Jin, S. Waydo, E.B. Wildanger, M. Lammers, H. Scholze, P. Foley, D. Held, and R.M. Murray. MVWT-II: The second generation Caltech multi-vehicle wireless testbed. In *Proceedings of the 2004 American Control Conference*, volume 6, pages 5321–5326. IEEE, 2004.

19. A. Kashyap, T. Başar, and R. Srikant. Correlated jamming on mimo Gaussian fading channels. *IEEE Transactions on Information Theory*, 50(9):2119–2123, 2004.

20. Y.W. Law, M. Palaniswami, L.V. Hoesel, J. Doumen, P. Hartel, and P. Havinga. Energy-efficient link-layer jamming attacks against wireless sensor network MAC protocols. *ACM Transactions on Sensor Networks (TOSN)*, 5(1):6, 2009.

21. H. Li, L. Lai, and R.C. Qiu. A denial-of-service jamming game for remote state monitoring in smart grid. In *2011 45th Annual Conference on Information Sciences and Systems (CISS)*, pages 1–6. IEEE, 2011.

22. Y. Li, L. Shi, P. Cheng, J. Chen, and D.E. Quevedo. Jamming attack on cyber-physical systems: A game-theoretic approach. In *2013 IEEE 3rd Annual International Conference on Cyber Technology in Automation, Control and Intelligent Systems (CYBER)*, pages 252–257. IEEE, 2013.

23. Y. Liu, P. Ning, and M.K. Reiter. False data injection attacks against state estimation in electric power grids. *ACM Transactions on Information and System Security (TISSEC)*, 14(1):13, 2011.

24. Y. Mo, E. Garone, A. Casavola, and B. Sinopoli. False data injection attacks against state estimation in wireless sensor networks. In *2010 49th IEEE Conference on Decision and Control*, pages 5967–5972. IEEE, 2010.

25. Y. Mo, T.H.J. Kim, K. Brancik, D. Dickinson, H. Lee, A. Perrig, and B. Sinopoli. Cyber–physical security of a smart grid infrastructure. *Proceedings of the IEEE*, 100(1):195–209, 2012.

26. Y. Mo and B. Sinopoli. Secure control against replay attacks. In *2009 47th Annual Allerton Conference on Communication, Control, and Computing*, pages 911–918, 2009.

27. Y. Mo and B. Sinopoli. Integrity attacks on cyber-physical systems. In *Proceedings of the International Conference on High Confidence Networked Systems*, pages 47–54, 2012.

28. J. Nash. Non-cooperative games. *Annals of Mathematics*, 54(2):286–295, 1951.

29. D.E. Quevedo, A. Ahlén, and J. Østergaard. Energy efficient state estimation with wireless sensors through the use of predictive power control and coding. *IEEE Transactions on Signal Processing*, 58(9):4811–4823, 2010.

30. S. Roy, C. Ellis, S. Shiva, D. Dasgupta, V. Shandilya, and Q. Wu. A survey of game theory as applied to network security. In *2010 43rd Hawaii International Conference on System Sciences (HICSS)*, pages 1–10. IEEE, 2010.

31. L. Shi, P. Cheng, and J. Chen. Sensor data scheduling for optimal state estimation with communication energy constraint. *Automatica*, 47(8):1693–1698, 2011.

32. L. Shi, M. Epstein, and R. M. Murray. Kalman filtering over a packet-dropping network: a probabilistic perspective. *IEEE Transactions on Automatic Control*, 55(3):594–604, 2010.

33. L. Shi, Q.S. Jia, Y. Mo, and B. Sinopoli. Sensor scheduling over a packet-delaying network. *Automatica*, 47(5):1089–1092, 2011.

34. L. Shi, K.H. Johansson, and L. Qiu. Time and event-based sensor scheduling for networks with limited communication resources. In *World Congress of the International Federation of Automatic Control (IFAC)*, 2011.

35. L. Shi and L. Xie. Optimal sensor power scheduling for state estimation of gauss–markov systems over a packet-dropping network. *IEEE Transactions on Signal Processing*, 60(5):2701–2705, 2012.

36. B. Sinopoli, L. Schenato, M. Franceschetti, K. Poolla, M.I. Jordan, and S.S. Sastry. Kalman filtering with intermittent observations. *IEEE Transactions on Automatic Control*, 49(9):1453–1464, 2004.

37. S. Sun and Z. Deng. Multi-sensor optimal information fusion Kalman filter. *Automatica*, 40(6):1017–1023, 2004.

38. H. Zhang, P. Cheng, L. Shi, and J. Chen. Optimal DoS attack policy against remote state estimation. In *2013 IEEE 52nd Annual Conference on Decision and Control (CDC), Florence, Italy*, 2013.

39. M. Zuba, Z. Shi, Z. Peng, and J.H. Cui. Launching denial-of-service jamming attacks in underwater sensor networks. In *Proceedings of the Sixth ACM International Workshop on Underwater Networks*, page 12. ACM, 2011.

Chapter 2

Secure State Estimation against Stealthy Attack

Yilin Mo

School of Electrical and Electronic Engineering, Nanyang Technological University

CONTENTS

2.1 Introduction

In this chapter, we consider the problem of inferring the system state from sensory data in an adversarial environment, where some of the sensors are compromised by an adversary and can report arbitrary readings. Our goal is to construct a secure state estimator, whose estimation error is bounded regardless of the adversary's action. This chapter is organized as follows: In Section 2.2, we consider the estimation problem in a static setting, where we try to estimate the state using the measurements collected at a single time step. The problem of estimating the state of a linear time-invariant system from sequential sensory data is further considered in Section 2.3.

2.1.1 Notations

■ For a discrete set $\mathcal{I}$, let $|\mathcal{I}|$ be the cardinality of the set.

■ Denote $\mathbb{Z}$ as the set of integers and $\mathbb{N}$ as the set of non-negative integers. Denote $\mathbb{R}$ as the set of real numbers and $\mathbb{C}$ as the set of complex numbers. For any $x \in \mathbb{C}$, denote its real part as $\mathfrak{Re}(x)$ and its absolute value as $|x|$.

■ For a matrix $M \in \mathbb{R}^{n \times m}$, denote M_i to be its ith row. Furthermore, let $\mathcal{I} = \{i_1, \ldots, i_l\} \subseteq \{1, \ldots, m\}$ be an index set. We define $M_{\mathcal{I}}$ as

$$M_{\mathcal{I}} \triangleq \begin{bmatrix} M_{i_1} \\ \vdots \\ M_{i_l} \end{bmatrix}.$$

■ Let $x \in \mathbb{R}^n$ be a vector; then $\|x\|_2$ is the 2-norm of x. $\|x\|_\infty$ is the infinity norm of x. $\|x\|_0$ is the zero norm of x, that is, the number of non-zero entries of x.

■ We denote restriction of an infinite sequence $\{x(t)\}_{t \in \mathbb{N}}$ to its first T elements with $x(0 : T)$, that is,

$$x(0 : T) \triangleq (x(0), \ldots, x(T)).$$

To avoid confusion, we will denote the infinite sequence $\{x(t)\}_{t \in \mathbb{N}}$ as $x(:) = x(0 : \infty)$ and its value at a given time t as $x(t)$.

■ The infinity norm of the finite sequence $x(0 : T)$ is defined as

$$\|x(0 : T)\|_\infty \triangleq \max_{0 \leq t \leq T} \max_i |x_i(t)|.$$

The infinity norm of an infinite sequence $x(:)$ is defined as

$$\|x(:)\|_\infty \triangleq \sup_{t \in \mathbb{N}} \max_i |x_i(t)|.$$

We denote l_∞^n as the space of infinite sequences of n-dimensional vectors with bounded infinity norm. We will write l_∞ when there is no confusion on dimension of the vector.

■ For an infinite sequence $\{x(t)\}_{t \in \mathbb{N}}$, let $\mathrm{supp}(x) \triangleq \{i : \exists t \text{ s.t. } x_i(t) \neq 0\}$ and we define $\|x(:)\|_0 = |\mathrm{supp}(x)|$.

■ For any matrix $A \in \mathbb{R}^{m \times n}$, define $\sigma(A)$ as the spectral radius of A, that is, the maximum over the absolute values of all its eigenvalues. We denote its norm induced by infinity norm as

$$\|A\|_\infty \triangleq \sup_{x \neq 0} \frac{\|Ax\|_\infty}{\|x\|_\infty} = \max_i \sum_j |a_{ij}|,$$

and its norm induced by 2-norm as

$$\|A\|_2 \triangleq \sup_{x \neq 0} \frac{\|Ax\|_2}{\|x\|_2} = \sqrt{\sigma(AA^T)}.$$

■ All comparisons between matrices in this chapter are in the positive semi-definite sense.

2.2 Static Estimation

2.2.1 System and Attack Model

The goal is to estimate the state $x \in \mathbb{R}^n$ from a vector

$$y \triangleq \begin{bmatrix} y_1 & \cdots & y_m \end{bmatrix}^T \in \mathbb{R}^m$$

consisting of m sensor measurements $y_i \in \mathbb{R}$, where i belongs to $\mathcal{S} \triangleq \{1, \ldots, m\}$, with the caveat that some of the sensors can potentially be compromised and their measurements can be arbitrarily manipulated by an adversary. For the good sensors, we assume that they measure a linear combination of the state x. As a result, the relationship between the state x and sensor measurements y can be characterized by the following equation:

$$y = Cx + Dw + a, \tag{2.1}$$

where $w \in \mathbb{R}^m$ is the sensor noise, which is assumed to be 2-norm bounded, that is, $\|w\|_2 \leq \delta$. $D \in \mathbb{R}^{m \times m}$ is assumed to be full rank. The vector a is the bias injected by the attacker. The non-zero entries of a indicate the set of compromised sensors. If $a = 0$, then we will say that the system is *not* under attack. In this chapter, we assume that the attacker can only manipulate up to l sensors. As a result, $\|a\|_0 \leq l$.

Remark 2.1   The parameter l can also be interpreted as a design parameter for the system operator. In other words, the system operator needs to choose how many compromised sensors can be tolerated by the estimation algorithm. In general, increasing l will increase the security of the estimator under attack. However, a large l could result in higher computational complexity and performance degradation during normal operation when no sensor is compromised. Therefore, there exists a trade-off between security and efficiency, which can be tuned by choosing a suitable parameter l.

An estimator is a function $f : \mathbb{R}^m \to \mathbb{R}^n$, which maps the sensor measurements to the state estimation, which is defined as $\hat{x} = f(y)$. The estimation error e can be defined as

$$e \triangleq x - \hat{x} = x - f(y). \tag{2.2}$$

It is clear that the estimation error e is a function of the true state x, the noise w, the bias a injected by the adversary, and our choice of the estimator f. In this chapter, we are concerned with the worst-case estimation error defined as

$$e^*(f) = \sup_{x \in \mathbb{R}^n, \|w\|_2 \leq \delta, \|a\|_0 \leq l} \|e\|_2. \tag{2.3}$$

We will call an estimator a secure estimator if the corresponding e^* is finite. The problems of interests are

1. *Fundamental limit:* Under what conditions does there not exist a secure estimator?

2. *Estimator design:* Can we construct a secure estimator and analyze its performance?

2.2.2   Preliminary

Before continuing on, we need to introduce the concept of Chebyshev center of a set. A ball $B(x, r) \subset \mathbb{R}^n$ is defined as

$$B(x, r) \triangleq \{x' \in \mathbb{R}^n : \|x' - x\|_2 \leq r\}.$$

Consider a bounded set $S \subseteq \mathbb{R}^n$. A ball $B(x, r)$ covers S if and only if $S \subseteq B(x, r)$. For any point $x \in \mathbb{R}^n$, define the radius of the minimum covering ball as

$$\rho(x, S) \triangleq \inf\{r \in \mathbb{R}^+ : S \subseteq B(x, r)\}.$$

For a bounded set S, define its radius $r(S) \in \mathbb{R}^+$ and Chebyshev center $c(S) \in \mathbb{R}^n$ of S to be

$$r(S) \triangleq \inf_{x \in \mathbb{R}^n} \rho(x, S), \ c(S) \triangleq \arg\min_{x \in \mathbb{R}^n} \rho(x, S).$$

In essence, $B(c(S), r(S))$ is the smallest ball that covers S. We further define the diameter $d(S)$ of the set S as

$$d(S) \triangleq \sup_{x \in S} \rho(x, S) = \sup_{x, x' \in S} \|x - x'\|_2.$$

Notice that in general, $d(S) \neq 2r(S)$. For example, for an equilateral triangle with side length 1, we have $d(S) = 1$, while $r(S) = 1/\sqrt{3}$. In general, the following relation between $r(S)$ and $d(S)$ holds:

Theorem 2.1

Let $S \subset R^n$ be a non-empty and bounded set; then the following inequalities hold on $r(S)$ and $d(S)$:

$$\frac{d(S)}{2} \leq r(S) \leq \sqrt{\frac{n}{2n+2}} d(S) \leq \frac{1}{\sqrt{2}} d(S).$$

Proof 2.1 The first inequality is due to the fact that $S \subseteq B(c(S), r(S))$, which implies that

$$d(S) \leq d[B(c(S), r(S))] = 2r(S).$$

The second inequality is from Jung's theorem [2]. The third inequality is trivial. □

2.2.3 Fundamental Limit

This subsection is devoted to the derivation of the fundamental limit of the static estimation problem. To this end, let us define the set $\mathbb{Y}$ as the set of all possible "manipulated" measurements, that is,

$$\mathbb{Y} \triangleq \{y \in \mathbb{R}^m : \exists x, \|w\|_2 \leq \delta, \|a\|_0 \leq l, \text{ such that } y = Cx + Dw + a\}. \quad (2.4)$$

For any $y \in \mathbb{Y}$, we can define the set $\mathbb{X}(y)$ as the set of feasible x that can generate y, that is,

$$\mathbb{X}(y) \triangleq \{x \in \mathbb{R}^n : \exists \|w\|_2 \leq \delta, \|a\|_0 \leq l, \text{ such that } y = Cx + Dw + a\}. \quad (2.5)$$

Thus, from the definition of the sets $\mathbb{Y}$ and $\mathbb{X}(y)$, we can rewrite the worst-case estimation error as

$$e^*(f) = \sup_{y \in \mathbb{Y}} \left(\sup_{x \in \mathbb{X}(y)} \|f(y) - x\|_2 \right) = \sup_{y \in \mathbb{Y}} \rho(f(y), \mathbb{X}(y)) \geq \sup_{y \in \mathbb{Y}} r(\mathbb{X}(y)). \quad (2.6)$$

Therefore, if the radius of $\mathbb{X}(y)$ is not uniformly bounded, then the worst-case estimation error is infinity, and hence no secure estimator will exist. This observation is used in the following theorem to provide a fundamental limit on the static estimation problem:

Theorem 2.2
If there exists an index set $\mathcal{K} \subset \mathcal{S}$ with cardinality $|\mathcal{K}| = m - 2l$ such that $C_{\mathcal{K}}$ is not full column rank, then $r(\mathbb{X}(y))$ is not uniformly bounded and thus no secure estimator exists.

Proof 2.2   If there exists a $|\mathcal{K}| = m - 2l$, such that $C_{\mathcal{K}}$ is not full column rank, we can find index set $\mathcal{I}, \mathcal{J}$, such that $|\mathcal{I}| = |\mathcal{J}| = m - l$ and $\mathcal{I} \cap \mathcal{J} = \mathcal{K}$. Furthermore, we know that there exists $x \neq 0$ such that

$$C_{\mathcal{K}} x = 0.$$

As a result, let us choose

$$a_i = \begin{cases} 0 & i \in \mathcal{I} \\ -C_i x & i \notin \mathcal{I} \end{cases}, \quad a_i' = \begin{cases} 0 & i \in \mathcal{J} \\ C_i x & i \notin \mathcal{J} \end{cases}.$$

One can verify that

$$Cx + a = a'.$$

Denote $y = Cx + a = a'$. Clearly, $y \in \mathbb{Y}$ and both x and 0 belong to $\mathbb{X}(y)$. Therefore, the radius of the set $\mathbb{X}(y)$ is at least $\|x\|_2 / 2$. Moreover, by linearity, we know that

$$\alpha x \in \mathbb{X}(\alpha y) \text{ and } 0 \in \mathbb{X}(\alpha y), \forall \alpha \in \mathbb{R}.$$

Hence, the radius of $\mathbb{X}(y)$, $y \in \mathbb{Y}$, is not uniformly bounded and no secure estimator exists. □

2.2.4 Secure Estimator Design and Performance Analysis

In this subsection, we provide an estimation algorithm based on semi-definite programming and analyze its worst-case error. Due to Theorem 2.2, throughout this section, we make the following assumption:

Assumption 2.1 $C_{\mathcal{K}}$ is full column rank for any $\mathcal{K} \subset \mathcal{S}$ with cardinality $m - 2l$.

It is clear from the definition of the Chebyshev center and (2.6), that the optimal estimator (in terms of the worst-case estimation error e^*) is given by

$$f^*(y) \triangleq c(\mathbb{X}(y)), \tag{2.7}$$

with worst-case error

$$e^*(f^*) = \sup_{y \in \mathbb{Y}} r(\mathbb{X}(y)).$$

To numerically compute the Chebyshev center and analyze the estimation error, we need to characterize the shape of $\mathbb{X}(y)$. To this end, let us define the following set:

$$\mathbb{X}_{\mathcal{I}}(y) \triangleq \{x \in \mathbb{R}^n : \exists w, a, \text{ such that } \|w\|_2 \le \delta, a_{\mathcal{I}} = 0 \text{ and } y = Cx + Dw + a\}.$$

Hence, $\mathbb{X}_{\mathcal{I}}(y)$ represents all possible states that can generate measurement y when the sensors in $\mathcal{I}$ are good and the sensors in $\mathcal{I}^c$ are (potentially) compromised. By enumerating all possible index sets, it is easy to see that $\mathbb{X}(y)$ can be written as

$$\mathbb{X}(y) = \bigcup_{|\mathcal{I}|=m-l} \mathbb{X}_{\mathcal{I}}(y).$$

For any index set $\mathcal{I}$, define the matrix

$$R^{\mathcal{I}} \triangleq D_{\mathcal{I}} D_{\mathcal{I}}^T.$$

Since D is full rank, $D_{\mathcal{I}}$ is full row rank, which implies that $R^{\mathcal{I}}$ is full rank. By Assumption 2.1 for any index set $\mathcal{I}$ with cardinality $|\mathcal{I}| \ge m - 2l$, $C_{\mathcal{I}}$ is full column rank. Thus, the following matrices are well defined:

$$P^{\mathcal{I}} \triangleq \left(C_{\mathcal{I}}^T \left(R^{\mathcal{I}} \right)^{-1} C_{\mathcal{I}} \right)^{-1}, K^{\mathcal{I}} \triangleq P^{\mathcal{I}} C_{\mathcal{I}}^T \left(R^{\mathcal{I}} \right)^{-1},$$

$$U^{\mathcal{I}} \triangleq (I - C_{\mathcal{I}} K^{\mathcal{I}})^T \left(R^{\mathcal{I}} \right)^{-1} (I - C_{\mathcal{I}} K^{\mathcal{I}}).$$

To characterize the shape of $\mathbb{X}^{\mathcal{I}}(y)$, we need the following lemma:

Lemma 2.1
Define the function $V^{\mathcal{I}}(x) : \mathbb{R}^n \to \mathbb{R}$ as the solution of the following optimization problem:

$$\underset{w \in \mathbb{R}^m}{\text{minimize}} \qquad \|w\|_2^2$$

$$\text{subject to} \qquad D_{\mathcal{I}} w = y_{\mathcal{I}} - C_{\mathcal{I}} x. \qquad (2.8)$$

Then $V^{\mathcal{I}}(x)$ is given by

$$V^{\mathcal{I}}(x) = (x - \hat{x}^{\mathcal{I}}(y))^T \left(P^{\mathcal{I}} \right)^{-1} (x - \hat{x}^{\mathcal{I}}(y)) + \varepsilon^{\mathcal{I}}(y), \qquad (2.9)$$

where

$$\hat{x}^{\mathcal{I}}(y) = K^{\mathcal{I}} y_{\mathcal{I}}, \qquad (2.10)$$

and

$$\varepsilon^{\mathcal{I}}(y) = y_{\mathcal{I}}^T U^{\mathcal{I}} y_{\mathcal{I}}. \tag{2.11}$$

Proof 2.3 Consider the constraint of the optimization problem (2.8):

$$y_{\mathcal{I}} - C_{\mathcal{I}} x = D_{\mathcal{I}} w. \tag{2.12}$$

As D is full row rank, $D_{\mathcal{I}}$ is also full row rank. Consider the singular value decomposition of $D_{\mathcal{I}}$; we get

$$D_{\mathcal{I}} = Q_l \begin{bmatrix} \Lambda & \mathbf{0} \end{bmatrix} Q_r,$$

where Q_l, Q_r are orthogonal matrices with proper dimensions and Λ is an invertible and diagonal matrix. Hence, (2.12) implies that

$$\Lambda^{-1} Q_l^T y_{\mathcal{I}} - \Lambda^{-1} Q_l^T C_{\mathcal{I}} x = \begin{bmatrix} I & \mathbf{0} \end{bmatrix} v, \tag{2.13}$$

where $v = Q_r w$ and $\|v\|_2 = \|w\|_2$. By projecting $\Lambda^{-1} Q_l^T y_{\mathcal{I}}$ into the subspace $\text{span}(\Lambda^{-1} Q_l^T C_{\mathcal{I}})$, we have

$$\left[\Lambda^{-1} Q_l^T y_{\mathcal{I}} - \Lambda^{-1} Q_l^T C_{\mathcal{I}} \hat{x}^{\mathcal{I}}(y) \right] + \Lambda^{-1} Q_l^T C_{\mathcal{I}} \left[x - \hat{x}^{\mathcal{I}}(y) \right] = \begin{bmatrix} I & \mathbf{0} \end{bmatrix} v. \tag{2.14}$$

The first term on the left-hand side (LHS) of (2.14) is perpendicular to the second term. Thus, (2.14) is equivalent to

$$\varepsilon^{\mathcal{I}}(y) + (x - \hat{x}^{\mathcal{I}}(y))^T P_{\mathcal{I}}^{-1}(x - \hat{x}^{\mathcal{I}}(y)) = \left\| \begin{bmatrix} I & \mathbf{0} \end{bmatrix} v \right\|_2^2 \leq \|v\|_2^2 = \|w\|_2^2.$$

Clearly, the equality holds when $v = \begin{bmatrix} v_1, \ldots, v_{|\mathcal{I}|}, 0, \ldots, 0 \end{bmatrix}$. Hence,

$$V^{\mathcal{I}}(x) = \varepsilon^{\mathcal{I}}(y) + (x - \hat{x}^{\mathcal{I}}(y))^T P_{\mathcal{I}}^{-1}(x - \hat{x}^{\mathcal{I}}(y)).$$

□

By Lemma 2.1, we immediately have the following result:

Lemma 2.2
If $\varepsilon^{\mathcal{I}}(y) > \delta^2$, then $\mathbb{X}_{\mathcal{I}}(y)$ is an empty set. Otherwise, $\mathbb{X}_{\mathcal{I}}(y)$ is an ellipsoid given by

$$\mathbb{X}_{\mathcal{I}}(y) = \{x : (x - \hat{x}_{\mathcal{I}}(y))^T P_{\mathcal{I}}^{-1}(x - \hat{x}_{\mathcal{I}}(y)) \leq \delta^2 - \varepsilon^{\mathcal{I}}(y)\}. \tag{2.15}$$

Proof 2.4 By definition, $x \in \mathbb{X}_{\mathcal{I}}(y)$ is equivalent to the existence of w, such that $\|w\|_2 \leq \delta$ and

$$y_{\mathcal{I}} = C_{\mathcal{I}} x + D_{\mathcal{I}} w.$$

Hence, the lemma holds by Lemma 2.1.

□

We will call $\mathcal{I}$ to be an *admissible* index set if the corresponding $\mathbb{X}_{\mathcal{I}}(y)$ is not empty. Let us denote the collection of all admissible index sets as

$$\mathcal{L} \triangleq \{\mathcal{I} \subset \mathcal{S} : |\mathcal{I}| = m - l, \ \delta^2 \geq \varepsilon^{\mathcal{I}}(y)\}. \tag{2.16}$$

Since $\mathbb{X}(y) = \bigcup_{|\mathcal{I}|=l} \mathbb{X}_{\mathcal{I}}(y) = \bigcup_{\mathcal{I} \in \mathcal{L}} \mathbb{X}_{\mathcal{I}}(y)$, we know that $\mathbb{X}(y)$ is a union of ellipsoids. To check if a ball covers a union of ellipsoids, we have the following theorem:

Theorem 2.3
A ball $B(x,r)$ covers $\mathbb{X}(y)$ if and only if for every index set $\mathcal{I} \in \mathcal{L}$, there exists $\alpha^{\mathcal{I}} \geq 0$ such that

$$\alpha^{\mathcal{I}} \Omega^{\mathcal{I}} \geq \begin{bmatrix} I & -x & 0 \\ -x^T & r^2 & x^T \\ 0 & x & -I \end{bmatrix}, \tag{2.17}$$

where $\Omega^{\mathcal{I}}$ is defined as

$$\Omega^{\mathcal{I}} = \begin{bmatrix} \left(P^{\mathcal{I}}\right)^{-1} & -\left(P^{\mathcal{I}}\right)^{-1}\hat{x}^{\mathcal{I}}(y) & 0 \\ -\hat{x}^{\mathcal{I}}(y)^T \left(P^{\mathcal{I}}\right)^{-1} & \hat{x}^{\mathcal{I}}(y)^T \left(P^{\mathcal{I}}\right)^{-1}\hat{x}^{\mathcal{I}}(y) + \varepsilon^{\mathcal{I}}(y) - \delta^2 & 0 \\ 0 & 0 & 0 \end{bmatrix}.$$

Proof 2.5   This theorem can be proved by Proposition 2.7 and Lemma 2.8 in [3]. □

Therefore, we can derive the optimal state estimate $\hat{x}$ as the solution of the following semi-definite programming problem:

$$\underset{\hat{x} \in \mathbb{R}^n, \varphi \geq 0, \alpha^{\mathcal{I}} \geq 0}{\text{minimize}} \qquad \varphi \tag{2.18}$$

$$\text{subject to} \qquad \alpha^{\mathcal{I}} \Omega^{\mathcal{I}} \geq \begin{bmatrix} I & -\hat{x} & 0 \\ -\hat{x}^T & \varphi & \hat{x}^T \\ 0 & \hat{x} & -I \end{bmatrix}, \forall \mathcal{I} \in \mathcal{L},$$

where the radius of the Chebyshev ball is $r = \sqrt{\varphi}$. In conclusion, the optimal state estimation can be computed via the following algorithm:

1. Enumerate all possible $|\mathcal{I}| = m - l$. Compute $\hat{x}^{\mathcal{I}}(y)$ and $\varepsilon^{\mathcal{I}}(y)$ via (2.10) and (2.11).

2. Check whether $\varepsilon^{\mathcal{I}}(y)$ is no greater than δ^2. Compute the index set $\mathcal{L}$ via (2.16).

3. Solve the optimization problem (2.8).

We now analyze the performance of the optimal estimator (2.7), which is given by the following theorem:

Theorem 2.4
If for all $|\mathcal{K}| = m - 2l$, $C_{\mathcal{K}}$ is full column rank, then for all possible $y \in \mathbb{Y}$, we have

$$\sup_{y \in \mathbb{Y}} d(\mathbb{X}(y)) = 2\delta \max_{|\mathcal{K}|=m-2l} \sqrt{\sigma(P^{\mathcal{K}})}. \tag{2.19}$$

Therefore, e^* satisfies

$$\max_{|\mathcal{K}|=m-2l} \delta\sqrt{\sigma(P^{\mathcal{K}})} \le e^* \le \max_{|\mathcal{K}|=m-2l} \delta\sqrt{2\sigma(P^{\mathcal{K}})}. \tag{2.20}$$

Before proving Theorem 2.4, we need the following lemma:

Lemma 2.3
If $\mathcal{K}_1 \subseteq \mathcal{K}_2 \subseteq \mathcal{S}$ and $C_{\mathcal{K}_1}$ is full column rank, then the following inequality holds:

$$P^{\mathcal{K}_1} \ge P^{\mathcal{K}_2}.$$

Proof 2.6 To prove $P^{\mathcal{K}_1} \ge P^{\mathcal{K}_2}$, we only need to prove that

$$C_{\mathcal{K}_1}^T \left(R^{\mathcal{K}_1}\right)^{-1} C_{\mathcal{K}_1} \le C_{\mathcal{K}_2}^T \left(R^{\mathcal{K}_2}\right)^{-1} C_{\mathcal{K}_2}. \tag{2.21}$$

Without loss of generality, let us assume that

$$C_{\mathcal{K}_2} = \begin{bmatrix} C_{\mathcal{K}_1} \\ C_{\mathcal{K}_2 \backslash \mathcal{K}_1} \end{bmatrix}, \quad D_{\mathcal{K}_2} = \begin{bmatrix} D_{\mathcal{K}_1} \\ D_{\mathcal{K}_2 \backslash \mathcal{K}_1} \end{bmatrix}. \tag{2.22}$$

From definition of $R^{\mathcal{I}}$, we can write $R^{\mathcal{K}_2}$ as

$$R^{\mathcal{K}_2} = \begin{bmatrix} R_{11} & R_{12} \\ R'_{12} & R_{22} \end{bmatrix},$$

where $R_{11} = R^{\mathcal{K}_1}$. Using Schur complements, we have

$$\left(R^{\mathcal{K}_2}\right)^{-1} = \begin{bmatrix} R_{11}^{-1} & 0 \\ 0 & 0 \end{bmatrix} + \begin{bmatrix} R_{11}^{-1}R_{12} \\ I \end{bmatrix} \left(R_{22} - R'_{12}R_{11}^{-1}R_{12}\right)^{-1} \begin{bmatrix} R'_{12}R_{11}^{-1} & I \end{bmatrix}.$$

Combining with (2.22), we can prove (2.21). □

We are now ready to prove Theorem 2.4:

Proof of Theorem 2.4   We first prove (2.19). Suppose that for all $\mathcal{K} \subset \mathcal{S}$ with cardinality $m - 2l$, $C_{\mathcal{K}}$ is full column rank. Let us consider a pair of set $\mathcal{I}, \mathcal{J}$ with cardinality $m - l$. Define $\mathcal{K}$ as

$$\mathcal{K} = \mathcal{I} \bigcap \mathcal{J}.$$

Clearly, $|\mathcal{K}| \geq m - 2l$ and $C_{\mathcal{K}}$ is also full column rank. Now for any point $x \in \mathbb{X}_{\mathcal{I}}(y)$ and $x' \in \mathbb{X}_{\mathcal{J}}(y)$, we have:

$$Cx + Dw + a = Cx' + Dw' + a' = y, \tag{2.23}$$

where $a_{\mathcal{I}} = 0$ and $a'_{\mathcal{J}} = 0$. Therefore, $a_{\mathcal{K}} = a'_{\mathcal{K}} = 0$, which implies that

$$C_{\mathcal{K}}x + D_{\mathcal{K}}w = C_{\mathcal{K}}x' + D_{\mathcal{K}}w'. \tag{2.24}$$

Thus,

$$x - x' = K^{\mathcal{K}} D_{\mathcal{K}}(w' - w). \tag{2.25}$$

By the fact that $\|w - w'\|_2 \leq 2\delta$, we have

$$\|x - x'\|_2 \leq 2\|K^{\mathcal{K}} D_{\mathcal{K}}\|_2 \delta = 2\delta\sqrt{\sigma(P^{\mathcal{K}})}.$$

Therefore, by Lemma 2.3, we have

$$\sup_{y \in \mathbb{Y}} d(\mathbb{X}(y)) \leq 2\delta \max_{|\mathcal{K}| \geq m - 2l} \sqrt{\sigma(P^{\mathcal{K}})} = 2\delta \max_{|\mathcal{K}| = m - 2l} \sqrt{\sigma(P^{\mathcal{K}})}.$$

Now we need to prove that the equality of (2.19) holds. Suppose that we find x, x' and $\|w\|_2, \|w'\|_2 \leq \delta$ that satisfies (2.24) and $\|x - x'\|_2 = 2\sqrt{\sigma(P^{\mathcal{K}})}$. We know that

$$C_{\mathcal{K}}x + D_{\mathcal{K}}w = C_{\mathcal{K}}x' + D_{\mathcal{K}}w'. \tag{2.26}$$

Therefore, let us create a y, such that

$$y_{\mathcal{K}} = C_{\mathcal{K}}x + D_{\mathcal{K}}w, \qquad\qquad y_{\mathcal{I}\backslash\mathcal{K}} = C_{\mathcal{I}\backslash\mathcal{K}}x + D_{\mathcal{I}\backslash\mathcal{K}}w,$$

$$y_{\mathcal{S}\backslash(\mathcal{I}\cup\mathcal{J})} = 0, \qquad\qquad y_{\mathcal{J}\backslash\mathcal{K}} = C_{\mathcal{J}\backslash\mathcal{K}}x' + D_{\mathcal{J}\backslash\mathcal{K}}w'.$$

Thus,

$$y_{\mathcal{K}} - C_{\mathcal{K}}x = D_{\mathcal{K}}w, y_{\mathcal{I}\backslash\mathcal{K}} - C_{\mathcal{I}\backslash\mathcal{K}}x = D_{\mathcal{I}\backslash\mathcal{K}}w,$$

which implies that $x \in \mathbb{X}_{\mathcal{I}}(y)$. On the other hand,

$$y_{\mathcal{K}} - C_{\mathcal{K}}x' = C_{\mathcal{K}}(x - x') + D_{\mathcal{K}}w = D_{\mathcal{K}}w', y_{\mathcal{J}\backslash\mathcal{K}} - C_{\mathcal{J}\backslash\mathcal{K}}x' = D_{\mathcal{J}\backslash\mathcal{K}}w',$$

which implies that $x' \in \mathbb{X}_\mathcal{J}(y)$. Therefore, (2.19) holds. Equation (2.20) can be proved by applying Theorem 2.1. □

It is worth noticing that the computational complexity for solving (2.18) is quite high, due to the fact that one needs to enumerate all possible $\binom{m}{l}$ index sets $\mathcal{I}$ with cardinality $m - l$ in order to compute the set $\mathcal{L}$ and the matrix $\Omega^\mathcal{I}$.

One possible way to reduce the complexity is to use any point in $\mathbb{X}(y)$ as an estimate of x. To be specific, we can construct a suboptimal estimator by solving the following optimization problem:

$$\underset{\hat{x} \in \mathbb{R}^n, w, a \in \mathbb{R}^m}{\text{minimize}} \quad \|w\|_2 \tag{2.27}$$

$$\text{subject to} \quad y = C\hat{x} + Dw + a$$

$$\|w\|_2 \leq \delta, \|a\|_0 \leq l.$$

Such an estimator will not be optimal in terms of the worst-case error e^*. However, it can be easily seen that the estimation error cannot exceed the diameter of $\mathbb{X}(y)$. As a result, by Theorem 2.4, the e^* of this suboptimal estimator satisfies

$$e^* \leq 2\delta \max_{|\mathcal{K}|=m-2l} \sqrt{\sigma(P^\mathcal{K})}. \tag{2.28}$$

which is at most two times the e^* of the optimal estimator. On the other hand, one may be able to use techniques such as l_1 relaxation of l_0 norm to solve (2.27) efficiently and greatly reduce the computational complexity.

To conclude this section, we provide a numerical example to illustrate our estimator design. We assume that $n = 2$, $m = 4$ and one sensor is compromised. The noise is assumed to satisfy $\|w\|_2 \leq \delta = 1$. We further assume that

$$C = \begin{bmatrix} 1 & 0 \\ 0 & 1 \\ 1 & 1 \\ 1 & -1 \end{bmatrix}, \quad y = \begin{bmatrix} -0.851 \\ 2.753 \\ 0.5257 \\ 0 \end{bmatrix}, \quad D = I.$$

The optimal state estimate $\hat{x}$ and radius of the corresponding Chebyshev radius is given by

$$\hat{x} = \begin{bmatrix} -0.851 \\ 1.376 \end{bmatrix}, \quad r(\mathbb{X}(y)) = 1.618.$$

The set $\mathbb{X}(y)$ and the corresponding Chebyshev ball is illustrated in Figure 2.1.

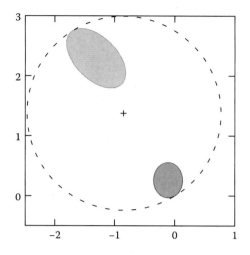

Figure 2.1: The performance of the optimal state estimator. The larger ellipse corresponds to $\mathbb{X}_{\{1,3,4\}}(y)$ and the smaller ellipse corresponds to $\mathbb{X}_{\{1,2,3\}}(y)$. The sets $\mathbb{X}_{\{2,3,4\}}(y)$ and $\mathbb{X}_{\{1,2,4\}}(y)$ are empty. The + is the optimal state estimate, while the black dashed line is the boundary of the Chebyshev ball for $\mathbb{X}(y)$.

2.3 Dynamic Estimation

2.3.1 System and Attack Model

In this subsection, we introduce the dynamic version of the static estimation problem discussed in Section 2.2.1. To this end, we assume that the evolution of the state $x(k)$ can be modeled as a linear time-invariant system, that is,

$$x(t+1) = Ax(t) + Bw(t), \qquad (2.29)$$

where $x(t) \in \mathbb{R}^n$ is the state at time t and $w(t) \in \mathbb{R}^p$ is the noise, which is assumed to be an l_∞ function, that is, $\|w(:)\|_\infty \leq \delta$. Throughout this section, we make the assumptions that the initial state $x(0) = 0$. Notice that this assumption can be easily relaxed to the case where $x(0)$ is bounded. Similar to the static case, the sensory model is given by

$$y(t) = Cx(t) + Dw(t) + a(t), \qquad (2.30)$$

where $y(t) \triangleq \begin{bmatrix} y_1(t) & \dots & y_m(t) \end{bmatrix}^T$ is the measurement from sensor 1 to m and $a(t)$ is the bias injected by the adversary. We assume that the adversary can only manipulate up to l sensors, which implies that $\|a(:)\|_0 \leq l$. We further assume that the matrix $\begin{bmatrix} B \\ D \end{bmatrix}$ is full row rank.

A dynamic estimator (predictor) is an infinite sequence of functions $f = (f_0, f_1, \ldots)$, where each f_t is a mapping from the past measurements $y(0 : t - 1)$ to the state estimate $\hat{x}(t)$. Define the estimation error at time t to be

$$e(t) \triangleq x(t) - \hat{x}(t) = x(t) - f_t(y(0 : t - 1)). \tag{2.31}$$

Clearly, the estimation error $e(t)$ defined in (2.31) depends on the noise process w, the bias a injected by the adversary, and the estimator f. As a result, we can write it as $e(w, a, f, t)$. However, we will simply write $e(t)$ when there is no confusion. In this subsection, we consider designing the estimator against the worst w and a. To this end, let us define the worst-case estimation performance as

$$e^*(f) \triangleq \sup_{\|w\|_\infty \le \delta, \|a\|_0 \le l, t} \|e(w, a, f, t)\|_\infty. \tag{2.32}$$

An estimator f is called a secure estimator if $e^*(f)$ is bounded. Similar to the static case, the problems of interests are

1. *Fundamental limit:* Under what conditions does there not exists a secure estimator?

2. *Estimator design:* Can we construct a secure estimator and analyze its performance?

The diagram of the dynamic estimation problem is illustrated in Figure 2.2.

Before continuing on, we first introduce some preliminary results on the l_1 norm of a linear system.

2.3.2 Preliminary

Consider a linear time-invariant (LTI) system:

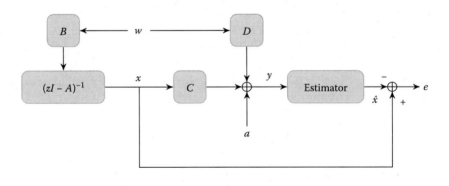

Figure 2.2: Diagram of the dynamic estimation problem in adversarial environment.
z^{-1} **is the unit delay.**

$$x(t+1) = Ax(t) + Bw(t), x(0) = 0. \tag{2.33}$$

$$y(t) = Cx(t) + Dw(t).$$

Define the following function $H : \mathbb{N} \to \mathbb{R}^{m \times p}$:

$$H(t) \triangleq \begin{cases} D & t = 0 \\ CA^{t-1}B & t \geq 1 \end{cases}. \tag{2.34}$$

Hence, $y = H * w$. With slight abuse of notation, denote

$$H \triangleq \left[\begin{array}{c|c} A & B \\ \hline C & D \end{array} \right]. \tag{2.35}$$

If A is strictly stable, H is a bounded operator with its l_∞ induced norm being [1]:

$$\|H\|_1 \triangleq \sup_{\|w(:)\|_\infty \neq 0} \frac{\|y(:)\|_\infty}{\|w(:)\|_\infty} = \max_{1 \leq i \leq n} \sum_{j=1}^{p} \sum_{t=0}^{\infty} |h_{ij}(t)|.$$

This induced norm is called the l_1 norm of the linear system H. For any $\|w(:)\|_\infty \leq \delta$, we have

$$\|y(:)\|_\infty \leq \|H\|_1 \delta. \tag{2.36}$$

Now consider the estimation problem; if (A, C) is detectable, then we know there exists a $K \in \mathbb{R}^{n \times m}$, such that $A + KC$ is strictly stable. Now consider the following linear estimator:

$$\hat{x}(t+1) = A\hat{x}(t) - K(y(t) - C\hat{x}(t)), \hat{x}(0) = 0. \tag{2.37}$$

Define the corresponding estimation error and the residue vector as

$$e(t) \triangleq x(t) - \hat{x}(t), r(t) \triangleq y(t) - C\hat{x}(t). \tag{2.38}$$

The following lemma provides bounds on $e(t)$ and $r(t)$:

Lemma 2.4
For the estimator defined in (2.37) with $A + KC$ strictly stable, the following inequalities hold:

$$\|e(:)\|_\infty \leq \|E(K)\|_1 \delta, \tag{2.39}$$

$$\|r(:)\|_\infty \leq \|G(K)\|_1 \delta, \tag{2.40}$$

where

$$E(K) \triangleq \left[\begin{array}{c|c} A+KC & B+KD \\ \hline I & 0 \end{array}\right], G(K) \triangleq \left[\begin{array}{c|c} A+KC & B+KD \\ \hline C & D \end{array}\right].$$

Proof 2.7   Manipulating (2.37), we have

$$e(t+1) = (A+KC)e(t) + (B+KD)w(t), e(0) = 0,$$

$$r(t) = Ce(t) + Dw(t).$$

Therefore, (2.39) and (2.40) can be derived from (2.36). □

The following lemma characterizes the relationship between the measurements $y(:)$ and the state $x(:)$:

Lemma 2.5
Consider system (2.33) with a detectable pair of (A,C) and $\|w(:)\|_\infty \le \delta$. If $y(0:T) = 0$, then

$$\|x(0:T)\|_\infty \le \inf_{K:A+KC \text{ strictly stable}} \|E(K)\|_1 \delta \tag{2.41}$$

Proof 2.8   The assumption that (A,C) is detectable implies the existence of K such that $A+KC$ is strictly stable. For such a stabilizing K, we construct a state estimator from (2.37). The condition $y(0:T) = 0$ implies that $\hat{x}(0:T) = 0$. Therefore, by Lemma 2.4 we have

$$\|x(0:T)\|_\infty = \|x(0:T) - \hat{x}(0:T)\|_\infty$$

$$= \|e(0:T)\|_\infty \le \|e(:)\|_\infty \le \|E(K)\|_1 \delta. \tag{2.42}$$

Since (2.42) holds for all stabilizing K, we can take the infimum over all such K and get (2.41). □

2.3.3   Fundamental Limit

This subsection is devoted to establishing a nonexistence proof of a secure estimator under certain conditions. To this end, we need the following lemma:

Lemma 2.6
There does not exist a secure estimator if there exist infinite sequences x, x', w, w', a, a', y, and y' of proper dimensions such that the following conditions hold:

1. x, a, w, y satisfy

$$x(t+1) = Ax(t) + Bw(t), x(0) = 0,$$

$$y(t) = Cx(t) + Dw(t) + a(t),$$

with $\|w(:)\|_\infty \leq \delta$ and $\|a(:)\|_0 \leq l$.

2. x', a', w', y' satisfy

$$x'(t+1) = Ax'(t) + Bw'(t), x'(0) = 0,$$

$$y'(t) = Cx'(t) + Dw'(t) + a'(t),$$

with $\|w'(:)\|_\infty \leq \delta$ and $\|a'(:)\|_0 \leq l$.

3. $y(t) = y'(t)$ for all t and $\sup_t \|x(t) - x'(t)\|_\infty = \infty$.

Proof 2.9   Let f be an arbitrary estimator. By the definition of the estimation error, we have

$$e(w, a, f, t) = x(t) - f_t(y(0:t-1)), e(w', a', f, t) = x'(t) - f_t(y'(0:t-1)).$$

Since $y(t) = y'(t)$ for all t, we know that $f_t(y(0:t-1)) = f_t(y'(0:t-1))$. Therefore,

$$e(w, a, f, t) - e(w', a', f, t) = x(t) - x'(t).$$

By triangular inequality,

$$\|e(w, a, f, t)\|_\infty + \|e(w', a', f, t)\|_\infty \geq \|x(t) - x'(t)\|_\infty.$$

Since $\sup_t \|x(t) - x'(t)\|_\infty = \infty$, at least one of the following equalities holds:

$$\sup_t \|e(w, a, f, t)\|_\infty = \infty, \sup_t \|e(w', a', f, t)\|_\infty = \infty.$$

Therefore, by the definition of $e^*(f)$, we know that $e^*(f) = \infty$ for all f, which implies the nonexistence of a secure estimator.   □

We are now ready to state the main theorem on the nonexistence of a secure dynamic estimator:

Theorem 2.5

 There does not exist a secure dynamic estimator if $(A, C_\mathcal{K})$ is not detectable for some set $\mathcal{K} \subset \mathcal{S}$ with cardinality $|\mathcal{K}| = m - 2l$.

Proof 2.10   Let $\mathcal{K}$ be a subset of $\mathcal{S}$ with cardinality $m - 2l$ such that $(A, C_\mathcal{K})$ is not detectable. We can find two index sets $\mathcal{K}_1, \mathcal{K}_2 \subset \mathcal{S} \backslash \mathcal{K}$ such that

$$|\mathcal{K}_1| = |\mathcal{K}_2| = l, \mathcal{K} \cup \mathcal{K}_1 \cup \mathcal{K}_2 = \mathcal{S}.$$

Since $(A, C_{\mathcal{K}})$ is not detectable, there exists an unstable and unobservable eigenvector $z \in \mathbb{C}^n$ and eigenvalue $\lambda \in \mathbb{C}$ of the matrix A such that

$$Az = \lambda z, \, C_{\mathcal{K}}z = 0, \, |\lambda| \geq 1.$$

Since we assume that $\begin{bmatrix} B \\ D \end{bmatrix}$ is full row rank, we can find $w_* \in \mathbb{C}^n$, such that

$$\begin{bmatrix} B \\ D \end{bmatrix} w_* = \begin{bmatrix} z \\ 0 \end{bmatrix}.$$

Without loss of generality, we can scale z such that each entry of w_* has its absolute value to be no greater than δ, that is, $|w_{*,i}| \leq \delta$ for all i. Now let us consider the following noise process:

$$w(t) = \mathfrak{Re}\left(\frac{\lambda^t}{|\lambda|^t} w_*\right).$$

One can verify that

$$\|w(:)\|_\infty \leq \sup_{t,i} \left| \frac{\lambda^t}{|\lambda|^t} \times w_{*,i} \right| \leq \delta.$$

Since A is real, the corresponding sequence of the state x generated by w is given by

$$x(t) = \begin{cases} 0 & t = 0 \\ \mathfrak{Re}\left[\lambda^{t-1}\left(1 + \frac{1}{|\lambda|} + \cdots + \frac{1}{|\lambda|^{t-1}}\right)z\right] & t > 0 \end{cases}.$$

One can verify that $\|x(:)\|_\infty = \infty$ for both of the cases where $|\lambda| > 1$ and $|\lambda| = 1$. Now by the fact that C, D matrices are real, we have

$$C_{\mathcal{K}}x(t) + D_{\mathcal{K}}w(t) = 0.$$

In order to use Lemma 2.6 to give a nonexistence proof, we need to construct sequences a, y, x', w', y', and a'. First, let us define a as

$$a_i(t) = \begin{cases} 0 & i \in \mathcal{K} \cup \mathcal{K}_2 \\ -C_i x(t) & i \in \mathcal{K}_1 \end{cases}.$$

Clearly $\|a(:)\|_0 \leq |\mathcal{K}_1| = l$. The corresponding $y(t)$ satisfies

$$y_i(t) = \begin{cases} 0 & i \in \mathcal{K} \cup \mathcal{K}_1 \\ C_i x(t) & i \in \mathcal{K}_2 \end{cases}.$$

Now let us consider another sequence of noise $w' = 0$. Therefore, the corresponding $x' = 0$. Let us construct the bias a' injected by the adversary as

$$a_i'(t) = \begin{cases} 0 & i \in \mathcal{K} \cup \mathcal{K}_1 \\ C_i x(t) & i \in \mathcal{K}_2 \end{cases}.$$

One can verify that $\|a'(:)\|_0 \le |\mathcal{K}_2| = l$. The corresponding $y'(t)$ satisfies

$$y_i'(t) = \begin{cases} 0 & i \in \mathcal{K} \cup \mathcal{K}_1 \\ C_i x(t) & i \in \mathcal{K}_2 \end{cases}.$$

Therefore, we have $y = y'$ and $\sup_t \|x(t) - x'(t)\|_\infty = \infty$, which implies the nonexistence of any secure estimator by Lemma 2.6. □

2.3.4 Secure Estimator Design and Performance Analysis

This subsection is devoted to the derivation of a secure estimator with performance guarantees. By Theorem 2.5, we know that the following assumption is necessary for the existence of the secure estimator:

Assumption 2.2 $(A, C_\mathcal{K})$ is detectable for any $\mathcal{K} \subset S$ with cardinality $m - 2l$.

Therefore, we will assume that Assumption 2.2 holds for the rest of this subsection. Let $\mathcal{I} = \{i_1, \cdots, i_{m-l}\} \subset S$ be an index set with cardinality $n - l$. Denote the collection of all such index sets as

$$\mathcal{L}(-1) \triangleq \{\mathcal{I} \subset S : |\mathcal{I}| = m - l\}.$$

For any $\mathcal{I} \in \mathcal{L}(-1)$, Assumption 2.2 implies the existence of a $K^\mathcal{I}$ such that $A + K^\mathcal{I} C_\mathcal{I}$ is strictly stable. Therefore, we can construct a stable "local" estimator, which only uses the truncated measurement $y_\mathcal{I}(t)$ to compute the state estimate

$$\hat{x}^\mathcal{I}(t+1) = A\hat{x}^\mathcal{I}(t) - K^\mathcal{I}(y_\mathcal{I}(t) - C_\mathcal{I}\hat{x}^\mathcal{I}(t)), \qquad (2.43)$$

with initial condition $\hat{x}^\mathcal{I}(0) = 0$. For each local estimator, let us define the corresponding error and residue vector as

$$e^\mathcal{I}(t) \triangleq x(t) - \hat{x}^\mathcal{I}(t), \, r^\mathcal{I}(t) \triangleq y_\mathcal{I}(t) - C_\mathcal{I}\hat{x}^\mathcal{I}(t), \qquad (2.44)$$

and the corresponding linear operators as follows:

$$E^\mathcal{I}(K^\mathcal{I}) \triangleq \left[\begin{array}{c|c} A + K^\mathcal{I}C_\mathcal{I} & B + K^\mathcal{I}D_\mathcal{I} \\ \hline I & 0 \end{array} \right], \qquad (2.45)$$

$$G^\mathcal{I}(K^\mathcal{I}) \triangleq \left[\begin{array}{c|c} A + K^\mathcal{I}C_\mathcal{I} & B + K^\mathcal{I}D_\mathcal{I} \\ \hline C_\mathcal{I} & D_\mathcal{I} \end{array} \right]. \qquad (2.46)$$

By Lemma 2.4, we know that if $a_\mathcal{I}(t) = 0$ for all t, that is, if $\mathcal{I}$ does not contain any compromised sensors, then the following inequality holds:

$$\|r^\mathcal{I}(:)\|_\infty \le \|G^\mathcal{I}(K^\mathcal{I})\|_1 \delta. \qquad (2.47)$$

As a result, we will assign each local estimator a detector, which checks if the following inequality holds at each time t:

$$\|r^{\mathcal{I}}(0:t)\|_\infty \leq \|G^{\mathcal{I}}(K^{\mathcal{I}})\|_1 \delta. \tag{2.48}$$

If (2.48) fails to hold, then we know the set $\mathcal{I}$ contains at least one compromised sensor, and hence the local estimate $\hat{x}^{\mathcal{I}}(t)$ is corrupted by the adversary. On the other hand, we call $\hat{x}^{\mathcal{I}}(t)$ an *admissible* local estimate at time t if (2.48) holds at time t. We further define the set $\mathcal{L}(t)$ of all admissible index sets $\mathcal{I}$ at time t as

$$\mathcal{L}(t) \triangleq \{\mathcal{I} \in \mathcal{S} : (2.48) \text{ holds at time } t\}. \tag{2.49}$$

Remark 2.2 Notice that (2.48) is only a necessary condition for the index set $\mathcal{I}$ to not contain compromised sensors. Therefore, one can potentially design better detectors to check if there exist compromised sensors in the index set $\mathcal{I}$ to provide better performance. However, the detector based on (2.48) is sufficient for us to design a secure estimator.

We will then fuse all the *admissible* local estimations $\hat{x}^{\mathcal{I}}(t)$ at time t to generate the state estimate $\hat{x}(t)$. Since we are concerned with the infinite norm of the estimation error, the following equation will be used to compute each entry of $\hat{x}(t)$:

$$\hat{x}_i(t) = \frac{1}{2}\left(\min_{\mathcal{I}\in\mathcal{L}(t)} \hat{x}_i^{\mathcal{I}}(t) + \max_{\mathcal{I}\in\mathcal{L}(t)} \hat{x}_i^{\mathcal{I}}(t)\right). \tag{2.50}$$

We now provide an upper bound on the worst-case performance $e^*(f)$ for our estimator design, which is given by the following theorem:

Theorem 2.6
Under Assumption 2.2, the state estimator (2.50) is a secure estimator. Furthermore, the following inequality on $e^*(f)$ holds:

$$e^*(f) \leq \max_{\mathcal{I},\mathcal{J}\in\mathcal{L}}\left(\|E^{\mathcal{I}}(K^{\mathcal{I}})\|_1 + \frac{1}{2}\alpha^{\mathcal{I}\cap\mathcal{J}}[\beta^{\mathcal{I}}(K^{\mathcal{I}}) + \beta^{\mathcal{J}}(K^{\mathcal{J}})]\right)\delta, \tag{2.51}$$

where $\alpha^{\mathcal{K}}$ is defined as

$$\alpha^{\mathcal{K}} \triangleq \inf_{K:A+KC_{\mathcal{K}} \text{ strictly stable}} \left\| \begin{bmatrix} A+KC_{\mathcal{K}} & \begin{bmatrix} I & K \end{bmatrix} \\ I & 0 \end{bmatrix} \right\|_1,$$

and $\beta^{\mathcal{I}}(K^{\mathcal{I}})$ is defined as

$$\beta^{\mathcal{I}}(K^{\mathcal{I}}) \triangleq \max(\|K^{\mathcal{I}}\|_\infty, 1)\|G^{\mathcal{I}}(K^{\mathcal{I}})\|_1.$$

Remark 2.3 It is worth noticing that if $\mathcal{I}$ does not contain compromised sensors, then minimizing the infinite norm of the local estimation error $e^{\mathcal{I}}(t)$ is equivalent to minimizing $\|E^{\mathcal{I}}(K^{\mathcal{I}})\|_1$. The second term on the right-hand side (RHS) of (2.51) exists since the estimator does not know which local estimate can be trusted at the beginning.

Several intermediate results are needed before proving Theorem 2.6. We first prove the following lemma to bound the divergence of the local estimates:

Lemma 2.7
For any two index sets $\mathcal{I}, \mathcal{J} \in \mathcal{L}(T)$, the following inequality holds:

$$\|\hat{x}^{\mathcal{I}}(0:T) - \hat{x}^{\mathcal{J}}(0:T)\|_\infty \le \alpha^{\mathcal{I} \cap \mathcal{J}}[\beta^{\mathcal{I}}(K^{\mathcal{I}}) + \beta^{\mathcal{J}}(K^{\mathcal{J}})]\delta. \qquad (2.52)$$

Proof 2.11 By (2.43), we have

$$\hat{x}^{\mathcal{I}}(t+1) = A\hat{x}^{\mathcal{I}}(t) - K^{\mathcal{I}}r^{\mathcal{I}}(t), \hat{x}^{\mathcal{I}}(t) = 0,$$

$$y_{\mathcal{I}}(t) = C_{\mathcal{I}}\hat{x}^{\mathcal{I}}(t) + r^{\mathcal{I}}(t). \qquad (2.53)$$

Let us define $\mathcal{K} = \mathcal{I} \cap \mathcal{J}$; we know that

$$y_{\mathcal{K}}(t) = C_{\mathcal{K}}\hat{x}^{\mathcal{I}}(t) + P_{\mathcal{K},\mathcal{I}}r^{\mathcal{I}}(t),$$

where $P_{\mathcal{K},\mathcal{I}} \in \mathbb{R}^{|\mathcal{K}| \times |\mathcal{I}|}$ is the unique matrix that satisfies:

$$y_{\mathcal{K}} = P_{\mathcal{K},\mathcal{I}}y_{\mathcal{I}}, \forall y \in \mathbb{R}^m.$$

Now let us define $\phi^{\mathcal{I}}(t) \triangleq -K^{\mathcal{I}}r^{\mathcal{I}}(t)$ and $\varphi^{\mathcal{I}}(t) \triangleq P_{\mathcal{K},\mathcal{I}}r^{\mathcal{I}}(t)$. If $t \le T$, then by (2.48), we have

$$\|r^{\mathcal{I}}(t)\|_\infty \le \|G^{\mathcal{I}}(K^{\mathcal{I}})\|_1\delta.$$

As a result,

$$\left\| \begin{bmatrix} \phi^{\mathcal{I}}(t) \\ [3pt]\varphi^{\mathcal{I}}(t) \end{bmatrix} \right\|_\infty \le \left\| \begin{bmatrix} -K^{\mathcal{I}} \\ P_{\mathcal{K},\mathcal{I}} \end{bmatrix} \right\|_\infty \times \|r^{\mathcal{I}}(t)\|_\infty = \beta^{\mathcal{I}}(K^{\mathcal{I}})\delta,$$

where we use the fact that each row of $P_{\mathcal{K},\mathcal{I}}$ is a canonical basis vector in $\mathbb{R}^{|\mathcal{I}|}$. Therefore, we have

$$\hat{x}^{\mathcal{I}}(t+1) = A\hat{x}^{\mathcal{I}}(t) + \begin{bmatrix} I & 0 \end{bmatrix} \begin{bmatrix} \phi^{\mathcal{I}}(t) \\ \varphi^{\mathcal{I}}(t) \end{bmatrix}, \hat{x}^{\mathcal{I}}(t) = 0,$$

$$y_{\mathcal{K}}(t) = C_{\mathcal{K}}\hat{x}^{\mathcal{I}}(t) + \begin{bmatrix} 0 & I \end{bmatrix} \begin{bmatrix} \phi^{\mathcal{I}}(t) \\ \varphi^{\mathcal{I}}(t) \end{bmatrix}. \qquad (2.54)$$

Similarly, the following equations hold:

$$\hat{x}^{\mathcal{J}}(t+1) = A\hat{x}^{\mathcal{J}}(t) + \begin{bmatrix} I & 0 \end{bmatrix} \begin{bmatrix} \phi^{\mathcal{J}}(t) \\ \varphi^{\mathcal{J}}(t) \end{bmatrix}, \hat{x}^{\mathcal{J}}(t) = 0,$$

$$y_{\mathcal{K}}(t) = C_{\mathcal{K}}\hat{x}^{\mathcal{J}}(t) + \begin{bmatrix} 0 & I \end{bmatrix} \begin{bmatrix} \phi^{\mathcal{J}}(t) \\ \varphi^{\mathcal{J}}(t) \end{bmatrix}, \tag{2.55}$$

with

$$\left\| \begin{bmatrix} \phi^{\mathcal{J}}(t) \\ \varphi^{\mathcal{J}}(t) \end{bmatrix} \right\| \leq \beta^{\mathcal{J}}(K^{\mathcal{J}})\delta, \forall t \leq T.$$

Now let us consider $\Delta\hat{x}(t) = \hat{x}^{I}(t) - \hat{x}^{\mathcal{J}}(t)$. By (2.54) and (2.55), we know that

$$\Delta\hat{x}(t+1) = A\Delta\hat{x}(t) + \begin{bmatrix} I & 0 \end{bmatrix} \begin{bmatrix} \phi^{\mathcal{I}}(t) - \phi^{\mathcal{J}}(t) \\ \varphi^{\mathcal{I}}(t) - \varphi^{\mathcal{J}}(t) \end{bmatrix}, \Delta\hat{x}(t) = 0,$$

$$0 = C_{\mathcal{K}}\Delta\hat{x}(t) + \begin{bmatrix} 0 & I \end{bmatrix} \begin{bmatrix} \phi^{\mathcal{I}}(t) - \phi^{\mathcal{J}}(t) \\ \varphi^{\mathcal{I}}(t) - \varphi^{\mathcal{J}}(t) \end{bmatrix}.$$

Now, Assumption 2.2 implies that $(A, C_{\mathcal{K}})$ is detectable. Hence, (2.52) can be proved by Lemma 2.5. □

Lemma 2.8
Let $r_1, \ldots, r_l$ be real numbers. Define

$$r = \frac{1}{2}\left(\max_i r_i + \min_i r_i \right).$$

Then for any i, we have

$$|r - r_i| \leq \frac{1}{2}\max_j |r_j - r_i|. \tag{2.56}$$

Proof 2.12 Without loss of generality, we assume that r_1 and r_2 are the largest and the smallest number among all r_i's, respectively. Therefore,

$$r - r_i = \frac{1}{2}(r_1 - r_i) - \frac{1}{2}(r_i - r_2).$$

Therefore, if $r_1 - r_i \geq r_i - r_2$, then

$$|r - r_i| = r - r_i \leq \frac{1}{2}(r_1 - r_i) = \frac{1}{2}\max_j |r_j - r_i|.$$

Similarly, one can prove that (2.56) holds when $r_1 - r_i < r_i - r_2$. □

Now we are ready to prove Theorem 2.6.

Proof 2.13 Let $\mathcal{G} \in \mathcal{L}$ be the true set of good sensors. By Lemma 2.4, we know that

$$\|x(t) - \hat{x}^{\mathcal{G}}(t)\|_\infty \leq \|E^{\mathcal{G}}(K^{\mathcal{G}})\|_1 \delta.$$

Furthermore, $\mathcal{G}$ must be admissible at all time, that is, $\mathcal{G} \in \mathcal{L}(t)$ for all t. At any given time t, assume that the index set $\mathcal{J}$ also belongs to $\mathcal{L}(t)$. By Lemma 2.7, we have

$$\|\hat{x}^{\mathcal{G}}(t) - \hat{x}^{\mathcal{J}}(t)\|_\infty \leq \alpha^{\mathcal{G} \cap \mathcal{J}} [\beta^{\mathcal{G}}(K^{\mathcal{G}}) + \beta^{\mathcal{J}}(K^{\mathcal{J}})] \delta.$$

Therefore, by Lemma 2.8, we know that

$$\|\hat{x}(t) - \hat{x}^{\mathcal{G}}(t)\|_\infty \leq \frac{1}{2} \max_{\mathcal{J} \in \mathcal{L}} \alpha^{\mathcal{G} \cap \mathcal{J}} [\beta^{\mathcal{G}}(K^{\mathcal{G}}) + \beta^{\mathcal{J}}(K^{\mathcal{J}})] \delta.$$

By triangular inequality, we have

$$\|e(t)\|_\infty \leq \|E^{\mathcal{G}}(K^{\mathcal{G}})\|_1 \delta + \frac{1}{2} \max_{\mathcal{J} \in \mathcal{L}} \alpha^{\mathcal{G} \cap \mathcal{J}} [\beta^{\mathcal{G}}(K^{\mathcal{G}}) + \beta^{\mathcal{J}}(K^{\mathcal{J}})] \delta. \qquad (2.57)$$

Thus, by taking the supremum over all possible "good" sensor sets $\mathcal{G}$, we can prove (2.51). □

Combining Theorems 2.5 and Theorem 2.6, we have the following corollary:

Corollary 2.1 A necessary and sufficient condition for the existence of a secure estimator is that $(A, C_\mathcal{K})$ is detectable for any index set $\mathcal{K} \subset \mathcal{S}$ with cardinality $2l$.

Finally, we provide a numerical example to illustrate our secure estimator design. We choose the following parameters for the system:

$$A = 1, C = \begin{bmatrix} 1 \\ 1 \\ 1 \end{bmatrix}, B = \begin{bmatrix} 1 & 0 & 0 & 0 \end{bmatrix}, D = \begin{bmatrix} 0 & I \end{bmatrix}.$$

We further assume that $\delta = 1$ and $l = 1$. For our secure estimator design, we need to compute the gains $K^{\{1,2\}}$, $K^{\{2,3\}}$, $K^{\{3,1\}}$. Due to symmetry, we will only consider the gain of the following form:

$$K^{\{1,2\}} = K^{\{2,3\}} = K^{\{3,1\}} = \begin{bmatrix} \mu & \mu \end{bmatrix}.$$

One can check that $\alpha^{\mathcal{K}} = 2$ for $\mathcal{K} = \{1\}, \{2\}, \{3\}$ and

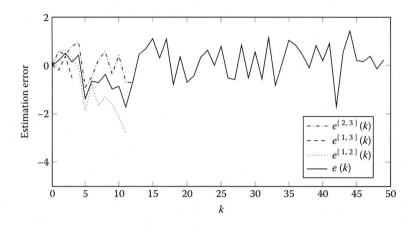

Figure 2.3: The estimation error of the secure estimator. The dash-dotted, dashed, and dotted lines correspond to the estimation errors for three local estimators, while the solid line is the estimation error after fusion. The dashed and dotted lines terminate at time 4 and time 12, respectively, when the corresponding detector detects a violation of (2.48). Hence, the dash-dotted and the solid lines coincide starting from time 13.

$$\|E^{\mathcal{I}}(K^{\mathcal{I}})\|_1 = \frac{1+|2\mu|}{1-|1+2\mu|}, \|G^{\mathcal{I}}(K^{\mathcal{I}})\| = 1 + \frac{1+|2\mu|}{1-|1+2\mu|},$$

where $\mu \in (-1,0)$ to ensure the stability of the local estimator. As a result,

$$\beta^{\mathcal{I}}(K^{\mathcal{I}}) = 1 + \frac{1+|2\mu|}{1-|1+2\mu|}.$$

Hence, the upper bound on the worst-case estimation error is

$$e^*(f) \leq 2 + 3 \times \frac{1+|2\mu|}{1-|1+2\mu|}. \tag{2.58}$$

The optimal μ that minimizes the RHS of (2.58) is $\mu = -0.5$, and the corresponding upper bound is 8.

Figure 2.3 illustrates the performance of the secure estimator. The noise $w(k)$ is randomly generated from a uniform distribution on the set $\|w(k)\|_\infty \leq 1$. We assume that the adversary compromises the first sensor and adds an increasing bias $a(t) = \begin{bmatrix} 0.5t & 0 & 0 \end{bmatrix}^T$. The trajectory of the estimation error of the secure estimator is plotted in Figure 2.3.

One can see that our secure estimator will detect that the index sets $\{1,2\}$ and $\{1,3\}$ contain the compromised sensor and hence discard the corresponding local estimates. As a result, the estimation error remains bounded.

2.4 Summary

■ This chapter is concerned with estimating the state x $(x(t))$ from sensory data y $(y(t))$, where we assume that the sensors measure a linear function of the state, subject to bounded noise w $(w(t))$ and sparse (potentially unbounded) bias vector a $(a(t))$ injected by the adversary. The estimate is either done as a single snapshot (static estimation problem) or sequentially (dynamic estimation problem).

■ An estimator is called a secure estimator if its estimation error is bounded regardless of the true state, noise vector, and bias vector.

■ The fundamental problems of interests in secure estimation are (1) whether a secure estimator exists and (2) how to design a secure estimator and analyze its performance.

■ For the static estimation problem, a secure estimator exists if and only if after removing an arbitrary set of $2l$ sensors, the corresponding $C_{\mathcal{K}}$ matrix is full column rank.

■ For the dynamic estimation problem, a secure estimator exists if and only if after removing an arbitrary set of $2l$ sensors, the corresponding $(A, C_{\mathcal{K}})$ is detectable.

■ The optimal static estimator is the Chebyshev center of $\mathbb{X}(y)$, which is a union of ellipsoids. The Chebyshev center can be computed as the solution of a semidefinite programming problem.

■ The proposed secure dynamic estimator design in this chapter consists of (1) local estimators (2.43), (2) local detectors (2.48), and (3) global fusion (2.50).

References

1. M. A. Dahleh and I. J. Diaz-Bobillo. *Control of Uncertain Systems: A Linear Programming Approach.* Prentice-Hall, Upper Saddle River, NJ, 1994.

2. L. Danzer, B. Grünbaum, and V. Klee. Helly's theorem and its relatives. In: *Proceedings of Symposia in Pure Mathematics: Convexity.* American Mathematical Society, Providence, RI, 1963.

3. E. Yildirim. On the minimum volume covering ellipsoid of ellipsoids. *SIAM Journal on Optimization*, 17(3):621–641, 2006.

Chapter 3

Secure State Estimation in Industrial Control Systems

Arash Mohammadi

Department of Electrical and Computer Engineering, University of Toronto

Konstantinos N. Plataniotis

Department of Electrical and Computer Engineering, University of Toronto

CONTENTS

Motivated by recent evolution of cutting-edge sensor technologies with complex-valued measurements, this chapter analyzes attack models and diagnostic solutions for monitoring industrial control systems against complex-valued cyber attacks. By capitalizing on the knowledge that the existing detection and closed-loop estimation algorithms ignore the full second-order statistical properties of the received measurements, we show that an adversary can attack the system by maximizing the correlations between the real and imaginary parts of the reported measurements. Consequently, the adversary can pass the conventional attack detection methodologies and change the underlying system beyond repair. In the rest of the chapter, the first section surveys recent developments in secure closed-loop state estimation methodologies, and then reviews the fundamentals of complex-valued signals and their applications. The second section highlights the drawbacks of the state-of-the-art estimation methodologies and illustrates their vulnerability to cyber attacks. In the third section, we first review the existing attack models and then introduce the noncircular attack model. The fourth section surveys the state-of-the-art attack detection diagnostics and shows how to transform the cyber-attack detection problem into a problem of comparing statistical distance measures between probability distributions. The fifth section provides illustrative examples, followed by future research directions and conclusions.

3.1 Introduction

Recent advancements in communication and sensor technologies have paved the way for deployment of a large number of sensor nodes in industrial control systems [1, 2], resulting in an exceptional growth of such practical implementations and opportunistic applications of such systems. The rapid growth of industrial control systems and the fact that their applications are typically safety critical have increased a recent surge of interest in investigating their security issues [3–9]. Potential cyber attacks could lead to a variety of severe consequences, such as destruction of critical infrastructures and endangerment of human lives. This makes identification, detection, and prevention of cyber attacks of significant practical importance. This is the focus of the chapter.

In particular and motivated by recent evolution of cutting-edge sensor technologies with complex-valued measurements in industrial control systems [10, 11], the chapter investigates and exposes complex-valued data injection attacks in this context. Due to the transformation in sensor technologies, complex-valued signal processing techniques [12–15] have found several applications of practical engineering importance [16]. However, despite the fact that complex-valued signals are ubiquitous in several real-world control applications, their vulnerabilities to cyber attacks have not yet been fully investigated. The emerging growth of complex-valued signals in industrial control systems, on the one hand, and the absence of proper diagnostic solutions, on the other hand, calls for an urgent quest to identify potential complex-valued attacks and devise reliable detection diagnostics against them. By capitalizing on the knowledge that the existing estimation algorithms do not incorporate the full second-order statistics of the received measurements, an adversary can attack the system by maximizing the correlations between real and imaginary parts of the reported measurements, that is, making the reported measurements highly noncircular. In contrary to this imminent threat, all the existing attack models and monitoring and detection solutions are restricted to real-valued signals and have not yet been developed for complex-valued statistics. It is therefore essential and critical to identify possible vulnerabilities and design smart solutions to detect potential complex-valued attacks. The rest of the chapter further opens this compelling research field, but first we review recent developments in this context.

3.1.1 Classification of Cyber Attacks

In general, the existing methodologies proposed to solve security problems in industrial control systems can be classified as follows:

1. *Information security:* The primarily focus of the methodologies belonging to this category is to ensure the integrity and availability of information in control systems; therefore, they are mainly focused on problems such as encryption, data security, and time synchronization [17]. Such approaches are, however, typically implemented without using the physical model of the underlying system. This drawback limits the ability of information security to predict and prevent attacks on the physical system. From an information security point of view, for instance, a reasonable solution is to disconnect the compromised sensors; however, if implemented poorly, this could result in the instability of the entire system.

2. *Control system security:* The main focus of approaches belonging to this category is to analyze and identify different types of failures and design robust algorithms that can withstand such attacks. The classical examples

of this category, such as robust estimation and control and fault detection algorithms, typically assume the failures to be random; however, a smart attack is deliberately designed based on the vulnerabilities of the underlying physical system. To design countermeasures against such attacks, the classical robust estimation and control algorithms require reexaminations.

In this chapter, we focus on the second category and consider an abstract model of an industrial control system consisting of the physical model and the control unit, as shown in Figure 3.1. The latter typically consists of two main blocks: (1) a feedback estimation mechanism that is used to dynamically track the state of the physical system, and (2) a control algorithm that computes the control commands based on the current state estimate of the system. Attacks on the system can be generally classified as physical attacks, where the attack is implemented directly on the plant itself (e.g., A5 in Figure 3.1), and cyber attacks (e.g., A1–A4 in Figure 3.1). Although the results of a physical attack could be more devastating, cyber attacks are more attractive to a risk-averse adversary as they are less detectable, less physically dangerous, and can be implemented remotely.

One can further classify cyber attacks based on the following three parameters [4]: (1) *a priori system model knowledge*, where the adversary has access to the parameters of the system; (2) *disclosure power*, where the attacker has access to real-time state information, and (3) *disruption resources*, where the attacker has the capability to disturb the system. For example, to implement a DoS attack [18], only the third parameter is required, while to design a deception attack, typically two or all of these parameters are needed. The focus of the chapter is on the deception category of cyber attacks, as shown in Figure 3.2, which can be roughly classified into the following categories:

■ *Replay attacks*, which are implemented by hijacking sensor measurements, record their readings for a certain period of time and repeat such readings while injecting the attack signal into the physical system.

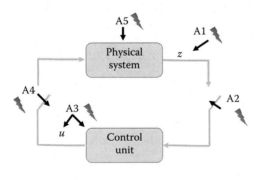

Figure 3.1: Abstract model of an industrial control systems.

■ *Covert attacks* are closed-loop replay attacks where the output attack is chosen to cancel out the effect on the measurements of the state attack.

■ *Complex-valued attacks*, which are the focus of this chapter, are designed by maximizing the correlations between real and imaginary parts of the reported complex-valued measurements.

Classification of Cyber Attacks

Cyber attacks on the control unit can be classified into the following two broad categories:

1. *Denial-of-service (DoS) attacks*, where the adversary prevents the control unit from receiving sensor readings. Such attacks can be accomplished, for example, by jamming the communication channels, compromising the observation nodes, or attacking the routing protocols. For example, in Figure 3.1, the adversary prevents the controller from receiving sensor measurements (A2) or prevents the control commands from reaching the physical system (A4).

2. *Deception attacks (false data injection)*, where the adversary compromises the integrity of sensor measurements or control commands. Such attacks can be accomplished, for example, by sending false information from one or more sensors to the control unit. In Figure 3.1, for instance, the adversary can launch deception attacks by compromising some sensors (A1) or controllers (A3).

As a motivating example for the last category, let us consider utilization of complex-valued measurements in state estimation problems of the next-generation power grids. Incorporation of phasor measurements from evolving smart metering devices, the phasor measurement units (PMUs) [10, 11], offers a new paradigm for state estimation in a power grid's monitoring system. As shown in Figure 3.3, PMUs report complex numbers representing the magnitude and phase angle of voltage and current waveforms and are considered to be the measurement technology that makes the dream of next-generation power systems, smart grids, a reality. To ensure security of smart grids with new meter technologies, recently there has been a surge of interest in identifying and preventing false data injection attacks [6], for example, in the context of static state estimation, [8] and [9] introduced false data injection attacks and showed that such attacks can efficiently bypass the existing residue-based bad data detection algorithms and modify the state estimates in a predicted manner. In the context of dynamic state estimation, for example, [19] considered false data injection attacks and proposed a Kalman filter (KF) estimator by forming vector from real and imaginary parts of the voltage phasors; that is, the possible correlations

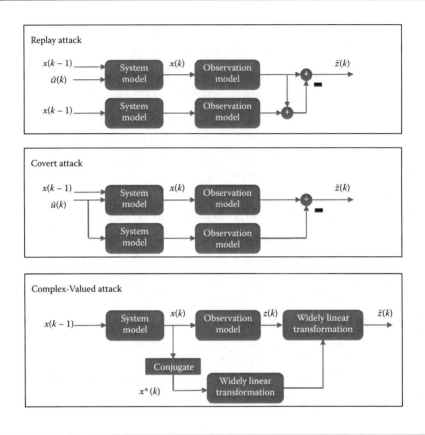

Figure 3.2: Deception attack models, where k denotes the iteration index, $x(\cdot)$ is the state vector, $\tilde{u}(\cdot)$ is the attack vector, and $\tilde{z}(\cdot)$ is the attacked measurement.

between the real and imaginary parts of the signal are not considered. This is a common trend in industrial control systems where the attack models and monitoring and detection solutions [3–7] are limited to real-valued statistics.

As complex-valued measurements play a critical role for effective state estimation and control of industrial control systems, investigating their vulnerabilities becomes of paramount importance; therefore, the chapter investigates and exposes complex-valued data injection attacks as briefly outlined below:

1. *Complex-valued attack model:* Our first objective is to investigate drawbacks of the existing state-of-the-art estimation and control methodologies commonly used in the industrial control systems, and then identify their potential vulnerabilities to cyber attacks. In particular, our objective is to investigate the vulnerability of the industrial control systems in scenarios where the measurements are complex-valued. After reviewing special statistical properties of complex-valued signals, we show how

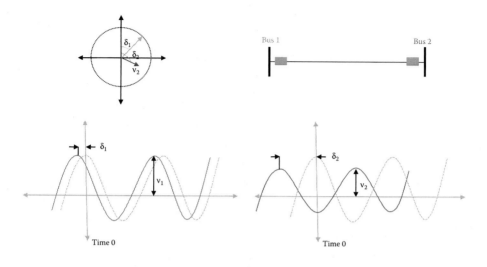

Figure 3.3: Phasor representation of power grid. Top left figure shows phasor representation, while bottom plots show corresponding sinusoidal waveforms.

an adversary can utilize these properties to manipulate the system. More specifically, we develop a complex-valued attack model that modifies the measurements in such a way that the attacked innovation sequence becomes second-order noncircular. In other words, the attacked innovation sequence has the same conventional covariance matrix as the original sequence, but its pseudocovariance matrix is no longer zero. As the convectional diagnostic is designed based on a specific model that ignores the full second-order statistics of the measurements (pseudocovariances), this attack will remain undetected.

2. *Detection via statistical distance measure:* Our second objective is to analyze properties of the existing state-of-the-art attack detection and diagnostic solutions. In particular, we show that the conventional metrics used for attack detection in industrial control systems use a predefined set of statistics corresponding to the innovation sequence (some of which are precalculated). The chapter then argues that relying on the statistics of the innovation sequence instead of using the whole distribution is structurally destructive. Intuitively speaking, the probability distribution of the innovation sequence is the sole source that contains all the required information for diagnosing and monitoring the system; therefore, the detection criterion should be defined in terms of the distribution itself and not its precalculated statistics. Capitalizing on this intuition, we propose to monitor the statistical distance between the probability distribution of the innovation sequence under attack against the distribution of the innovation sequence

with an optimal profile. Essentially, attack detection becomes the problem of comparing the distance between two probability distributions, which increases the deepness of diagnostics and, therefore, increases the robustness of the control process in the presence of cyber attacks.

Next, we review fundamentals of complex-Valued signals.

3.2 Complex-Valued Signals

A complex-valued signal is typically characterized by its second-order statistical properties represented by the conventional covariance matrix and the complementary covariance matrix (pseudocovariance). While the former captures the information corresponding to the total power of the signal, the latter characterizes the correlations between the real and imaginary parts of the signal. Traditionally, complex-valued processing algorithms ignore the complementary covariance matrix and solely consider the conventional covariance matrix, making them applicable only to proper (circular) signals that are characterized by a vanishing complementary covariance matrix. On the contrary, several real-world signals are improper and do not have statistical properties similar to real signals. A complex-valued signal is improper (noncircular) as long as it is correlated with its complex conjugate; that is, the complementary covariance matrix (pseudocovariance) does not vanish. Noncircularity of the received observations in control systems could originate from either the modulation scheme or the coding method. Several important digital modulation schemes [28–30] produce noncircular complex signals under various circumstances; for example, noncircular signal constellations are binary phase-shift keying (BPSK), offset quadrature phase-shift keying (OQPSK), minimum-shift keying (MSK), *M*-array amplitude-shift keying (ASK), and simplex signals in general. The space-time block coding (STBC) scheme [31] is an example that results in noncircularity of the received observations. Finally, imbalance between the in-phase and quadrature components of a communicated signal can also result in noncircularity.

Intuitively speaking, for an improper signal, the processing model should depend on both the signal itself and its complex conjugate. Such a methodology is referred to as widely linear processing [20] and has been utilized to develop several estimation [21], filtering [22], and detection [28] algorithms. Widely linear processing can potentially improve the performance of the estimation algorithm when the complementary covariance matrix exists and can be exploited. In the context of feedback state estimation, complex-valued KFs [23] were initially developed by inherently assuming that the state and observation noises as well as the initial conditions are circular. Recently [24–26], however, there has been a surge of interest in developing widely linear implementations of the KF that fully utilize the complete available second-order statistics of the complex signals. For improper, complex-valued, and nonlinear dynamical systems, widely linear processing has also been used to develop augmented complex extended Kalman filters (EKFs) [24], complex-valued unscented Kalman filters [25], and augmented complex particle filters [26]. Below and before talking about widely linear models and augmented KFs, first we briefly outline the notions and then review properties of complex Gaussian distributions.

3.2.1 Notation

Throughout the chapter, the following notations are used: nonbold letter z denotes a scalar variable, lowercase bold letter $\mathbf{z}$ represents a vector, and capital bold letter $\mathbf{Z}$ denotes a matrix. The real and complex domains are represented by $\mathbb{R}$ and $\mathbb{C}$, while $\Re\{z\}$ and $\Im\{z\}$ represent the real and imaginary parts of z. The complex conjugate, transpose, and Hermitian of vector $\mathbf{z}$ are, respectively, denoted by $\mathbf{z}^*$, $\mathbf{z}^T$, and $\mathbf{z}^H$. The augmented version of a complex vector $\mathbf{z}$ is represented by an underlined letter $\underline{\mathbf{z}}$ and is defined as $\underline{\mathbf{z}} = [\mathbf{z}^T, \mathbf{z}^H]^T$. An augmented matrix (a 2×2 block matrix) is denoted by an underlined capital bold letter, for example,

$$\underline{R} = \begin{bmatrix} R & \tilde{R} \\ \tilde{R}^* & R^* \end{bmatrix}, \tag{3.1}$$

and satisfies a particular structure (block pattern); that is, the lower-left block is the conjugate of the upper-right block, and the lower-right block is the conjugate of the upper-left block. We refer to the upper-left block of an augmented matrix as the *conventional block* and denote it by a capital bold letter (e.g., R), whereas the upper-right block of an augmented matrix is referred to as the *pseudoblock* and is denoted by a capital bold letter with a tilde (e.g., $\tilde{R}$). Finally, term $\mathcal{N}(a; b, c)$ denotes the probability at a of a normal distribution with mean b and covariance c, and term $\mathbb{E}\{\cdot\}$ denotes the expectation operator. Next, we describe properties of complex signals and complex-valued Gaussian distributions.

3.2.2 Complex-Valued Gaussian Distribution

Generally speaking, three closely related vectors can be constructed to represent an $n_\mathbf{z}$-dimensional complex-valued vector $\mathbf{z} \in \mathbb{C}^{n_\mathbf{z}}$ as follows:

1. The conventional complex vector representation, that is,

$$\mathbf{z} = \Re\{\mathbf{z}\} + j\Im\{\mathbf{z}\} \in \mathbb{C}^{n_\mathbf{z}}. \tag{3.2}$$

2. The real composite representation, which is a $2n_\mathbf{z}$-dimensional vector, obtained by stacking the real and imaginary parts of $\mathbf{z}$. A complex-valued vector in real composite form is represented by $w_\mathbf{z}$ and is given by

$$w_\mathbf{z} = \begin{bmatrix} \Re\{\mathbf{z}\} \\ \Im\{\mathbf{z}\} \end{bmatrix} \in \mathbb{R}^{2n_\mathbf{z}}. \tag{3.3}$$

3. The complex augmented vector where the bottom entries are the complex conjugates of the top entries. An augmented vector is represented by an underlined letter $\underline{z}$ and is defined as follows:

$$\underline{z} = \begin{bmatrix} \mathbf{z} \\ \mathbf{z}^* \end{bmatrix} \in \mathbb{C}^{2n_z}. \tag{3.4}$$

The real composite vector w_z, is related to the augmented complex vector $\underline{z}$ based on the following linear transformation:

$$w_z = \frac{1}{2} \underbrace{\begin{bmatrix} I_{n_z} & jI_{n_z} \\ I_{n_z} & -jI_{n_z} \end{bmatrix}}_{T_{n_z}} \underline{z}, \tag{3.5}$$

where I_{n_z} is the $n_z \times n_z$ identity matrix. In this chapter, we focus on the properties of the augmented representations of a complex-valued Gaussian distribution. A general (possibly improper) complex-valued Gaussian distribution is defined in augmented format as follows:

Definition 3.1   The probability density function $f(\mathbf{z})$ corresponding to a complex Gaussian vector $\mathbf{z}$ is defined as follows:

$$\mathcal{N}(\underline{x}; \underline{\mu}_z, \underline{P}_z) = \frac{1}{\pi^{n_z}\sqrt{|\underline{P}_z|}} \exp\left\{ -\frac{1}{2}(\underline{z} - \underline{\mu}_z)^H \underline{P}_z^{-1}(\underline{z} - \underline{\mu}_z) \right\}, \tag{3.6}$$

where the augmented mean vector $\underline{\mu}_z = [\mu_z^T, \mu_z^H]^T$ is the expected value of $\underline{z}$, and the augmented covariance matrix $\underline{P}_z$ is given by

$$\underline{P}_z = \mathbb{E}\left\{ (\underline{z} - \underline{\mu}_z)(\underline{z} - \underline{\mu}_z)^H \right\} = \begin{bmatrix} P_z & \tilde{P}_z \\ \tilde{P}_z^* & P_z^* \end{bmatrix}, \tag{3.7}$$

where covariance P_z and pseudocovariance $\tilde{P}_z$ blocks are defined as

Covariance: $\quad P_z = \mathbb{E}\left\{ (\mathbf{z} - \mu_z)(\mathbf{z} - \mu_z)^H \right\}, \tag{3.8}$

Pseudocovariance: $\quad \tilde{P}_z = \mathbb{E}\left\{ (\mathbf{z} - \mu_z)(\mathbf{z} - \mu_z)^T \right\}. \tag{3.9}$

The complex-valued Gaussian probability distribution function (PDF) in (3.6) depends algebraically on the augmented vector $\underline{z}$ and is interpreted as the joint PDF of the real and imaginary parts of $\mathbf{z}$.

Noncircular (Improper) Signals

In general, both covariance (P_z) and pseudocovariance ($\tilde{P}_z$) matrices are required for a complete second-order characterization of complex-valued signal $\mathbf{z}$. A complex-valued Gaussian vector $\mathbf{z}$ is called proper or circularly symmetric (proper) if $\tilde{P}_z = 0$ and noncircular (improper) when $\tilde{P}_z \neq 0$.

In order to evaluate noncircularity of a given signal, we define the so-called *noncircularity matrix*, which allows the estimator to examine each measurement and how far away it is from circularity. In order to mitigate the dependency on the power of **z**, the noncircularity matrix A_z is designed as

$$A_z \triangleq P_z^{-\frac{1}{2}} \tilde{P}_z P_z^{-\frac{T}{2}},$$ (3.10)

which only conveys information about the impropriety of **z**. In cases where z is scalar, we refer to the noncircularity matrix as the *noncircularity coefficient* denoted by α_z, which is computed as follows:

$$\alpha_z = \frac{\delta_z^2}{\tilde{\delta}_z^2},$$ (3.11)

where δ_z^2 and $\tilde{\delta}_z^2$ are, respectively, the conventional and pseudovariances of z. Signal z is circular if $\alpha_z = 0$ and maximally noncircular if $\alpha_z = 1$. Figure 3.4 illustrates scatter plots of circular and noncircular complex Gaussian random variables with different noncircularity coefficients to provide a better insight into their differences. When the noncircularity coefficient is close to zero, the pseudoblocks can safely be disregarded, while ignoring them when the noncircularity coefficient is close to 1 results in loss of critical second-order statistical information, which leaves the system susceptible to complex-valued attacks. Next, we propose a circularization approach that transforms a noncircular signal into its proper counterpart, which is useful in scenarios where widely linear processing is not applicable.

3.2.3   Circularization

As stated previously, conventionally, the term *complex Gaussian distribution* implicitly assumed circularity; that is, the pseudocovariance is ignored ($\tilde{P}_z = 0$).

> Intuitively speaking, in this section we have the following problem: given a noncircular complex-valued random vector and its corresponding augmented covariance matrix, how can we form a circular version of the given signal?

The following definition presents the density function of a circular complex-valued Gaussian distribution.

Definition 3.2   The circular (proper) complex-valued Gaussian density $f_c(z)$ is defined as follows:

$$f_c(\mathbf{z}) = \frac{1}{\pi^{n_z}\sqrt{|\underline{P}_z|}} \exp\left\{-(\underline{z}-\underline{\mu})^H \underline{P}_z^{-1}(\underline{z}-\underline{\mu})\right\},$$ (3.12)

where $\underline{P}_z = \begin{bmatrix} P_z & 0 \\ 0 & P_z \end{bmatrix}$.

When dealing with an improper Gaussian distribution, one approach is to ignore the pseudocovariance blocks and assume it is circular. This approach may result in severe performance degradation. An alternative solution is to circularize the Gaussian distribution. Circularization can be performed to simplify an improper complex-valued Gaussian distribution and therefore reduces the computational complexity associated with it.

A noncircular signal has two specific properties: (1) the power of the signal is not equally distributed between its real and imaginary parts, and (2) there exists a correlation between its real and imaginary parts.

Circularizing the signal can therefore be interpreted as equally distributing the power of the signal between the real and imaginary parts and removing the correlations between these parts. Consequently, in order to circularize a signal, we

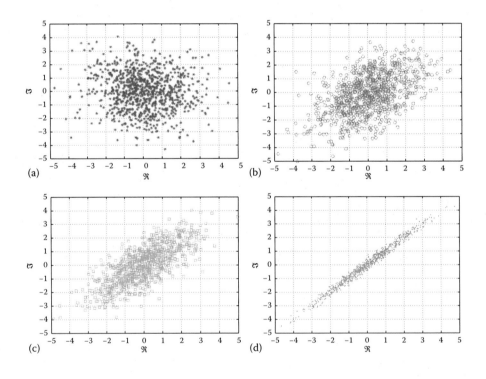

Figure 3.4: (a–d) Scatter plots of complex Gaussian random variables when $\alpha_z = 0$, $\alpha_z = 0.25$, $\alpha_z = 0.5$, and $\alpha_z = 0.99$.

have to first rotate the coordinate system of the signal to remove the correlation between real and imaginary parts. The eigenvectors of the augmented covariance matrix of the underlying signal can be used as the new coordinate system. The new coordinates are then scaled separately by the inverse of the square root of the corresponding eigenvalues to get equal power on real and complex domains. In effect, the circularization process changes the projection space. Intuitively speaking, the circularization process is alignment of the coordinates and whitening of the space therefore, it can be seen as a particular application of principal component analysis (PCA) or a singular value decomposition (SVD) type of operator for this particular problem. Based on the above intuition, we design the following circularizing filter (CF).

Definition 3.3   Given an improper complex-valued vector $\mathbf{z} \in \mathbb{C}^{n_z}$, the circularizing filter converts $\mathbf{z}$ into a proper random vector

$$\tilde{\mathbf{z}} = \mathbf{D}_{\mathbf{z}}\mathbf{z} + \tilde{\mathbf{D}}_{\mathbf{z}}\mathbf{z}^*, \tag{3.13}$$

where $\mathbf{D}_{\mathbf{z}}, \tilde{\mathbf{D}}_{\mathbf{z}} \in \mathbb{C}^{n_z \times n_z}$ are selected such that the pseudocovariance matrix of $\tilde{\mathbf{z}}$ vanishes, that is, $\tilde{\mathbf{P}}_{\tilde{\mathbf{z}}} = 0$.

The first step to construct the CF is to compute terms $\mathbf{D}_{\mathbf{z}}$ and $\tilde{\mathbf{D}}_{\mathbf{z}}$. Since the CF converts the improper covariance matrix, it is reasonable to derive matrices $\mathbf{D}_{\mathbf{z}}$ and $\tilde{\mathbf{D}}_{\mathbf{z}}$ based on the augmented covariance matrix $\underline{\mathbf{P}}_{\mathbf{z}}$. The following lemma is used to achieve this goal.

Lemma 3.1
An eigendecomposition of the augmented covariance matrix $\underline{\mathbf{P}}_{\mathbf{z}}$ corresponding to a complex-valued vector $\mathbf{z} \in \mathbb{C}^{n_z \times n_z}$ is given by $\underline{\mathbf{P}}_{\mathbf{z}} = \mathbf{\Psi}_{\mathbf{z}}\mathbf{\Omega}_{\mathbf{z}}\mathbf{\Psi}_{\mathbf{z}}^H$, where

$$\mathbf{\Psi}_{\mathbf{z}} = \begin{bmatrix} \mathbf{\Psi}_{\mathbf{z},1} & \mathbf{\Psi}_{\mathbf{z},2} \\ \mathbf{\Psi}_{\mathbf{z},1}^* & \mathbf{\Psi}_{\mathbf{z},2}^* \end{bmatrix}, \mathbf{\Omega}_{\mathbf{z}} = \begin{bmatrix} \mathbf{\Omega}_{1,n_z} & 0 \\ 0 & \mathbf{\Omega}_{n_z+1,2n_z} \end{bmatrix}. \tag{3.14}$$

Proof 3.1   Suppose $\mathbf{u} = [\mathbf{u}_1^T, \mathbf{u}_2^T]^T \in \mathbb{C}^{2n_z}$ is an eigenvector of $\underline{\mathbf{P}}_{\mathbf{z}} \in \mathbb{W}^{2n \times 2n}$ corresponding to the eigenvalue $\lambda \in \mathcal{R}$. It follows that

$$\begin{bmatrix} \mathbf{P}_{\mathbf{z}} & \tilde{\mathbf{P}}_{\mathbf{z}} \\ \tilde{\mathbf{P}}_{\mathbf{z}}^* & \mathbf{P}_{\mathbf{z}}^* \end{bmatrix} \begin{bmatrix} \mathbf{u}_1 \\ \mathbf{u}_2 \end{bmatrix} = \lambda \begin{bmatrix} \mathbf{u}_1 \\ \mathbf{u}_2 \end{bmatrix}. \tag{3.15}$$

By taking the complex conjugate of both sides in (3.16) and reordering the variables we will have the following equation:

$$\begin{bmatrix} \mathbf{P}_{\mathbf{z}} & \tilde{\mathbf{P}}_{\mathbf{z}} \\ \tilde{\mathbf{P}}_{\mathbf{z}}^* & \mathbf{P}_{\mathbf{z}}^* \end{bmatrix} \begin{bmatrix} \mathbf{u}_2^* \\ \mathbf{u}_1^* \end{bmatrix} = \lambda \begin{bmatrix} \mathbf{u}_2^* \\ \mathbf{u}_1^* \end{bmatrix}, \tag{3.16}$$

thus $[\boldsymbol{u}_2^H, \boldsymbol{u}_1^H]^T$ is also an eigenvector of $\underline{\boldsymbol{P}}_{\mathbf{z}}$ corresponding to the eigenvalue λ. Assuming that the eigenvalues of $\boldsymbol{P}_{\mathbf{z}}$ have no multiplicity, we have $[\boldsymbol{u}_2^H \boldsymbol{u}_1^H]^T = c[\boldsymbol{u}_1^T, \boldsymbol{u}_2^T]^T$, where c is a real-valued constant; therefore, $\boldsymbol{u}_1 = \boldsymbol{u}_2$. Finally, matrix $\boldsymbol{\Psi}_{\mathbf{z}}$, which contains all the eigenvectors of $\underline{\boldsymbol{P}}_{\mathbf{z}}$, takes the form in (3.19). $\qquad\square$

Using Lemma 3.1, the following proposition provides a solution for the CF.

Proposition 3.1   The circularizing filter's matrices, $\boldsymbol{D}_{\mathbf{z}}$ and $\tilde{\boldsymbol{D}}_{\mathbf{z}}$, are computed as follows:

$$\boldsymbol{D}_{\mathbf{z}} = \boldsymbol{\Omega}_{1,n_z}^{-\frac{1}{2}} \boldsymbol{\Psi}_{\mathbf{z},1}^H + j\boldsymbol{\Omega}_{n_z+1,2n_z}^{-\frac{1}{2}} \boldsymbol{\Psi}_{\mathbf{z},2}^H, \tag{3.17}$$

$$\tilde{\boldsymbol{D}}_{\mathbf{z}} = \boldsymbol{\Omega}_{1,n_z}^{-\frac{1}{2}} \boldsymbol{\Psi}_{\mathbf{z},1}^T + j\boldsymbol{\Omega}_{n_z+1,2n_z}^{-\frac{1}{2}} \boldsymbol{\Psi}_{\mathbf{z},2}^T, \tag{3.18}$$

where $\boldsymbol{\Omega}_{\mathbf{z}}$ and $\boldsymbol{\Psi}_{\mathbf{z}}$ are obtained from Lemma 3.1.

Proof 3.2   Let $\tilde{\mathbf{z}}_k = \boldsymbol{D}\mathbf{z}_k + \tilde{\boldsymbol{D}}\mathbf{z}_k^*$ where $\boldsymbol{D}$ and $\tilde{\boldsymbol{D}}$ are given by (3.17) and (3.18). The augmented circularized observation $\tilde{\underline{\mathbf{z}}}_k$ is computed from $\underline{\mathbf{z}}_k$ as

$$\tilde{\underline{\mathbf{z}}} = \boldsymbol{T}_{n_z}^H \boldsymbol{\Psi}_{\mathbf{z}} \underline{\boldsymbol{P}}_{\mathbf{z}}^{-\frac{1}{2}} \underline{\mathbf{z}}; \tag{3.19}$$

therefore, $\underline{\boldsymbol{P}}_{\tilde{\mathbf{z}}} = 4\boldsymbol{T}_{n_z}^H \underline{\boldsymbol{P}}_{\mathbf{z}}^{-\frac{1}{2}} \underline{\boldsymbol{P}}_{\mathbf{z}} \underline{\boldsymbol{P}}_{\mathbf{z}}^{-\frac{1}{2}} \boldsymbol{T}_{n_z} = 4\boldsymbol{I}$, which implies that $\tilde{\underline{\mathbf{z}}}$ is a proper random vector. $\qquad\square$

Alternatively, $\boldsymbol{D}_{\mathbf{z}}$ and $\tilde{\boldsymbol{D}}_{\mathbf{z}}$ can be designed such that the initial covariance matrix of the improper signal is preserved, that is,

$$\boldsymbol{D}_{\mathbf{z}} = \boldsymbol{P}_{\mathbf{z}}^{\frac{1}{2}} \left(\boldsymbol{\Omega}_{1,n_z}^{-\frac{1}{2}} \boldsymbol{\Psi}_{\mathbf{z},1}^H + j\boldsymbol{\Omega}_{n_z+1,2n_z}^{-\frac{1}{2}} \boldsymbol{\Psi}_{\mathbf{z},2}^H \right), \tag{3.20}$$

$$\tilde{\boldsymbol{D}}_{\mathbf{z}} = \boldsymbol{P}_{\mathbf{z}}^{\frac{1}{2}} \left(\boldsymbol{\Omega}_{1,n_z}^{-\frac{1}{2}} \boldsymbol{\Psi}_{\mathbf{z},1}^T + j\boldsymbol{\Omega}_{n_z+1,2n_z}^{-\frac{1}{2}} \boldsymbol{\Psi}_{\mathbf{z},2}^T \right). \tag{3.21}$$

Incorporating Proposition 3.1, the new mean value is given by

$$\tilde{\boldsymbol{\mu}}_{\mathbf{z}} = \boldsymbol{D}_{\mathbf{z}}\boldsymbol{\mu}_{\mathbf{z}} + \tilde{\boldsymbol{D}}_{\mathbf{z}}\boldsymbol{\mu}_{\mathbf{z}}^*. \tag{3.22}$$

The circularization procedure leads to the following circular complex Gaussian density function:

$$p_c(\tilde{\mathbf{z}}) = \frac{1}{\pi^{n_z} |\boldsymbol{P}|} \exp\left\{ -(\mathbf{z} - \tilde{\boldsymbol{\mu}}_{\mathbf{z}})^H \boldsymbol{P}^{-1} (\mathbf{z} - \tilde{\boldsymbol{\mu}}_{\mathbf{z}}) \right\}. \tag{3.23}$$

This completes our introduction to complex-valued signals. Next, we describe the conventional approach for state estimation and attack detection in industrial control systems and highlight their drawbacks.

3.3 Conventional Estimation and Monitoring Approach

In this section, first we present the conventional methodology used by the state-of-the-art state estimation and monitoring modules in the industrial control systems, and then we investigate its drawbacks to see how an adversary can manipulate the system incorporating these drawbacks and utilizing complex-valued statistics.

> In order to detect cyber attacks, one needs to perform the following three steps: (1) model the physical system, (2) develop an estimation algorithm based on the selected model, and (3) monitor the system and run diagnostics for attack detection.

The conventional approach typically consists of the following steps:

1. *State-space model*: Typically, the evolution of the states is represented by a time-invariant linear state-space model.

2. *Estimation algorithm*: The conventional strictly linear KF is commonly used to dynamically compute an estimate of the internal state of the system. Although the KF uses a time-varying gain, in its optimal operation scenario this gain will converge to its steady-state value relatively fast; therefore, it is a common practice in control systems [19] to use the steady-state form of the KF.

3. *Diagnostics/monitoring*: The conventional attack detection solution is the residue-based approaches [3] where the detection algorithm performs hypothesis test by checking the power of the residuals computed based on state estimates.

Below, we further elaborate on these conventional steps in more details.

3.3.1 *Strictly Linear State-Space Model and KF*

We focus on the state estimation problem based on a dynamic state model that is considered to be linear with Gaussian statistics. Consider a feedback control system represented by a set of n_x state variables denoted by $\mathbf{x}_k$, where ($k \geq 0$) is the time/iteration index. The evolution of the states (state dynamics) is represented by a linear state-space model, that is,

$$\mathbf{x}_k = \boldsymbol{F}\mathbf{x}_{k-1} + \boldsymbol{w}_k, \tag{3.24}$$

where $\boldsymbol{w}_k$ models the uncertainties in the state model, which is considered to be zero-mean normal distribution with known covariance matrix, that is, $\boldsymbol{w}_k \sim \mathcal{N}(\mathbf{0}, \boldsymbol{Q})$. In smart grids, for instance, the state vector includes voltage phasors (i.e., voltage magnitudes and voltage angles) at different bus nodes across the grid. The objective of the estimation framework is to determine the optimal value of the state vector $\mathbf{x}_k$ over time given the measurements provided by the observation module. We consider a scenario where complex-valued measurements obtained from a set of n_z sensors or meters are deployed for state estimation and to monitor the state variables dynamically over time. The relationship between the measurement vector, denoted by $\mathbf{z}_k$, and the state vector $\mathbf{x}_k$ is represented by the observation model given by

$$\mathbf{z}_k = \boldsymbol{H}\mathbf{x}_k + \mathbf{v}_k, \tag{3.25}$$

where $\mathbf{v}_k \sim \mathcal{N}(\mathbf{0}, \boldsymbol{R})$ models the uncertainties in the observation model. The KF is commonly used to dynamically compute an estimate of the internal state of the systems, which consists of prediction and update steps. In the KF, the conditional mean of the state variables during the prediction step at iteration k given observations up to time $(k-1)$ is denoted by

$$\hat{\mathbf{x}}_{k|k-1} \triangleq \mathbb{E}\left\{\mathbf{x}_k | \boldsymbol{Z}_{k-1}\right\}, \tag{3.26}$$

where $\boldsymbol{Z}_{k-1} = [\mathbf{z}_1^T, \ldots, \mathbf{z}_{k-1}^T]^T$ is the record of measurements. The conditional covariance matrix associated with the predictive estimate $\hat{\mathbf{x}}_{k|k-1}$ is defined as

$$\boldsymbol{P}_{k|k-1} \triangleq \mathbb{E}\left\{\left(\mathbf{x}_k - \hat{\mathbf{x}}_{k|k-1}\right)\left(\mathbf{x}_k - \hat{\mathbf{x}}_{k|k-1}\right)^H | \boldsymbol{Z}_{k-1}\right\}. \tag{3.27}$$

As for the prediction step, the following notation is associated with the conditional mean and covariance of the estimated state variables:

$$\hat{\mathbf{x}}_{k|k} \triangleq \mathbb{E}\{\mathbf{x}_k | \boldsymbol{Z}_k\}, \tag{3.28}$$

$$\boldsymbol{P}_{k|k} \triangleq \mathbb{E}\left\{\left(\mathbf{x}_k - \hat{\mathbf{x}}_{k|k}\right)\left(\mathbf{x}_k - \hat{\mathbf{x}}_{k|k}\right)^H\right\}. \tag{3.29}$$

Iteration k of the KF starts with the previous state estimate $\hat{\mathbf{x}}_{k-1|k-1}$ and its associated error covariance matrix denoted by $\boldsymbol{P}_{k-1|k-1}$, and performs the prediction step as follows:

$$\hat{\mathbf{x}}_{k|k-1} = \boldsymbol{F}\hat{\mathbf{x}}_{k-1|k-1}, \tag{3.30}$$

$$\boldsymbol{P}_{k|k-1} = \boldsymbol{F}\boldsymbol{P}_{k-1|k-1}\boldsymbol{F}^H + \boldsymbol{Q}. \tag{3.31}$$

Based on the predicted state estimates $\hat{\mathbf{x}}_{k|k-1}$, the KF computes the mean of the innovation sequence $\boldsymbol{r}_k$, that is, the current measurement $\mathbf{z}_k$ minus the projection of the states to the measurement space, and its corresponding innovation covariance matrix as follows:

$$\boldsymbol{r}_k = \mathbf{z}_k - \boldsymbol{H}\hat{\mathbf{x}}_{k|k-1}, \tag{3.32}$$

$$\boldsymbol{S}_k = \boldsymbol{H}\boldsymbol{P}_{k|k-1}\boldsymbol{H}^H + \boldsymbol{R}, \tag{3.33}$$

which results in the following definition of the innovation sequence

$$\text{Innovation sequence} \quad \sim \quad \mathcal{N}(r_k, S_k). \tag{3.34}$$

Using the innovation sequence, the Kalman gain is computed as $\mathcal{K}_k = P_{k|k-1} H^H S_k^{-1}$ and the state estimate and its corresponding covariance matrix are updated as follows:

$$\hat{x}_{k|k} = \hat{x}_{k|k-1} + \mathcal{K}_k r_k, \tag{3.35}$$

$$P_{k|k} = P_{k|k-1} - \mathcal{K}_k H^H P_{k|k-1}. \tag{3.36}$$

Although the KF uses a time-varying gain $\mathcal{K}_k$, in its optimal operation scenario this gain will converge to its steady-state value relatively fast. In this case, a finite steady-state covariance matrix $P \triangleq \lim_{k \to \infty} P_{k|k-1}$ exists and can be pre-computed when the pair $\{F, H\}$ is completely observable and the pair $\{F, Q\}$ is completely controllable. Consequently, the conventional approach uses precalculated statistics of the KF for further processing and attack detection purposes.

3.3.2 Residual-Based Detection Algorithms

Intuitively speaking, the innovation sequence and its error covariance matrix are the only elements of the KF-based estimators that can be used to diagnose its operations over cyber attacks. The mean of the innovation sequence can be computed at each filtering iteration and has to be close to zero. In theory, during the normal operation of the filter, the innovation sequence is a white type random process with zero mean. Therefore, the real question at this point is what happens to the covariance matrix of the innovation sequence. The covariance of the innovation can be precalculated using the steady-state Kalman gain; this is the theoretical covariance of the innovations. When the Kalman gain is precalculated, the second moment does not depend on the measurements and only depends on the characteristics of the noise. Essentially, two filters (the optimal filter and its implemented counterpart) produce two Gaussian distributions: one with the optimal profile and one with the operation profile. The objective is essentially to monitor the probability distribution of the innovation sequence against its optimal version. The only reasonable option to detect whether the estimator is in its normal operation state or under attack is to calculate the statistical distance between Gaussian distributions in the innovation space.

For linear systems with Gaussian statistics and when the filter is in its normal operation, the optimal innovation sequence, denoted by v_k^o, is a white Gaussian process with zero mean and known precomputed covariance S, where the steady-state covariance of the innovation sequence is precomputed as $S \triangleq HPH^H + R$. Therefore, when the KF has the optimal profile, the optimal innovation sequence is defined as

$$\text{Optimal innovation sequence:} \quad v_k^o \sim \mathcal{N}(\mathbf{0}, \mathbf{S}) \tag{3.37}$$

The χ^2-detectors, also referred to as residue-based detectors [3], are the most commonly used detection algorithms in industrial control systems. The detection criterion in the χ^2-detectors is given by

$$g_k = r_k^H \mathbf{S} r_k, \tag{3.38}$$

which is the power of the residuals r_k normalized by the steady-state innovation covariance matrix $\mathbf{S}$. The χ^2-detector at iteration k is therefore given by

$$g_k \underset{H_1}{\overset{H_0}{\lessgtr}} \tau \tag{3.39}$$

where the null hypotheses H_0 mean that the system is operating normally, whereas the alternative hypotheses H_1 mean that the system is under attack. The threshold τ is selected to satisfy a specific false alarm probability.

The conventional diagnostic metrics typically use some function, either determinant or trace, of the innovation sequence. For example, the χ^2-test uses the trace of the innovation sequence normalized by the steady-state innovation covariance matrix. The off-diagonal elements are not provided by the trace or determinant; therefore, such metrics lap the off-diagonal elements together with diagonal ones ignoring critical statistical information. Besides, while the distribution of the innovation sequence contains all the required information for diagnosing the system, metrics such as the χ^2-test only check the system against an assumed statistical model. Later, we use this intuitive concept to show the vulnerability of the control systems to complex-valued attacks, where the innovation sequence is circular under the optimal profile, while the adversary transforms it into a highly noncircular sequence. An adversary therefore creates an innovation sequence that eventually looks like the optimal sequence when the pseudocovariances of the now noncircular observations are ignored. The above conventional modeling and detection framework suffers from the following drawbacks when the underlying signals are complex-valued:

1. The conventional estimation or monitoring framework ignores the information regarding the complex nature of the underlying measurements; that is, the pseudocovariances are ignored and only the conventional covariances are considered.

2. When the Kalman gain is precalculated, the covariance of the innovation sequence does not depend on the measurements and only depends on the characteristics of the observation noise. Therefore, precomputing the innovation's covariance results in significant loss of information.

3. The conventional detection metrics, such as χ^2-detector, only check the system against an assumed statistical model. Such metrics lap the off-diagonal elements together with diagonal ones, ignoring critical statistical information.

In summary, as the conventional methods do not take into consideration the complex nature of the measurements, an attacker can modify the measurement model in such a way that the attacked innovation sequence becomes second-order noncircular. In other words, the attacked innovation sequence has the same conventional covariance matrix as the original sequence, but its pseudocovariance matrix is no longer vanishing. As the convectional diagnostic is designed based on a specific model that ignores the pseudocovariance statistics, this attack remain undetected. Later, we show that not only do the attacks remain undetected, but also they can drastically change the state of the underlying system beyond repair. The first step in designing a detection and prevention mechanism is to form the mathematical model of a potential cyber attack, which will be covered next.

3.4 Complex-Valued Attacks

In this section, we design a complex-valued attack model, extend the conventional linear state-space model to its widely linear counterpart in order to incorporate the full second-order statistical properties of complex-valued signals, present an augmented version of the KF using the widely linear model, and develop a noncircularity test as a simple test to check the measurements against noncircular attacks.

3.4.1 Complex-Valued Attack Model

The conventional model assumes measurement $\mathbf{z}_k$ to be circular. A circular signal is rotation invariant; that is, $\mathbf{z}_k$ and $e^{j\theta}\mathbf{z}_k$ have the same distribution for all $\theta \in \mathbb{R}$. Therefore, the conventional model is assumed to be rotation invariant. To provide more insight into the nature of circularity, consider the scalar case $z = m_z e^{j\theta_z}$, where m_z and θ_z are its magnitude and phase, respectively. As z is circular, m_z and θ_z are independent random variables. The PDF of z, denoted by $p(z)$, can then be computed in terms of the PDFs of its magnitude $p(m_z)$ and phase $p(\theta_z)$, that is, $p(z) = p(m_z, \theta_z) = p(\theta_z)p(m_z)$. To modify the measurements and, consequently, the innovation sequence, the attack model first manipulates the circularity of the measurements and then rotates them without being detected. The circularity of the observations is preserved under linear transformations; however, this is not the case with widely linear transformations. Therefore, the attack model can add noncircularity to the measurements via a widely linear transformation that is a linear combination of the original measurement $\mathbf{z}_k$ and its complex conjugate $\mathbf{z}_k^*$ as follows:

$$\mathbf{z}_k^a = \mathbf{D}_k\mathbf{z}_k + \tilde{\mathbf{D}}_k\mathbf{z}_k^*, \tag{3.40}$$

where $\mathbf{z}_k^a$ is the attacked observation, and matrices $\mathbf{D}_k, \tilde{\mathbf{D}}_k \in \mathbb{C}^{n_z \times n_z}$. Rearranging (3.40) we have

$$\mathbf{z}_k^a = \begin{bmatrix} \mathbf{D}_k\mathbf{H}_k & \tilde{\mathbf{D}}_k\mathbf{H}_k^* \end{bmatrix} \begin{bmatrix} \mathbf{x}_k \\ \mathbf{x}_k^* \end{bmatrix} + \begin{bmatrix} \mathbf{D}_k & \tilde{\mathbf{D}}_k \end{bmatrix} \begin{bmatrix} \mathbf{v}_k \\ \mathbf{v}_k^* \end{bmatrix}. \tag{3.41}$$

The processing model of the complex-valued attack is now widely linear (not strictly linear), as it depends on both the signal itself and its complex conjugate. This motivates the use of an augmented measurement vector $\underline{\mathbf{z}}_k^a$, which includes both $\mathbf{z}_k^a$ and its complex conjugate $(\mathbf{z}_k^a)^*$ as its elements, that is, $\underline{\mathbf{z}}_k^a = [(\mathbf{z}_k^a)^T, (\mathbf{z}_k^a)^H]^T$. The attack model in augmented format is therefore given by

$$\underline{\mathbf{z}}_k^a = \underline{\mathbf{H}}_k^a \, \underline{\mathbf{x}}_k + \underline{\mathbf{v}}_k^a, \tag{3.42}$$

where $\underline{\mathbf{H}}_k^a$ is the augmented observation model given by

$$\underline{\mathbf{H}}_k^a \triangleq \begin{bmatrix} \mathbf{H}_k^a & \tilde{\mathbf{H}}_k^a \\ [\tilde{\mathbf{H}}_k^a]^* & [\mathbf{H}_k^a]^* \end{bmatrix} = \begin{bmatrix} \mathbf{D}_k\mathbf{H}_k & \tilde{\mathbf{D}}_k\mathbf{H}_k^* \\ \tilde{\mathbf{D}}_k^*\mathbf{H}_k & \mathbf{D}_k^*\mathbf{H}_k^* \end{bmatrix}, \tag{3.43}$$

with $\mathbf{H}_k^a = \mathbf{D}_k\mathbf{H}_k$ and $\tilde{\mathbf{H}}_k^a = \tilde{\mathbf{D}}_k\mathbf{H}_k^*$ constituting its conventional and pseudoblocks, respectively. The observation noise $\underline{\mathbf{v}}_k^a = \mathcal{N}(0, \underline{\mathbf{R}}_k^a)$ of the attack model is a complex-valued Gaussian, but with different statistical properties in comparison to its original counterpart $\underline{\mathbf{v}}_k = \mathcal{N}(0, \underline{\mathbf{R}}_k)$. Specifically, its pseudocovariance $\tilde{\mathbf{R}}_k^a$ will be a nonzero matrix; therefore it carries critical statistical information for an adversary to change the state of the system.

In summary, the noncircular attack model modifies both the covariance and pseudocovariance of the innovation sequence; therefore, it injects critical statistical information that will remain hidden from the conventional monitoring algorithms. Such statistical information can result in significant degradation in the overall performance of the estimation algorithm, as will be shown later via Monte Carlo simulations.

As a complex-valued attack is the result of inducing noncircularity to the innovation sequence, it is logical to further analyze noncircularity detection tests. Remember, when the pseudocovariance matrix is zero, the covariance matrix $\mathbf{S}_k^a$ completely characterizes the second-order properties of the innovation sequence. In order to measure and adjust accordingly the noncircularity of an attack, the noncircularity matrix (introduced in Section 3.2.3) corresponding to the innovation covariance could be used as follows:

$$A_k \triangleq [\mathbf{S}_k^a]^{-\frac{1}{2}} \tilde{\mathbf{S}}_k^a [\mathbf{S}_k^a]^{-\frac{T}{2}}. \tag{3.44}$$

Innovation Sequence under Complex-Valued Attacks

A complex-valued attack model changes the statistical properties of the innovation sequence in such a way that the innovation covariance matrix S_k computed based on (3.33) is modified as follows:

$$S_k^a = [H_k^a]^H P_{k|k-1} H_k^a + [\tilde{H}_k^a]^T P_{k|k-1} [\tilde{H}_k^a]^* + R_k^a. \qquad (3.45)$$

The pseudocovariance of the innovation sequence, defined as follows,

$$\tilde{S}_k^a \triangleq \mathbb{E}\{r_k^a [r_k^a]^T\}, \qquad (3.46)$$

is zero under the assumptions of the conventional model; however, the pseudocovariance of the innovation sequence under attack is no longer zero, that is,

$$\tilde{S}_k^a = [H_k^a]^H P_{k|k-1} \tilde{H}_k^a + [\tilde{H}_k^a]^T P_{k|k-1} [H_k^a]^* + \tilde{R}_k^a, \qquad (3.47)$$

which results in the following innovation sequence.

Attacked innovation sequence: $\underline{v}_k^a \sim \mathcal{N}(\underline{r}_k^a, \underline{S}_k^a).$ $\qquad (3.48)$

The circularity coefficient can also be used as a test for detecting complex-valued attacks. Alternatively, a complex-valued generalized likelihood ratio test (GLRT) can be designed to test noncircularity of the observations themselves. The GLRT makes an attempt to isolate different failures by using knowledge of the different effects of failures on the system innovations. At iteration k, consider L observations denoted by $\mathbf{z}_{k-L:k}$ with augmented mean $\underline{\mu}$ and augmented covariance $\underline{R}$. The joint PDF of these samples is given by

$$p(\mathbf{z}_{k-L:k}) = \pi^{-Ln_z} (|\underline{R}|)^{-\frac{L}{2}} \exp\left\{ -\frac{1}{2} \sum_{m=0}^{L} (\mathbf{z}_{k-L+m} - \underline{\mu})^H \underline{R}^{-1} (\mathbf{z}_{k-L+m} - \underline{\mu}) \right\}$$

$$= \pi^{-Ln_z} (|\underline{R}|)^{-\frac{L}{2}} \exp\left\{ -\frac{L}{2} \mathrm{tr}(\underline{R}^{-1} \underline{\hat{R}}) \right\}, \qquad (3.49)$$

where $\underline{\hat{R}}$ is the sample augmented covariance matrix computed as follows:

$$\underline{\hat{R}} = \frac{1}{L} \sum_{m=0}^{L} (\mathbf{z}_{k-L+m} - \underline{\hat{\mu}})^H (\mathbf{z}_{k-L+m} - \underline{\hat{\mu}}) = \begin{bmatrix} \hat{R} & \widehat{\tilde{R}} \\ \widehat{\tilde{R}}^* & \hat{R}^* \end{bmatrix}, \qquad (3.50)$$

and $\underline{\hat{\mu}}$ is given by

$$\underline{\hat{\mu}} = \frac{1}{L} \sum_{m=0}^{L} \mathbf{z}_{k-L+m}. \qquad (3.51)$$

The detection purpose is to develop the following hypothesis test

$$\begin{cases} H_0 : \tilde{\pmb{R}} = 0, & \mathbf{z}_k \quad \text{is circular.} \\ H_1 : \tilde{\pmb{R}} \neq 0, & \mathbf{z}_k \quad \text{is noncircular.} \end{cases}$$

The following GLR, statistics can be used for testing the above two hypotheses:

$$\lambda = \frac{\underset{\pmb{R}}{\max}\ p(\mathbf{z})}{\underset{\pmb{R}}{\max}\ p(\mathbf{z})}. \tag{3.52}$$

Equation 3.52 provides the ratio between the constraint likelihood computed based on the assumption that $\tilde{\pmb{R}} = 0$ is circular and the unconstraint likelihood without any assumptions. The constraint maximum likelihood estimate $\underline{\pmb{R}}_c$ of the augmented covariance matrix $\underline{\pmb{R}}$ is given by the following diagonal matrix:

$$\underline{\hat{\pmb{R}}}_c = \begin{bmatrix} \hat{\pmb{R}} & 0 \\ 0 & \hat{\pmb{R}}^* \end{bmatrix}. \tag{3.53}$$

The unconstraint maximum likelihood estimate of $\underline{\pmb{R}}$ is given by (3.50). Therefore, the GLRT statistics in (3.52) can be simplified as follows:

$$l = \lambda^{2/L} = \frac{\det(\underline{\hat{\pmb{R}}})}{\det(\hat{\pmb{R}})^2} = \frac{\det(\hat{\pmb{R}} - \hat{\tilde{\pmb{R}}}\hat{\pmb{R}}^{-*}\hat{\tilde{\pmb{R}}}^*)}{\det(\hat{\pmb{R}})}. \tag{3.54}$$

Equation 3.54 expresses the GLR as the ratio of the determinant of the Schur complement of the estimated augmented covariance matrix to the determinant of the sample covariance matrix. Next, we develop the widely linear state-space representation as an alternative modeling solution to the conventional state-space model described in Section 3.3.1. Widely linear representation accounts for potential noncircularity of the observations; therefore, it is more appropriate for scenarios where the observations are complex-valued.

3.4.2 Widely Linear State-Space Model

Consider the following general discrete-time state-space model:

$$\mathbf{x}_k = \pmb{F}_k \mathbf{x}_{k-1} + \tilde{\pmb{F}}_k \mathbf{x}_{k-1}^* + \mathbf{w}_k, \tag{3.55}$$

$$\mathbf{z}_k = \pmb{H}_k \mathbf{x}_k + \tilde{\pmb{H}}_k \mathbf{x}_k^* + \mathbf{v}_k, \tag{3.56}$$

where $\pmb{F}_k$ and $\pmb{H}_k$ are, respectively, the state transition and observation matrices, and $\mathbf{w}_k$ and $\mathbf{v}_k$ are, respectively, the zero-mean state and observation noises with covariance matrices $\pmb{Q}_k$ and $\pmb{R}_k$ and pseudocovariances $\tilde{\pmb{Q}}_k$ and $\tilde{\pmb{R}}_k$. The above state-space model is typically [20] called *widely linear* since it linearly depends on both the state vector $\mathbf{x}_k$ and its conjugate $\mathbf{x}_k^*$. The conjugate matrices $\tilde{F}_k$ and $\tilde{H}_k$ determine whether (3.55) and (3.56) are widely linear or strictly linear, that is, (3.24) and (3.25). The above widely linear model can be expressed compactly as follows:

$$\underline{\mathbf{x}}_k = \underline{F}_k \underline{\mathbf{x}}_{k-1} + \underline{w}_k, \tag{3.57}$$

$$\underline{z}_k = \underline{H}_k \underline{\mathbf{x}}_k + \underline{v}_k, \tag{3.58}$$

where $\underline{\mathbf{x}}_k = [\mathbf{x}_k^T, \mathbf{x}_k^H]^T$, $\underline{z}_k = [z_k^T, z_k^H]^T$, and the augmented state and observation matrices are given by

$$\underline{F}_k = \begin{bmatrix} F_k & \tilde{F}_k \\ \tilde{F}_k^* & F_k^* \end{bmatrix} \quad \text{and} \quad \underline{H}_k = \begin{bmatrix} H_k & \tilde{H}_k \\ \tilde{H}_k^* & \tilde{H}_k^* \end{bmatrix}. \tag{3.59}$$

The augmented state and observation noise covariance matrices corresponding to mode *l*, for $(1 \leq l \leq n_f)$, are, respectively, given by

$$\underline{Q}_k = \mathbb{E}\{\underline{w}_k \underline{w}_k^H\} = \begin{bmatrix} Q_k & \tilde{Q}_k \\ \tilde{Q}_k^* & Q_k^* \end{bmatrix}, \tag{3.60}$$

$$\underline{R}_k = \mathbb{E}\{\underline{v}_k \underline{v}_k^H\} = \begin{bmatrix} R_k & \tilde{R}_k \\ \tilde{R}_k^* & R_k^* \end{bmatrix}, \tag{3.61}$$

where the diagonal blocks are the covariance matrices, while the off-diagonal blocks are pseudocovariances. A simple inspection of (3.57) through (3.61) reveals the following potential complex-valued scenarios:

1. *Strictly Linear:* In such cases, matrices $\tilde{F}(\cdot)$, $\tilde{H}(\cdot)$, $\tilde{Q}(\cdot)$, and $\tilde{R}(\cdot)$ are all zero matrices of appropriate dimensions. In other words, all the matrices in (3.57) and (3.58) are block-diagonal as

$$\underline{F}_k = \begin{bmatrix} F_k & 0 \\ 0 & F_k^* \end{bmatrix}, \underline{Q}_k = \begin{bmatrix} Q_k & 0 \\ 0 & Q_k^* \end{bmatrix},$$

$$\underline{H}_k = \begin{bmatrix} H_k & 0 \\ 0 & H_k^* \end{bmatrix}, \underline{R}_k = \begin{bmatrix} R_k & 0 \\ 0 & R_k^* \end{bmatrix},$$

where **0** is a matrix of appropriate dimension with zero elements. Furthermore, the initial distribution is circular.

2. *Improper (noncircular) initial conditions:* Similar to the previous scenario, all the off-diagonal blocks of matrices in (3.57) and (3.58) are zero blocks; however, the initial condition is improper:

$$\underline{Q}_0 = \mathbb{E}\{\underline{w}_0 \underline{w}_0^H\} = \begin{bmatrix} Q_0 & \tilde{Q}_0 \\ \tilde{Q}_0^* & Q_0^* \end{bmatrix}. \tag{3.62}$$

3. *Improper (noncircular) observation noises:* The observation noise is improper (noncircular); that is, $\tilde{R}$ is a nonzero matrix, while the state noise and the initial distribution are proper random variables. As stated previously, noncircular observations are more common in practice, as the modulation and coding procedures typically introduce noncircular complex noise to the communicated sensor observations. This necessitates incorporation of widely linear observation models and development of complex-valued estimators.

Note that the augmented state-space representation is superior over the linear state-space model, as it can accommodate Scenarios 2 and 3. For Scenario 1, the estimator can use the conventional strictly linear model, however, for Scenarios 2 and 3, the estimator has to use the widely linear model (Equations 3.57 and 3.58) instead of its conventional strictly linear counterpart. Table 3.1 outlines the update step of the structured complex-valued KF based on the state-space model defined in (3.57) and (3.58). At iteration k, the structured widely linear Kalman filter first performs the prediction step as follows:

$$P_{k|k-1} = F_k P_{k-1|k-1} F_k^H + \tilde{F}_k \tilde{P}_{k-1|k-1}^* F_k^H + F_k \tilde{P}_{k-1|k-1} \tilde{F}_k^H$$

$$+ \tilde{F}_k \tilde{P}_{k-1|k-1}^* \tilde{F}_k^H + Q_k, \tag{3.63}$$

$$\hat{\mathbf{x}}_{k|k-1} = F_k \hat{\mathbf{x}}_{k-1|k-1} + \tilde{F}_k \hat{\mathbf{x}}_{k-1|k-1}^*. \tag{3.64}$$

Table 3.1 Observation Update Step of Structured Complex Kalman Filter

Structure-Induced Widely Linear Kalman Filter
$\hat{\mathbf{x}}_{k|k} = \mathcal{K}_k r_k + \tilde{\mathcal{K}}_k r_k^*$
$r_k = \mathbf{z}_k - H_k \hat{\mathbf{x}}_{k|k-1} - \tilde{H}_k \hat{\mathbf{x}}_{k|k-1}^*$
$\mathcal{K}_k = \left(P_{k|k-1} H_k + \tilde{P}_{k|k-1} \tilde{H}_k^* \right) \mathrm{Sch}_k^{-1} - \left(P\tilde{H}_k + \tilde{P}H_k^* \right) S_k^{-*} \tilde{S}_k^* \mathrm{Sch}_k^{-1}$
$\tilde{\mathcal{K}}_k = \left(P\tilde{H}_k + \tilde{P}H_k^* \right) \mathrm{Sch}_k^{-*} - \left(PH_k + \tilde{P}\tilde{H}_k^* \right) S_k^{-1} \tilde{S}_k \mathrm{Sch}_k^{-*}$
$S_k = \begin{bmatrix} H_k & \tilde{H}_k \end{bmatrix} \underline{P}_{k|k-1} \begin{bmatrix} H_k^H \\ \tilde{H}_k^H \end{bmatrix} + R_k$
$\tilde{S}_k = \begin{bmatrix} H_k & \tilde{H}_k \end{bmatrix} \begin{bmatrix} \tilde{P}_{k|k-1} & P_{k|k-1} \\ P_{k|k-1}^* & \tilde{P}_{k|k-1}^* \end{bmatrix} \begin{bmatrix} H_k^T \\ \tilde{H}_k^T \end{bmatrix} + \tilde{R}_k$
$\mathrm{Sch}_k = S_k - \tilde{S}_k S_k^{-*} \tilde{S}_k^*$

Table 3.2   Prediction and Update Steps of the Augmented Kalman Filter

Complex-Valued Kalman Filter	
Prediction	$\underline{P}_{k\|k-1} = \underline{F}\,\underline{P}_{k-1\|k-1}\underline{F}^H + \underline{Q}_k$ $\hat{\underline{x}}_{k\|k-1} = \underline{F}\,\hat{\underline{x}}_{k-1\|k-1}$
Update	$\hat{\underline{x}}_{k\|k} = \hat{\underline{x}}_{k\|k-1} + \underline{\mathcal{K}}_k\underline{r}_k$ $\underline{P}_{k\|k} = \left[\underline{I} - \underline{\mathcal{K}}_k\underline{H}\right]\underline{P}_{k\|k-1}$
Innovation distribution : $\mathcal{N}(\underline{r}_k, \underline{S}_k)$	$\underline{r}_k = \begin{bmatrix} H\tilde{x}_k + \tilde{H}\tilde{x}_k^* + v_k \\ \tilde{x}_k^* H^* + \tilde{x}_k\tilde{H}^* + v_k^* \end{bmatrix}$ $\underline{S}_k = \underline{H}^H \underline{P}_{k\|k-1}\underline{H} + \underline{R}_k$
Kalman gain	$\underline{\mathcal{K}}_k = \underline{P}_{k\|k-1}\underline{H}\left[\underline{S}_k\right]^{-1}$

Table 3.2 outlines a simplified version of the complex-valued KF in terms of augmented variables. Note that the augmented variables in Table 3.2 are twice as big as the variables in Table 3.1. Next, we further analyze diagnostic and attack detection solutions to prevent complex-valued attacks.

3.5   Diagnostics and Monitoring

As stated previously, the conventional metrics used for attack detection in industrial control systems are functions of the statistics of the innovation sequence (some of which are precalculated). Intuitively speaking, the probability distribution of the innovation sequence is the sole source that contains all the required information for diagnosing and monitoring the system. Capitalizing on this intuition, we propose to monitor the statistical distance between the probability distribution of the innovation sequence under attack and the distribution of the innovation sequence with an optimal profile, and use this as the attack detection criterion. Later, we show that the χ^2-detector (the classical residue-based detection approach) is a special case of the proposed methodology, but first we briefly review properties of statistical distance measures.

3.5.1 Statistical Distance Measures

We are interested in computing the statistical distance or divergence between two Gaussian distributions using the F-divergence class. In probability theory, an F-divergence is a function that measures the difference between two probability distributions $f_1(\mathbf{x})$ and $f_2(\mathbf{x})$. This general notion has been introduced based on the intuitive fact that it is natural to measure the distance between two probability distributions f_1 and f_2 using the dispersion of the likelihood ratio (f_2/f_1) with respect to f_1. More rigorously, the F-divergence is defined as follows:

Definition 3.4   Let Φ be a continuous convex function and let Ψ be an increasing function. The class of F-divergence coefficients between two probabilities is given by

$$d_F(f_1, f_2) = \Psi\left[\mathbb{E}_{f_1}\left\{\Phi\left(\frac{f_2}{f_1}\right)\right\}\right], \tag{3.65}$$

where $\mathbb{E}_{f_1}\{\cdot\}$ is the expectation with respect to f_1.

Several well-known divergences, such as the Kullback–Leibler (KL) divergence, Bhattacharyya distance, and total variation distance, are special cases of the F-divergence class, each obtained based on a particular choice of Φ and Ψ. Any divergence, belonging to this class is nonnegative (i.e., $d_F(f_1, f_2) \geq 0$) and has the following joint convexity property:

Remark 3.1   If f, h_1, and h_2 are any PDFs over d-dimensions and $0 \leq w \leq 1$, the f-divergence functions satisfy the following inequalities:

$$d_F(wh_1 + \bar{w}h_2, f) \leq wd_F(h_1, f) + \bar{w}d_F(h_2, f), \tag{3.66}$$

$$d_F(f, wh_1 + \bar{w}h_2) \leq wd_F(f, h_1) + \bar{w}d_F(f, h_2), \tag{3.67}$$

where $\bar{w} = 1 - w$.

Next, we consider an important member of the F-divergence class used later for developing diagnostics for attack detection in industrial control systems.

3.5.1.1 Bhattacharyya Distance

The Bhattacharyya coefficient (BC), also known as Hellinger affinity, measures the amount of overlap between two statistical samples and is closely related to the Bhattacharyya distance (BD), which measures the similarity of two discrete or continuous probability distributions. Bhattacharyya distance is one of the most widely used techniques in several practical applications including (but not limited to) positioning in wireless local area networks [32] and image processing [33]. Specifically, the BD for Gaussian distributions is typically used for evaluating class separability in classification problems and feature extraction in pattern recognition. In general, feature extraction can be considered the process of transforming high-dimensional data into a low-dimensional feature space based on an optimization criterion. In other words, reducing dimensionality without a serious loss of class separability is the key to feature extraction. Dimensionality reduction and identification of relevant features are therefore important for the classification accuracy. In discriminant analysis, the Bayes error is the best criterion to evaluate feature sets, and posteriori probability functions are the ideal features. Unfortunately, the Bayes error is just too complex and useless as an analytical tool to extract features systematically. The BC/BD is directly related to the classification error and provides an upper bound on the Bayes error; therefore, it has been widely used in pattern recognition as an effective measure of the separability of two distributions. Bhattacharyya distance between two probability distributions ($f_1(\mathbf{x})$ and $f_2(\mathbf{x})$) is defined as follows:

$$d_{\mathrm{B}}(f_1, f_2) = -\log\left(\rho_{\mathrm{B}}(f_1, f_2)\right), \tag{3.68}$$

where the Bhattacharyya coefficient $\rho_{\mathrm{B}}(f_1, f_2)$ is defined as

$$\rho_{\mathrm{B}}(f_1, f_2) = \int_{\mathbb{R}^d} \sqrt{f_1(\mathbf{x})}\sqrt{f_2(\mathbf{x})}d\mathbf{x} = \int_{\mathbb{R}^d} f_2(\mathbf{x})\sqrt{\frac{f_1(\mathbf{x})}{f_2(\mathbf{x})}}d\mathbf{x}. \tag{3.69}$$

The Bhattacharyya distance can be obtained from (3.65) with $\Phi(\mathbf{x}) = -\sqrt{\mathbf{x}}$ and $\Psi(\mathbf{x}) = \log(-\mathbf{x})$. The Bhattacharyya is symmetric, that is, $d_{\mathrm{B}}(f_1, f_2) = d_{\mathrm{B}}(f_2, f_1)$, but it does not satisfy the triangle inequality. The following proposition provides a closed-form expression for the Bhattacharyya distance between two multivariate Gaussian distributions:

Proposition 3.2   Let $f_1(\mathbf{x})$ be the d-dimensional Gaussian PDF with mean vector μ_1 and positive definite covariance matrix P_1, and let $f_1(\mathbf{x})$ be the d-dimensional Gaussian PDF with mean vector μ_2 and positive definite covariance matrix P_2. The Bhattacharyya distance between $f_1(\mathbf{x})$ and $f_2(\mathbf{x})$ is

$$d_{\mathrm{B}}(f_1, f_2) = \frac{1}{8}(\mu_1 - \mu_2)\Gamma^{-1}(\mu_1 - \mu_2)^T + \frac{1}{2}\ln\left\{\frac{|\Gamma|}{\sqrt{|P_1||P_2|}}\right\}, \tag{3.70}$$

where

$$\Gamma = \frac{1}{2}(P_1 + P_2). \tag{3.71}$$

Next, BD is incorporated to develop a novel attack detection framework.

3.5.2 Detection via Statistical Distance Measure

In this section, we propose to compare the innovation sequence of the implemented filter with the one obtained theoretically, *in terms of the complex-valued Bhattacharyya distance between their probability distributions*. Essentially, two filters (the optimal filter and its implemented counterpart) produce two Gaussian distributions, one with the optimal profile and one with the operational profile. The only reasonable option to detect whether the estimator is in its normal operation state or under attack, is to calculate the statistical distance between Gaussian distributions in the innovation space. In other words, the attack detection becomes the problem of comparing the distance between two Gaussian representations of the innovation sequence. In order to show the vulnerability of industrial control systems to complex-valued attacks, first we extend Proposition 3.2 and derive a closed-form expression for computing the BD between the attacked complex-valued innovation sequence denoted by $\underline{v}_k^a = \mathcal{N}(\underline{r}_k^a, \underline{S}_k^a)$ and the optimal innovation sequence denoted by $\underline{v}^o = \mathcal{N}(\underline{r}^o, \underline{S}^o)$ as follows:

BD between complex-valued innovations :

$$d_{\mathrm{B}}(\underline{v}^o, \underline{v}_k^a) = \frac{1}{4}\left[\Re\left\{(\mathbf{r}^a - \mathbf{r}_k^o)^H \mathbf{\Lambda}_k (\mathbf{r}^a - \mathbf{r}_k^o)\right\}\right. \tag{3.72}$$

$$\left. + \Re\left\{(\mathbf{r}^a - \mathbf{r}_k^o)^H \tilde{\mathbf{\Lambda}}_k ([\mathbf{r}^a]^* - [\mathbf{r}_k^o]^*)\right\}\right] + \underbrace{\frac{1}{2}\ln\frac{|\mathbf{\Gamma}_k|}{\sqrt{|\underline{S}^a||\underline{S}_k^o|}}}_{\beta_k},$$

where

$$\mathbf{\Gamma}_k = \frac{1}{2}\left(\underline{S}^a + \underline{S}_k^o\right) \triangleq \begin{bmatrix} \mathbf{\Gamma}_k & \tilde{\mathbf{\Gamma}}_k \\ \tilde{\mathbf{\Gamma}}_k^* & \mathbf{\Gamma}_k^* \end{bmatrix} \tag{3.73}$$

and

$$\mathbf{\Lambda}_k \triangleq \mathbf{\Gamma}_k^{-1} = \begin{bmatrix} \mathbf{\Lambda}_k & \tilde{\mathbf{\Lambda}}_k \\ \tilde{\mathbf{\Lambda}}_k^* & \tilde{\mathbf{\Lambda}}_k^* \end{bmatrix}. \tag{3.74}$$

Equation 3.73 incorporates the complete second-order statistical properties of the attacked innovation sequence; therefore, it is a more descriptive criterion for attack detection purposes. Based on the statistical properties of the optimal innovation sequence (Equation 3.37), we have $\mathbf{r}^o = \mathbf{0}$ and $\underline{S}^o$ is a block-diagonal augmented matrix with the steady-state innovation covariance S on its upper-left block. The BD therefore reduces to

$$d_{\mathrm{B}}(\underline{v}^o, \underline{v}_k^a) = \frac{1}{4}\left[\Re\left\{(\mathbf{r}_k^a)^H \mathbf{\Lambda}_k (\mathbf{r}_k^a)\right\} + \Re\left\{(\mathbf{r}_k^a)^H \tilde{\mathbf{\Lambda}}_k (\mathbf{r}_k^a)^*\right\}\right] + \underbrace{\frac{1}{2}\ln\frac{|\mathbf{\Gamma}_k|}{|S|\sqrt{|\underline{S}_k^a|}}}_{\beta_k}. \tag{3.75}$$

The BD-detector at iteration k is finally given by

$$d_{\mathrm{B}}(\underline{\boldsymbol{v}}^o, \underline{\boldsymbol{v}}_k^a) \underset{H_1}{\overset{H_0}{\lessgtr}} \tau, \tag{3.76}$$

where

$$\begin{cases} H_0 : \text{ The system is normal.} \\ H_1 : \text{ The system is under attack.} \end{cases}$$

Note that the χ^2-detector is a special case of the BD-detector where only the first term is used. In ensuring the security of the estimation unit, all three terms are required; besides, it is essential to know which term is dominant, as this knowledge determines the type of attack and detection methodology that needs to be designed. The first and second terms on the right-hand side (RHS) of (3.75) measure the difference between the two distributions due to the distance between the means normalized by the covariance and pseudocovariance matrices, while the third term measures the distance solely due to the covariance and pseudocovariance shifts. The first two terms therefore provide the class separability due to the mean differences, while the third term provides the class separability due to the covariance and pseudocovariance differences. Note that the second term on the RHS of (3.75) appears due to the improperness of the injected innovation, that is, the presence of the pseudocovariance matrix $\tilde{\boldsymbol{\Lambda}}_k$. Similarly, the pseudo-covariance matrix plays an important role in the third term on the RHS of (3.75). The first and second terms disappear when $r_k^a = r$, and the third term disappears when $\underline{\boldsymbol{S}}_k^a = \underline{\boldsymbol{S}}$. Next, we further analyze when the third term dominates the distance between two innovations.

3.5.3 β-Dominance Attack

Up to this point, we have introduced a methodology for monitoring the innovation sequence using the BD. In this subsection, we investigate another potential attack on the estimation module that can only be observed based on the proposed BD-detector. A close look at (3.75) reveals that when the third term dominates the first two terms, the hypothesis calculation can become completely independent of the actual observations. We refer to this phenomenon as β-dominance [15] attack, which occurs when the determinant of $\underline{\boldsymbol{S}}_k^a$ is small. An adversary can attack the estimator by altering the innovation covariance matrix in such a way that its determinant is very small while the other terms maintain approximately the same size for both innovations.

We examine the dominance of β_k as a function of degrees of noncircularity of the underlying densities, that is, the true innovation sequence and its attacked counterpart, and derive an upper bound on its value. The proposed bound is significantly useful for identifying and avoiding the β-dominance problem. In particular, term β_k can be computed in terms of the eigenvalue decomposition of the noncircularity matrix (Equation 3.48). Since A_k is complex symmetric ($A_k = A_k^T$), there exists a special singular value decomposition (SVD), called Takagi's factorization [27], that is, $A_k = M_k K_k M_k^T$, where M_k is a unitary matrix and K_k is a diagonal matrix with the canonical correlations $1 \geq k_{\underline{S}_k^a}(1) \geq \ldots \geq k_{\underline{S}_k^a}(n_z) \geq 0$ on its diagonal [27]. The determinant of an augmented covariance matrix can be factorized in terms of these coefficients, for example,

$$|\underline{S}_k^a| = |S_k^a|^2 \prod_{i=1}^{n_z} \left(1 - k_{\underline{S}_k^a}^2(i)\right). \tag{3.77}$$

Using (3.77), β_k is factorized as

$$\beta_k = \ln \frac{|\Gamma_k|}{\sqrt{|S||S_k^a|}} + \frac{1}{2} \ln \prod_{i=1}^{n_z} \left(1 - k_{\underline{\Gamma}_k}^2(i)\right) - \frac{1}{4} \ln \prod_{i=1}^{n_z} \left(1 - k_{\underline{S}_k^a}^2(i)\right). \tag{3.78}$$

Note that the first term on the RHS of (3.78) represents the circular block of the underlying densities. For the special case of two scalar Gaussians, (3.78) simplifies to

$$\beta_k = \ln \frac{\Gamma_k}{\sqrt{ss_k^a}} + \ln \frac{1/2(1 - \alpha_{\Gamma_k}^2)}{\sqrt{1 - \alpha_{S_k^a}^2}},$$

where α_Γ is the circularity coefficient of Γ_k, and $\alpha_{S_k^a}$ is the circularity coefficient of S_k^a. Equation 3.78 highlights the difference between the BD computed based on the full second-order complex-valued statistics and its circular version obtained from the optimal operating profile of the estimator where the pseudocovariance blocks are zero. Although in the scalar case the circularity coefficient can be computed simply, for a general n_z-dimensional complex-valued Gaussian distribution, computing the circularity decomposition (i.e., computing $k_{P_*}(\cdot)$) could be computationally expensive. Developing upper and lower bounds on β_k in terms of the eigenvalues of the covariance and pseudocovariance matrices is of significant practical importance, as computing an eigenvalue decomposition is computationally less expensive. Consider the case where the eigenvalues of Γ_k and $\underline{S}_k^a$, denoted by $\lambda_\Gamma = [\lambda_{\Gamma,1}, \ldots, \lambda_{\Gamma,2n_z}]$ and $\lambda_{\underline{S}_k^a} = [\lambda_{\underline{S}_k^a,1}, \ldots, \lambda_{\underline{S}_k^a,n_z}]$, respectively, are known. Noting that $0 \leq \prod_{i=1}^{n_z} \left(1 - k_*^2(i)\right) \leq 1$, term β_k is upper bounded by

$$\beta_k \leq \frac{1}{4}\left[-\ln \prod_{i=1}^{n_z} \frac{4\lambda_{\underline{S}_k^a,i}\lambda_{\underline{S}_k^a,2n_z+1-i}}{(\lambda_{\underline{S}_k^a,i} + \lambda_{\underline{S}_k^a,2n_z+1-i})^2} + 2\ln \prod_{i=1}^{n_z} \frac{4\lambda_{\Gamma,2i}\lambda_{\Gamma,2i-1}}{(\lambda_{\Gamma,2i} + \lambda_{\Gamma,2i-1})^2} \right],$$

and lower bounded by

$$\beta_k \geq \prod_{i=1}^{n_z} \frac{4\lambda_{\Gamma,i}\lambda_{\Gamma,2n_z+1-i}}{(\lambda_{\Gamma,i} + \lambda_{\Gamma,2n_z+1-i})^2} - \prod_{i=1}^{n_z} \frac{4\lambda_{\underline{S}_k^a,2i}\lambda_{\underline{S}_k^a,2i-2}}{(\lambda_{\underline{S}_k^a,2i} + \lambda_{\underline{S}_k^a,2i-1})^2}. \tag{3.79}$$

In practical applications, the above bounds can be computed in a preprocessing step to check presence or absence of β-dominance. In the presence of β-dominance effect, one solution is to strip the β_k term and compute the BD only based on the first two terms. This completes our discussion on attack modeling and prevention, in the next section we perform experimental simulations.

3.6 Experiments

In this section, we present different simulation scenarios to evaluate the effects of complex-valued attacks on estimating a voltage signal using phasor measurements. The objective of the simulations is threefold. The first objective is to observe the effects of the noncircular attacks and test their capability to change the state of the system without being recognized by the conventional detectors. The second objective is to validate the effectiveness of the proposed BD-detector and noncircularity detection algorithms against that of the conventional χ^2-detector. The third objective is to evaluate the proposed circularization filter and test the performance of the χ^2-detector using circularized measurements.

3.6.1 State-Space Model

In the simulations, a voltage signal is estimated using noisy voltage phasor measurements. According to [16], a voltage signal is considered a sinusoidal wave, that is,

$$V_k = A_v^* \cos(\phi)cos(wk) - A_v \sin(\phi)\sin(wk), \tag{3.80}$$

where A_v is the voltage magnitude, ϕ is the voltage phase, and w is the angular velocity. Assuming that w is relatively constant over time, the state model is constructed as follows:

$$\begin{bmatrix} x_k^{(1)} \\ x_k^{(2)} \end{bmatrix} = \begin{bmatrix} 1 & 0 \\ 0 & 1 \end{bmatrix} \begin{bmatrix} x_{k-1}^{(1)} \\ x_{k-1}^{(2)} \end{bmatrix} + w_k, \tag{3.81}$$

where $x_k^{(1)} = A_v^* \cos(\phi)$ and $x_k^{(1)} = A_v \sin(\phi)$ are the state variables. In the experiments, we consider a sampling frequency of 2 KHz with $w = 60$ Hz, and $A_v = 1$. By considering a noisy version of the complex-valued voltage phasor $(z_k = A_v e^{j\phi})$, the observation model is constructed as

$$z_k = \begin{bmatrix} 1 & j \end{bmatrix} \mathbf{x}_k + v_k. \tag{3.82}$$

The state forcing term w_k, the observation noise v_k, and the initial conditions are assumed to be circular complex-valued Gaussian distributions with zero means. This completes the presentation of the state-space model used in the experiments.

3.6.2 Simulation Results

For a quantitative assessment of the performance, we present the results based on a Monte Carlo simulation of 1000 runs. Our comparison is based on two implementations of the KF, one based on the original observations and one based on the injected observations. The monitoring results are presented based on four detection algorithms: (1) the proposed BD-detector, (2) the proposed noncircularity detection algorithm, (3) the conventional χ^2-detector, and (4) the conventional χ^2-detector using the proposed circularization filter. We consider the following two scenarios:

Scenario 1: In this scenario, we design the attack model (Equation 3.40) in two steps: (1) a noncircular noise is added to the original measurements, that is, $z_k^a = z_k + v_k^a$, where v_k^a is a noncircular noise with a fixed $\alpha_{v_k^a} = 0.9$ degree of noncircularity, and (2) as the modified observation z_k^a is noncircular, it is no longer rotation invariant; therefore, in the second step of the attack, we rotate z_k^a to generate noncircular injected observations, that is, $z_k^a = e^{j\theta}(z_k + v^a)$, where $\theta \in \mathbb{R}$. Figure 3.5a shows an example of this attack where $\theta = \pi/1.5$ and the attack is injected at time $t = 0.04$. It is observed that the estimated voltage based on the true observations closely tracks the input signal. However, the state estimates are diverging from the input after the complex-valued attack is injected. Figure 3.5b shows (1) the BD between the innovation sequence obtained from the original measurements and its optimal counterpart, (2) the BD between the innovation sequence under attack and its optimal counterpart assuming circular observations, and (3) the proposed BD-detector, which incorporates the full second-order statistics of the observations. It is observed that the BD-detector

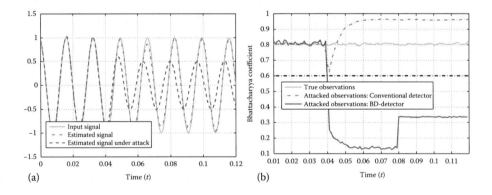

(a)

(b)

Figure 3.5: Complex-valued attack based on Scenario 1. (a) Estimated voltage. (b) BD-detector.

sharply detects the attack, whereas item 2, which ignores the pseudocovariances, provides false detection results.

Scenario 2: In this scenario, based on the attack model in (3.40), measurement is gradually modified as follows:

$$z_k^a = .5(\mathbf{z}_k + \mathbf{v}^a) + \left(\frac{2.5k}{N-1} - \frac{2.5}{N-1}\right)(\mathbf{z}_k + \mathbf{v}^a)^*, \tag{3.83}$$

where N is the number of filtering iterations. Figures 3.6 and 3.7 illustrates the effects of this attack based on a setup similar to that of Scenario 1. Figure 3.7a compares the root mean square error (RMSE) results based on the original observations with the one obtained from the attack observations. Based on the RMSE results, we can see that the estimation error resulting from this attack increases

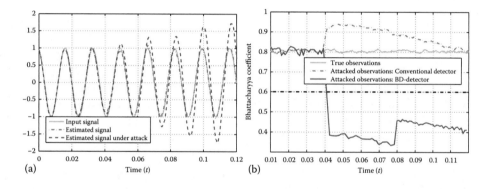

Figure 3.6: Complex-valued attack based on Scenario 2. (a) Estimated voltage. (b) BD-detector.

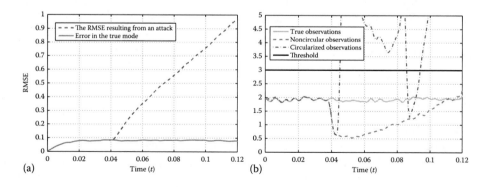

Figure 3.7: Complex-valued attack based on Scenario 2. (a) RMSE curves. (b) χ^2-detector.

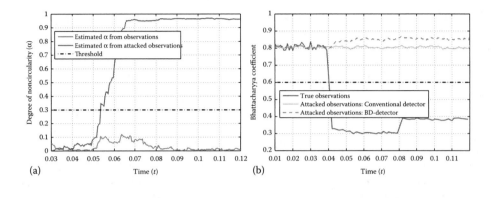

Figure 3.8: Performance of detection algorithms. (a) Noncircularity detector. (b) BD-detector.

unboundedly, which shows the destructive nature of the complex-valued attacks. Figure 3.7(b) shows (1) the χ^2-detector based on the original observations, (2) the χ^2-detector based on the injected observation, and (3) the χ^2-detector based on circularized observations obtained by applying the proposed CF. It is observed that while the conventional χ^2-detector misses the attack, its circularized version is capable of identifying the attack with a short delay. Finally, in Figure 3.8, we compare the performance of the noncircularity detection algorithm with the BD-detector. Both detection algorithms are capable of identifying complex-valued attacks; however, the BD-detector reacts faster to such attacks.

In summary, motivated by the emergence of sensor devices with complex-valued measurements, the chapter introduced a new class of deception attacks, referred to as noncircular attacks. Furthermore, (1) developed potential models for identifying and preventing complex-valued attacks, (2) introduced a new paradigm based on statistical distance measures to monitor the system against deception attacks, and (3) designed countermeasures to detect the noncircularity of the measurements and identify complex-valued attacks in real time.

3.7 Glossary

Bhattacharyya distance (BD): Measures the similarity of two discrete or continuous probability distributions and is closely related to the Bhattacharyya coefficient, which measures the amount of overlap between two statistical samples.

False data injection: An adversary compromises the integrity of sensor measurements or control commands to implement a cyber attack.

Innovation sequence: The difference between the observed value of state variables at a given time (e.g., k) and the optimal forecast of that value based on information available prior to that time.

noncircular process: A complex-valued signal is improper (noncircular) as long as it is correlated with its complex conjugate.

pseudocovariance: Characterizes the correlations between the real and imaginary parts of a complex-valued signal.

residue-based attack detection: The detection algorithm performs a hypothesis test by checking the power of the residuals computed based on state estimates.

widely linear processing: The processing model depends on both the signal itself and its complex conjugate.

References

1. J. Chen, X. Cao, P. Cheng, Y. Xiao, and Y. Sun, Distributed collaborative control for industrial automation with wireless sensor and actuator networks, *IEEE Transactions on Industrial Electronics*, vol. 57, no. 12, pp. 4219–4230, 2010.

2. F. Pasqualetti, F. Dorfler, and F. Bullo, Attack detection and identification in cyber-physical systems, *IEEE Transactions on Automatic Control*, vol. 58, no. 11, pp. 2715–2729, 2013.

3. S. Huang, K.K. Tan, and T.H. Lee, Fault diagnosis and fault-tolerant control in linear drives using the Kalman filter, *IEEE Transactions on Industrial Electronics*, vol. 59, no. 11, pp. 4285–4292, 2012.

4. A. Teixeira, I. Shames, H. Sandberg, and K.H. Johansson, A secure control framework for resource-limited adversaries, *Automatica*, vol. 51, pp. 135–148, 2015.

5. Y. Mo, R. Chabukswar, and B. Sinopoli, Detecting integrity attacks on SCADA systems, *IEEE Transactions on Control System Technology*, vol. 22, no. 4, pp. 1396–1407, 2014.

6. Z.H. Pang and G.P. Liu, Design and implementation of secure networked predictive control systems under deception attacks, *IEEE Transactions on Control System Technology*, vol. 20, no. 5, pp. 1334–1342, 2014.

7. Y. Mo, J. Hespanha, and B. Sinopoli, Resilient detection in the presence of integrity attacks, *IEEE Transactions on Signal Processing*, vol. 62, no. 1, pp. 31–43, 2014.

8. Q. Yang, J. Yang, W. Yu, D. An, N. Zhang, and W. Zhao, On false data-injection attacks against power system state estimation: Modeling and countermeasures, *IEEE Transactions on Parallel Distributed Systems*, vol. 25, no. 3, pp. 717–729, 2014.

9. Y. Liu, P. Ning, and M. K. Reiter, False data injection attacks against state estimation in electric power grids, *ACM Transactions on Information and System Security*, vol. 14, no. 1, pp. 13:1–13:33, 2011.

10. D. Ghosh, T. Ghose, and D.K. Mohanta, Communication feasibility analysis for smart grid with phasor measurement units, *IEEE Transactions on Industrial Informatics*, vol. 9, no. 3, pp. 1486–1496, 2013.

11. H. Morais, P. Vancraeyveld, A.H. Birger Pedersen, M. Lind, H. Johannsson, and J. Ostergaard, SOSPO-SP: Secure operation of sustainable power systems simulation platform for real-time system state evaluation and control, *IEEE Transactions on Industrial Informatics*, vol. 10, no. 4, pp. 2318–2329, 2014.

12. A. Mohammadi and K.N. Plataniotis, Improper complex-valued multiple-model adaptive estimation, *IEEE Transactions on Signal Processing*, vol. 63, no. 6, pp. 1528–1542, 2015.

13. A. Mohammadi and K.N. Plataniotis, Complex-valued Gaussian sum filter for nonlinear filtering of non-Gaussian/non-circular noise, *IEEE Signal Processing Letters*, vol. 22, no. 4, pp. 440–444, 2015.

14. A. Mohammadi and K.N. Plataniotis, Structure-induced complex Kalman filter for decentralized sequential Bayesian estimation, *IEEE Signal Processing Letters*, vol. 22, no. 9, pp. 1419–1423, 2015.

15. A. Mohammadi and K.N. Plataniotis, Improper complex-valued Bhattacharyya distance, *IEEE Transactions on Neural Networks and Learning Systems*, 2015, in press. [Epub ahead of print].

16. Y. Xia, S. Douglas, and D. Mandic, Adaptive frequency estimation in smart grid applications: Exploiting noncircularity and widely linear estimators, *IEEE Signal Processing Magazine*, vol. 29, no. 5, pp. 44–54, 2012.

17. J. He, J. Chen, P. Cheng, and X. Cao, Secure time synchronization in wireless sensor networks: A maximum consensus-based approach, *IEEE Transactions on Parallel and Distributed Systems,* vol. 25, no. 4, pp. 1055–1065, 2014.

18. H. Zhang, P. Cheng, L. Shi, and J. Chen, Optimal Denial-of-Service Attack Scheduling against Linear Quadratic Gaussian Control, In *Proceedings of American Control Conference*, Montreal. pp. 3996–4001, 2014.

19. K. Manandhar, X. Cao, F. Hu, and Y. Liu, Detection of faults and attacks including false data injection attack in smart grid using Kalman filter, *IEEE*

Transactions on Control of Network Systems, vol. 1, no. 4, pp. 370–379, 2014.

20. B. Picinbono and P. Chevalier, Widely linear estimation with complex data, *IEEE Transactions on Signal Processing*, vol. 43, no. 8, pp. 2030–2033, 1995.

21. P.J. Schreier and L.L. Scharf, *Statistical Signal Processing of Complex-Valued Data: The Theory of Improper and Noncircular Signals*, Cambridge: Cambridge University Press, 2010.

22. D.P. Mandic and S.L. Goh, *Complex Valued Nonlinear Adaptive Filters: Noncircularity, Widely Linear and Neural Models*, New York: Wiley, 2009.

23. P. Dash, A. Pradhan, and G. Panda, Frequency estimation of distorted power system signals using extended complex Kalman filter, *IEEE Transactions on Power Delivery*, vol. 14, no. 3, pp. 761–766, 1999.

24. D.H. Dini and D.P. Mandic, Class of widely linear complex Kalman filters, *IEEE Transactions on Neural Networks and Learning Systems*, vol. 23, no. 5, pp. 775–786, 2012.

25. W. Dang, and L.L. Scharf, Extensions to the theory of widely linear complex Kalman filtering, *IEEE Transactions on Signal Processing*, vol. 60, no. 12, pp. 6669–6674, 2012.

26. D.H. Dini, P.M. Djuric, and D.P. Mandic, The augmented complex particle filter, *IEEE Transactions on Signal Processing*, vol. 61, no. 17, pp. 4341–4346, 2013.

27. P.J. Schreier, Bounds on the degree of impropriety of complex random vectors, *IEEE Signal Processing Letters*, vol. 15, pp. 190–193, 2008.

28. A.S. Aghaei, K.N. Plataniotis, and S. Pasupathy, Widely linear MMSE receivers for linear dispersion space-time block-codes, *IEEE Transactions on Wireless Communications*, vol. 9, no. 1, pp. 8–13, 2010.

29. A.S. Aghaei, K.N. Plataniotis, and S. Pasupathy, Maximum likelihood detection in improper complex Gaussian noise, In *Proceedings of IEEE International Conference on Acoustics, Speech and Signal Processing*, Las Vegas. pp. 3209–3212, 2008.

30. A. Mirbagheri, K.N. Plataniotis, and S. Pasupathy, An enhanced widely linear CDMA receiver with OQPSK modulation, *IEEE Transactions on Communications*, vol. 54, no. 2, pp. 261–272, 2006.

31. Z. Gao, R.Q. Lai, and K.J. Ray Liu, Differential space-time network coding for multi-source cooperative communications, *IEEE Transactions on Communications*, vol. 59, pp. 3146–3157, 2011.

32. A. Kushki, K.N. Plataniotis, and A.N. Venetsanopoulos, Kernel-based positioning in wireless local area networks, *IEEE Transactions on Mobile Computing*, vol. 6, no. 6, pp. 689–705, 2007.

33. R. Lukac and K.N. Plataniotis, *Color Image Processing, Methods and Applications*, Boca Raton: CRC Press, 2007.

RESILIENT CONTROL THEORY

Chapter 4

Optimal Denial-of-Service Attack Policy against Wireless Industrial Control Systems

Heng Zhang

College of Control Science and Technology, Zhejiang University

Peng Cheng

College of Control Science and Technology, Zhejiang University

Ling Shi

Department of Electronic and Computer Engineering,
Hong Kong University of Science and Technology

Jiming Chen

College of Control Science and Technology, Zhejiang University

CONTENTS

Recently, a great deal of literature has been concerned with the security issues of wireless industrial control systems. However, there is still a lack of investigation on how the attacker should optimize its attack schedule in order to maximize the effect on the system performance due to the insufficiency of energy at the attacker side. This chapter fills in this gap from the aspect of control system performance. Especially, this chapter investigates the optimal denial-of-service (DoS) attack, which maximizes the linear-quadratic-Gaussian (LQG) control cost function under energy constraint. After analyzing the properties of the cost function under an arbitrary attack schedule, we derive the optimal jamming attack schedule and the corresponding cost function. System stability under this optimal attack schedule is also considered. Simulations are provided to demonstrate the effectiveness of the proposed optimal attack schedule.

4.1   Introduction

With the remarkable progress in the field of communication technology, industrial control systems nowadays are able to adopt wireless technology for data transmission and system control. Compared with the traditional wired industrial systems, the wireless industrial control systems have enormous advantages. For instance, the industrial control systems with wireless networks are highly flexible with low costs [2, 3]. However, wireless industrial control systems may be vulnerable to an increasing number of malicious attacks [18].

Various efforts have been devoted to studying the influence of specific malicious attacks, for example, denial-of-service (DoS) attacks [1, 22], replay attacks [23], and false data injection attacks [12], on wireless industrial control systems. Thereinto, *DoS attack*, which aims to prevent the communication between system components, has been widely studied, since this attack pattern is the most accomplishable and can result in serious consequences [1, 20, 21].

A typical DoS technique in wireless industrial control systems is the jamming attack, which can interfere with radio frequencies on the communication channels [13].

Recently researchers have studied the LQG problems under DoS attack [1, 5, 6]. A semidefinite programming-based solution was presented in [1] to find an optimal feedback controller that minimizes a cost function subject to safety and energy constraints in the presence of an attack with identical independent distributed actions. In [6], the optimal control law is designed against an intelligent jammer with limited actions. In [5], an event-triggered control strategy is derived in the presence of an energy-constrained periodic jamming attacker. The common characteristic of these related works is that they aim to find an optimal defensive control law. Our work, however, is from the viewpoint of the attacker; that is, we look for the optimal attack strategies to maximize the LQG cost function. This is equally important, as one can design an effective defensive control law only when he knows how the attacker behaves.

In almost all types of attacks, energy constraint is inherent and will affect an attacker's strategies [9, 11, 24]. Kashyap et al. [9] studied a zero-sum game on multiple-input multiple-output (MIMO) Gaussian Rayleigh fading channels between an intelligent DoS jammer and a decoder with bilateral power constraints. Li et al. [11] investigated the optimal jamming attack strategies by controlling the probability of jamming and the transmission range. Zuba et al. [24] studied the effect of a jamming attack on underwater wireless sensor networks and investigated the minimal energy consumption and the probability of detection in order to launch an effective DoS jamming attack.

In this chapter, we aim to design an optimal attack schedule to maximize the attacking effect on the wireless industrial control system. Specifically, we first consider a system where one sensor measures the system state and sends the data packets to a remote estimator through a wireless channel. The attacker has a limited energy budget in every active period and decides at each sampling time whether or not to jam the channel. The main contributions of this chapter that distinguish it from the related literature are summarized as follows:

1. We formulate a novel DoS attack problem and seek the optimal attack schedule that maximizes the LQG control cost function with energy constraint.

2. We investigate the analytical expression of the LQG cost function under an arbitrary attack schedule.

3. We provide the optimal attack schedule and analyze the system stability under this schedule.

4. We study the effectiveness of our proposed optimal attack schedules by software simulation and semiphysical simulation.

4.2 Problem Formulation

4.2.1 System Model

Consider the following discrete linear system:

$$x_{k+1} = Ax_k + Bu_k + w_k, \tag{4.1}$$

where $x_k \in \mathbb{R}^{n_x}$ is the state vector at time k, x_0 is the initial state vector, $u_k \in \mathbb{R}^{m_x}$ is the control input vector at time k, and $w_k \in \mathbb{R}^{n_x}$ is the zero-mean Gaussian with covariance $\mathrm{Cov}(w_i, w_j) = \delta_{ij} \Sigma_w$.* We assume that the pairs (A, B) and $(A, \Sigma_w^{\frac{1}{2}})$ are stabilizable [7].

The sensor measures the state x_k and sends it to the remote controller via a wireless channel (see Figure 4.1). The controller has a built-in estimator to estimate the state in case the packet is lost. The controller then generates a control packet u_k based on all received sensor measurements and historical control commands and sends u_k to the actuator through a reliable channel [16, 19].

We introduce $\theta_k = 1/0$ as the indicator function whether the data packet x_k is received or not by the controller at time k. Let $\mathcal{I}_k$ be the data set $\{\theta_1 x_1, \theta_2 x_2, \ldots, \theta_k x_k, u_1, u_2, \ldots, u_{k-1}\}$. We define $\hat{x}_k$ as the controller's minimum mean square error (MMSE) estimate of x_k at time k, that is,

$$\hat{x}_k = \mathbb{E}[x_k | \mathcal{I}_k], \tag{4.2}$$

where $\mathcal{I}_k$ is the controller's data at time k. The corresponding error covariance is

$$P_k = \mathbb{E}[(x_k - \hat{x}_k)(x_k - \hat{x}_k)' | \mathcal{I}_k]. \tag{4.3}$$

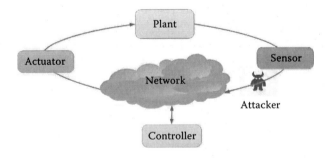

Figure 4.1: System architecture.

---

*δ_{ij} is the Kronecker delta function, that is, $\delta_{ij} = 1$ if $i = j$ and $\delta_{ij} = 0$ otherwise.

It is straightforward to obtain [16]

$$\hat{x}_k = \begin{cases} x_k, & \text{if } \theta_k = 1; \\ A\hat{x}_{k-1} + Bu_{k-1}, & \text{otherwise.} \end{cases} \tag{4.4}$$

We consider a linear static feedback controller of the form $u_k = L\hat{x}_k$, which is designed to minimize the following infinite-horizon LQG cost function [4, 12, 16]:

$$J = \lim_{N \to \infty} \frac{1}{N} \sum_{k=1}^{N} \mathbb{E}[x_k'Qx_k + u_k'Ru_k],$$

where $Q \geq 0$ and $R > 0$ are two weighting matrices and the expectation is taken over $\{w_k\}$. We will give the explicit form of L in Section 4.3. We assume that the controller does not know the existence of the attacker.

4.2.2 Attack Model

It has been summarized that there are three types of attack in a control system: integrity attacks, DoS attacks, and direct physical attacks to the process (see Figure 1 in [1]). In this chapter, we consider the case that the attacker performs a DoS attack with the objective of increasing the cost function J as much as possible subject to an energy constraint. We assume that the attacker can only attack the communication channel n times in a given active period T_{on}. After this period, he has to stop his attack actions and shift to an inactive period T_{off} to replenish the energy for the following attack period.

Let $\gamma(m) = (\gamma_{m,1}, \gamma_{m,2}, \ldots, \gamma_{m,T})$ be the attack schedule in the mth period, where $\gamma_{m,t}, t = 1, 2, \ldots, T_{\text{on}}$ is the attack decision variable in active time, that is, $\gamma_{m,t} = 1$ if he jams the transmission channel at time t of the mth active period, and $\gamma_{m,t} = 0$ otherwise, and $\gamma_{m,t} = 0$ for inactive time $i = T_{\text{on}}+1, T_{\text{on}}+2, \ldots, T$, with $T = T_{\text{on}} + T_{\text{off}}$. The consequence of attacking action $\gamma_{m,t} = 1$ is

$$\theta(\gamma_{m,t}) = \begin{cases} 1, & \text{with probability } 1 - \alpha; \\ 0, & \text{with probability } \alpha. \end{cases} \tag{4.5}$$

If he does not take action at time t, that is, $\gamma_{m,t} = 0$, the data from the sensor can be received by the controller, that is, $\theta(\gamma_{m,t}) = 1$.

From the attacker's viewpoint, it is of interest to design the optimal attack schedule that maximizes the expected cost function, that is,

Problem 4.1

$$\max_{\gamma \in \Theta} \mathbb{E}[J(\gamma)] \tag{4.6}$$

$$s.t. \ \| \gamma(m) \|_0 \leq n, \forall m \in \mathbb{N}, \tag{4.7}$$

where $\gamma = [\gamma(m)]_{m=1}^{\infty}$ is the attack schedule on the infinite time horizon $[1, \infty)$, and $\Theta = \{\gamma | \gamma_{m,t} \in \{0,1\}, \forall m \in \mathbb{N}, t \in \{1,2,\ldots,T_{\mathrm{on}}\}\}$ is the attack schedule space; $\| \cdot \|_0$ is the zero-norm operation.

4.3 System Control Performance under Given Attack Schedule

Before investigating the optimal attack schedule, we introduce some preliminary results when the system is under a given attack schedule.

From [12, 16], we can obtain the following conclusion:

Lemma 4.1

The optimal static feedback controller is given by

$$u_k = -(B'SB + R)^{-1}B'SA\hat{x}_k,$$

where S is the unique solution to the following equation [10]:

$$S = A'SA + Q - A'SB(B'SB + R)^{-1}B'SA,$$

and the minimal cost is

$$J^* = J_c + J_e,$$

where

$$J_c = \mathrm{Tr}[S\Sigma_w], \tag{4.8}$$

$$J_e = \lim_{N \to \infty} \frac{1}{N} \sum_{k=1}^{N} \mathrm{Tr}[MP_k], \tag{4.9}$$

with $M = A'SB(B'SB + R)^{-1}B'SA$.*

Now we present some properties of system performance under a given DoS attack schedule.

Lemma 4.2

Let $f(P) = \mathrm{Tr}(MP)$, where P is $n_x \times n_x$ positive definite. If positive definite matrices P_1 and P_2 are $n_x \times n_x$ and satisfy $P_1 \geq P_2$, we have $f(P_1) \geq f(P_2)$.

Proof 4.1   If $P_1 \geq P_2$, we have

$$\begin{aligned} f(P_1) - f(P_2) &= \mathrm{Tr}(MP_1) - \mathrm{Tr}(MP_2) \\ &= \mathrm{Tr}\left[M(P_1 - P_2)\right] \geq 0. \end{aligned}$$

□

---

*It can be seen that M is a positive semidefinite matrix.

Note that J_c is the fixed part of J and J_e is the variable part that is affected by an attack schedule γ. Thus, we only need to find an optimal attack schedule to maximize $\mathbb{E}[J_e(\gamma)]$. From Lemma 4.2, to maximize $\mathbb{E}[J_e(\gamma)]$ is equivalent to maximizing

$$\lim_{N \to \infty} \frac{1}{N} \sum_{k=1}^{N} \mathbb{E}[P_k(\gamma)].$$

Thus, we have to investigate $\mathbb{E}[P_k]$ under arbitrary attack schedule γ.

From (4.3) and (4.4) one can see that

$$P_{k+1} = \begin{cases} \Sigma_w, & \text{if } \theta_{k+1} = 0, P_k = 0; \\ AP_kA' + \Sigma_w, & \text{if } \theta_{k+1} = 0, P_k \neq 0; \\ 0, & \text{otherwise.} \end{cases}$$

Then we have

$$\mathbb{E}[P_{k+1}|P_k = 0] = \begin{cases} \alpha \Sigma_w, & \text{if } \gamma_{k+1} = 1; \\ 0, & \text{otherwise,} \end{cases}$$

and

$$\mathbb{E}[P_{k+1}|P_k \neq 0] = \begin{cases} \alpha(AP_kA' + \Sigma_w), & \text{if } \gamma_{k+1} = 1; \\ 0, & \text{otherwise.} \end{cases}$$

The following lemma shows how the consecutive attack actions affect the error covariance:

Lemma 4.3

For a given consecutive attack time interval $[s+1, s+t]$, that is, $\gamma_s = 0, \gamma_{s+1} = \gamma_{s+2} = \cdots = \gamma_{s+t} = 1, \gamma_{s+t+1} = 0$, we have

$$\mathbb{E}[P_{s+j}] = \sum_{i=0}^{j-1} (\alpha^i - \alpha^{i+1}) H_i + \alpha^j H_j, \tag{4.10}$$

where $H_i = \sum_{l=0}^{i-1} (A^l) \Sigma_w (A^l)'$ and $j = 1, 2, \ldots, t$.

Proof 4.2 Since $\gamma_s = 0, \gamma_{s+1} = 1$, we have $P_s = 0$, and $P_{s+1} = \mathbb{E}[w_s w_s'] = \Sigma_w$ with probability α, and $P_{s+1} = 0$ with probability $1 - \alpha$. By the recursive method for time $s+k$, one has

$$\Pr\{P_{s+k} = H_i\} = \begin{cases} \alpha^i - \alpha^{i+1}, & i = 0, 1, \ldots, k-1; \\ \alpha^k, & i = k, \end{cases}$$

where $H_0 = 0$.

Then we can see that

$$\mathbb{E}[P_{s+k}] = \sum_{i=0}^{k} H_i \Pr\{P_{s+k} = H_i\}$$

$$= \sum_{i=0}^{k-1} (\alpha^i - \alpha^{i+1}) H_i + \alpha^k H_k,$$

which finishes the proof. □

From Lemma 4.3, for the given consecutive attack time interval $[s+1, s+t]$, $\mathbb{E}[P_{s+j}], j = 1, 2, \ldots, t$, does not depend on s. Thus, we denote $\Gamma_\alpha(t)$ as the sum of expected error covariance with any t times consecutive attack in the given active attack period $[s+1, s+t]$. Using the same method, any attack strategy with consecutive attack sequence $t_1, t_2, \ldots, t_s$ in a given period will lead to the same sum of expected error covariance. Then, we denote $\Gamma_\alpha(t_1 \oplus t_2 \oplus \cdots \oplus t_s)$ as the sum of expected error covariance with consecutive attack sequence $t_1, t_2, \ldots, t_s$ in the given attack period. Note that $t_1, t_2, \ldots, t_s$ satisfy the commutative law in $\Gamma_\alpha(t_1 \oplus t_2 \oplus \cdots \oplus t_s)$.

In fact,

$$\Gamma_\alpha(t) = \sum_{j=1}^{t} \mathbb{E}[P_{s+j}]$$

$$= \sum_{j=1}^{t} \left[\sum_{i=0}^{j-1} (\alpha^i - \alpha^{i+1}) H_i + \alpha^j H_j \right]$$

$$= \sum_{i=0}^{t} \left[(t - i + 1)\alpha^i - (t - i)\alpha^{i+1} \right] H_i,$$

and

$$\Gamma_\alpha(t_1 \oplus t_2 \oplus \cdots \oplus t_s) = \sum_{i=1}^{s} \sum_{j_i=1}^{t_i} \mathbb{E}[P_{s_i+j_i}] = \sum_{i=1}^{s} \Gamma_\alpha(t_i).$$

Lemma 4.4

The following statements are true:

1. $\Gamma_\alpha(t_1) \leq \Gamma_\alpha(t_2)$, where $t_1 \leq t_2$.

2. $\Gamma_\alpha(t_1 \oplus t_2) \leq \Gamma_\alpha(t)$, where $t = t_1 + t_2$.

3. $\Gamma_\alpha(t_1 \oplus t_2 \oplus \cdots \oplus t_s) \le \Gamma_\alpha(t)$, where $t = t_1 + t_2 + \cdots + t_s$.

4. $\Gamma_\alpha(t_1 \oplus t_2) \le \Gamma_\alpha(t_3 \oplus t_4)$, where $t_1 + t_2 = t_3 + t_4$ and $\max\{t_1, t_2, t_3, t_4\}$ is t_3 or t_4.

Proof 4.3

1. We have

$$\Gamma_\alpha(t_2) - \Gamma_\alpha(t_1) = \sum_{k=0}^{t_1} (t_2 - t_1)(\alpha^k - \alpha^{k+1}) H_k$$

$$+ \sum_{k=t_1+1}^{t_2} [(t_2 - k)(\alpha^k - \alpha^{k+1}) + \alpha^k] H_k \ge 0.$$

2. It can be seen that

$$\Gamma_\alpha(t) - \Gamma_\alpha(t_1 \oplus t_2) = \sum_{k=1}^{t_2} \left[\alpha^k (H_{t_1+k} - H_k) + \sum_{i=0}^{k-1} (\alpha^i - \alpha^{i+1})(H_{t_1+i} - H_i) \right]$$

$$\ge 0.$$

3. Since the proof of this result is similar to that of 2, it is omitted here.

4. It is sufficient to prove

$$\Gamma_\alpha(t_1 \oplus t_2) \le \Gamma_\alpha\big((t_1 - 1) \oplus (t_2 + 1)\big),$$

with $t_1 \le t_2$. In fact,

$$\Gamma_\alpha\big((t_1 - 1) \oplus (t_2 + 1)\big) - \Gamma_\alpha(t_1 \oplus t_2) = \sum_{k=t_1}^{t_2} \alpha^{k+1} (H_{k+1} - H_k) \ge 0.$$

$\square$

From Lemma 4.4, one can see that more attack times (more energy) used in any given period lead to higher cost function. Thus, the constraint (4.7) in Problem 4.1 can be replaced by

$$\| \gamma(m) \|_0 = n, \forall m \in \mathbb{N}. \tag{4.11}$$

4.4 Optimal Attack Schedule Analysis

In this section, we first present the optimal attack schedule for Problem 4.1, and then analyze the system stability under the proposed attack schedule.

4.4.1 Optimal Attack Schedule

Theorem 4.1

The optimal attack schedule γ^* for Problem 4.1 is any n times consecutive attack at active periods, and the expected corresponding cost function can be calculated as follows:

$$\mathbb{E}[J(\gamma^*)] = J_c + \frac{1}{T}\mathrm{Tr}\big[M\Gamma_\alpha(n)\big]. \tag{4.12}$$

Proof 4.4   From Lemma 4.4, one can obtain that any n times consecutive attack at active periods is an optimal attack schedule.

Let $N = qT + m$ with $q, m \in \mathbb{N}$, and $0 \le m < T$. Then

$$\frac{1}{(q+1)T} < \frac{1}{N} \le \frac{1}{qT}.$$

If $0 \le m \le n$, we have

$$\sum_{k=1}^{N} \mathbb{E}[P_k] = qT\Gamma_\alpha(n).$$

If $n < m < T$,

$$\sum_{k=1}^{N} \mathbb{E}[P_k] = (qT + 1)\Gamma_\alpha(n).$$

Then, we have

$$\frac{1}{(q+1)T}\sum_{k=1}^{N} \mathbb{E}[P_k] < \frac{1}{N}\sum_{k=1}^{N} \mathbb{E}[P_k] \le \frac{1}{qT}\sum_{k=1}^{N} \mathbb{E}[P_k].$$

Taking the limit $N \to \infty$ to this inequality, we can obtain

$$\lim_{N\to\infty} \frac{1}{N}\sum_{k=1}^{N} \mathbb{E}[P_k] = \frac{1}{T}\big[\Gamma_\alpha(n)\big],$$

which finishes the proof. □

Example 4.1   Consider Problem 4.1 with $T_{on} = 5, n = 3$, and $T_{off} = 5$. From Theorem 4.1, it can be seen that in any active period the attack schedules $(1, 1, 1, 0, 0)$, $(0, 1, 1, 1, 0)$, and $(0, 0, 1, 1, 1)$ are optimal. The corresponding result on J can be calculated as

$$\mathbb{E}[J(\gamma^*)] = J_c + \frac{1}{10}\mathrm{Tr}\big[M\Gamma_\alpha(3)\big].$$

From Theorem 4.1, under an optimal attack schedule, the jamming instances are grouped together and this schedule does not depend on system parameters.

4.4.2 Stability Analysis under Optimal Attack Schedule

Stability is critical in an industrial control system. Before presenting results on the stability of system (4.12) under an optimal attack schedule, we give the definition of system stability formally as follows [7, 8]:

Definition 4.1 The system (4.12) is stable if the covariance of system state is bounded, that is, $\text{Var}(x_k) \leq C^*$, where C^* is a constant matrix.

Theorem 4.2
Consider system (4.12) under an optimal attack schedule γ^*. When $T_{\text{off}} > 0$, the system is stable in the sense of bounded covariance.

Proof 4.5 From Definition 4.1, one can see that

$$\text{Var}(x_k) = \mathbb{E}\left[\mathbb{E}\left((x_k - \mathbb{E}(x_k))(x_k - \mathbb{E}(x_k))' | \mathcal{I}_k\right)\right]$$
$$= \mathbb{E}\left[\mathbb{E}\left((x_k - \widehat{x}_k)(x_k - \widehat{x}_k)' | \mathcal{I}_k\right)\right] = \mathbb{E}\left[P_k\right].$$

Since $T_{\text{off}} > 0$, from (4.3) we have

$$\text{Var}(x_k) \leq \sum_{i=0}^{n-1}(\alpha^i - \alpha^{i+1})H_i + \alpha^n H_n = C^*;$$

that is, the covariance of state is bounded for a given n. □

Theorem 4.2 shows that the optimal DoS attack cannot change the system stability if $T_{\text{off}} > 0$. If $T_{\text{off}} = 0$, the attacker has unlimited energy, and he can always attack the wireless channel. Thus, $\theta_k, k = 1, 2, \ldots$ becomes an independent and identically distributed (i.i.d.) Bernoulli random variable sequence with $\mathbb{E}(\theta_k) = 1 - \alpha$. Similar to [17], the system is stable if and only if $\alpha < 1 - \gamma_c$, where $\gamma_c = \inf_\gamma \{\gamma | X = A'XA + \Sigma_w - \gamma A'XB(B'XB + R)^{-1}B'XA\}$.

4.5 Simulations

Security testbeds for industrial systems can be classified into three categories: physical testbeds, software simulation testbeds, and semiphysical simulation testbeds [14]. Physical testbeds, which employ the same experimental equipment as the real world to construct the security test platform, need a long cycle of construction and have cost. A typical one is the supervisory control and data acquisition (SCADA) test platform built by Idaho National Laboratory of the U.S. Department of Energy, which includes a full-scale power grid, a wireless testbed, and an Internet security testbed [14]. The software simulation testbeds include physical process simulation, network simulation, and attack simulation. One can set up such a testbed in a short time with little cost. The low-cost testbeds have a big gap from the actual system, so they are mainly used for academic research. The shortcoming of the software simulation testbeds is that the software cannot fully simulate the real environment. In order to overcome this drawback, semi-physical testbeds are designed to study the security issues in industrial control systems. This type of testbeds can save costs and simulate some real industrial environments.

In order to evaluate the effectiveness of our proposed attack schedules, we study the control performance under DoS attack by the software simulation approach and semiphysical simulation approach, respectively.

4.5.1 Software Simulation

We consider system (4.12) with

$$A = 2, B = 1, \Sigma_w = 1, Q = 1, R = 1.$$

Assume the length of the attacker's active period is $T_{on} = 80$, the inactive period is $T_{off} = 20$, and attack energy constraint is $n = 20$. With the help of Matlab, we illustrate the effects of different attack schedules on the expected cost function $\mathbb{E}(J)$ and the system stability under optimal attacks by the Monte Carlo method. The simulation system runs 10,000 times for each illustration.

4.5.1.1 Different Attack Schedules

Figure 4.2 shows the variation of the expected cost function under different attack schedules when the sensor uses a deterministic channel for transmission. In Figure 4.2, we examine the attack effect under different attack success probabilities from $\alpha = 0.01$ to $\alpha = 0.35$. The top curve of this figure stands for the performance under the attack schedule given by Theorem 4.1, that is, $\mathbb{E}[J(\gamma^*)]$, which maximizes the expected cost function. It can also be seen that the expected cost function increases rapidly with α. The second line from the top shows the performance under a common attack schedule with consecutive attack sequence $t_1 = t_2 = 10$. It can be seen that the expected cost function grows much slower than $\mathbb{E}[J(\gamma^*)]$. We also can find that the performance under a uniform attack with energy constraint in each active period is nearly the same as those without attack.

4.5.1.2 System Stability

Figures 4.3 and 4.4 plot the evolution of the system state with different attack parameters T_{on} and T_{off}. In Figure 4.3, the attacker is working with $n = 20$, $T_{on} = 80$, $T_{off} = 40$, and $\alpha = 0.2$. From Theorem 4.1, the attack schedule in Figure 4.3 (attacking at times $[81, 100]$, $[221, 240]$, $[361, 380]$, and so on) is an

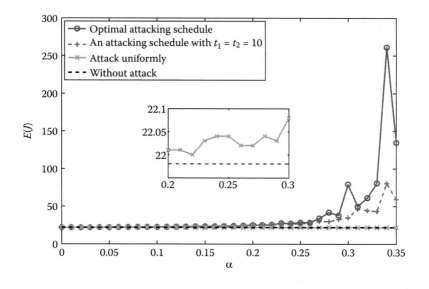

Figure 4.2: Example of attacking effect on expected cost function $E(J)$ with different attack schedules.

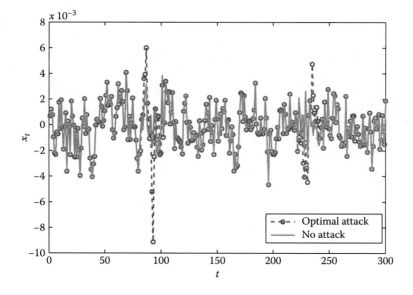

Figure 4.3: System state under optimal attack schedule with and without attack, respectively. Here the attacker can take actions $n = 20$ times in any active period $T_{on} = 100$ with successful probability $\alpha = 0.2$ ($T_{off} = 40$).

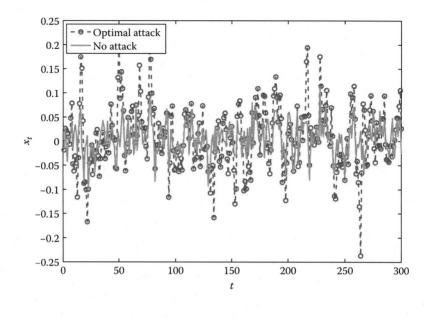

Figure 4.4: System state under optimal attack schedule with and without attack, respectively. Here the attacker can always take actions in any active period $T_{on} = 20$ with successful probability $\alpha = 0.2$ ($T_{off} = 1$).

optimal attack policy. In Figure 4.3, the dash line demonstrates that the system is still stable under this optimal attack schedule. The solid line shows the system's state without attack. In Figure 4.4, the attacker is working with $T_{on} = n = 20$ and $T_{off} = 1$. We can see that the state is still stable under this optimal attack.

Figures 4.5 and 4.6 show the effectiveness of the optimal attack schedule on the system state when there is no inactive period, that is, $T_{off} = 0$ and $T_{on} = n$. From Theorem 4.2, the stability depends on the attack successful probability α. In our example, the critical value can be calculated by $\gamma_c = 1 - \frac{1}{A^2} = 0.75$ [17]. From Figure 4.5, one can see that the system is still stable when it suffers from the consecutive attack with successful probability $\alpha = 0.2 < 1 - \gamma_c = 0.25$. The system will become unstable, however, if the successful probability is $\alpha = 0.8 > 1 - \gamma_c$, which has been demonstrated in Figure 4.6.

4.5.2 Semiphysical Simulation

We establish a semiphysical security testbed that consists of a virtual plant, physical controller, and communication process. The architecture of this testbed is given in Figure 4.7. The completed closed-loop control is described as follows: Real-time system states of the virtual plant are sent to the programmable logic controller (PLC) through wirelessly. Then, the control algorithm reads the system

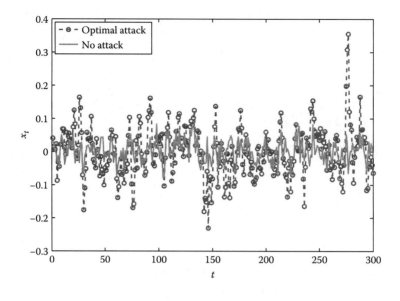

Figure 4.5: System state under optimal attack schedule with and without attack, respectively. Here the attacker can always take actions in any active period $T_{on} = 20$ with successful probability $\alpha = 0.2$ ($T_{off} = 0$).

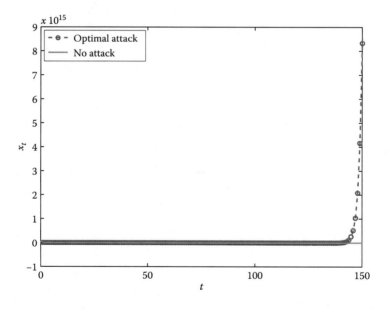

Figure 4.6: System state under optimal attack schedule with and without attack, respectively. Here the attacker can always take actions in any active period $T_{on} = 20$ with successful probability $\alpha = 0.8$ ($T_{off} = 0$).

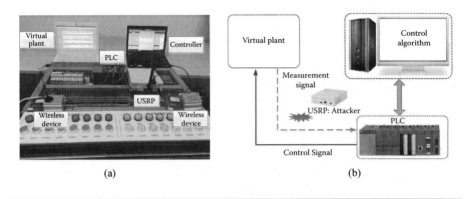

(a)　　　　　　　　　　　　　　　　(b)

Figure 4.7: Structure of semiphysical testbed.

states to calculate the control commands and writes them back to the PLC. The control commands are transmitted to the virtual plant in a wired channel.

We build an inverted pendulum control system for experiments, which is based on the system presented in [15]. The parameters are given as follows:

$$A = \begin{pmatrix} 1.001 & 0.005 & 0.000 & 0.000 \\ 0.350 & 1.001 & -0.135 & 0.000 \\ -0.001 & 0.000 & 1.001 & 0.005 \\ -0.375 & -0.001 & 0.590 & 1.001 \end{pmatrix}, B = \begin{pmatrix} 0.001 \\ 0.540 \\ -0.002 \\ -1.066 \end{pmatrix},$$

$$\Sigma_w = qq', q = \begin{pmatrix} 0.003 \\ 1.000 \\ -0.005 \\ -2.150 \end{pmatrix}, Q = \begin{pmatrix} 5 & 0 & 0 & 0 \\ 0 & 1 & 0 & 0 \\ 0 & 0 & 1 & 0 \\ 0 & 0 & 0 & 1 \end{pmatrix}, R = 2.$$

When the whole closed-loop control system runs without attack, the control performance and control command curve are shown in Figure 4.8.

With the help of Universal Software Radio Peripheral (USRP) N210, we implement a DoS jamming attack on the wireless channel. We verify the optimal attack strategies through real experiments based on our established semiphysical testbed. We set $T = 40, T_{on} = 30, T_{off} = 10$, and the attack times $n = 20$ for each period. This means that we can allocate the 20 times attack freely for the first 30 times in each period, and then the attacker has to pause to restore energy. In our experiment, we set the total attack period equal to 10. Although the total attack time in each period is the same, different attack time allocations will influence the attack effect.

In our experiment, we start the DoS jamming attack at time 220. USRP is exploited to generate jamming signals that have the same center frequency as those produced by the industrial wireless device. We can change the amplitude of the radio signal to modulate the jamming intensity. The effect of the DoS attack is determined by the signal frequency and attack intensity. When the DoS attack

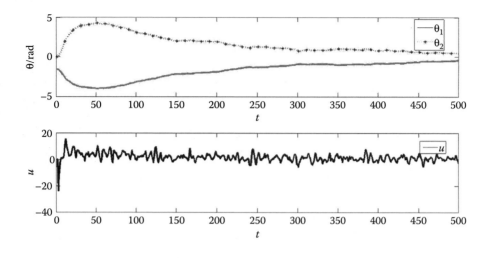

Figure 4.8: Variation of angles and control commands without attack.

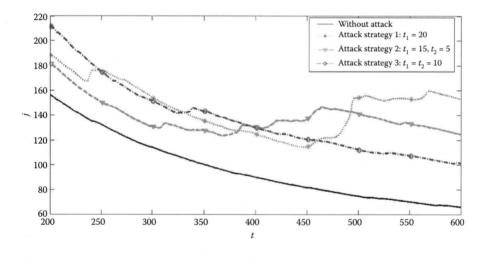

Figure 4.9: Variation of cost function J under different attack strategies.

is successful, the measurement cannot be received by PLC, and the controller has to run the control algorithm with the estimated state values.

In order to verify the optimal DoS jamming attack strategy, we implement attack with different strategies. These strategies are as follows:

1. Attack strategy 1: Attack sequence $t_1 = 20$

2. Attack strategy 2: Attack sequence $t_1 = 15, t_2 = 5$

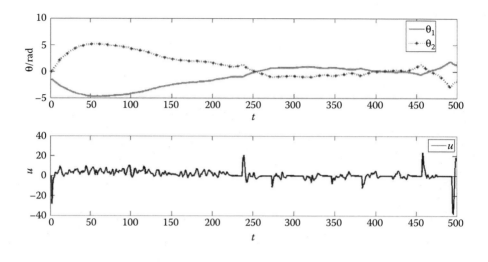

Figure 4.10: Variation of angles and control commands under attack strategy 1.

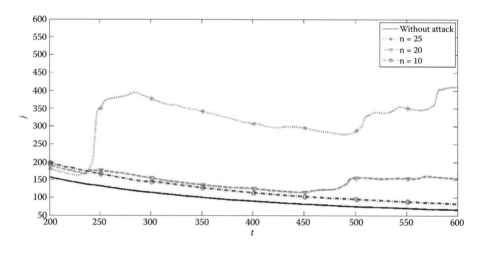

Figure 4.11: Variation of cost *J* under optimal attack with different *n*.

3. Attack strategy 3: Attack sequence $t_1 = t_2 = 10$

From Figure 4.9, one can see that after a long period of time, the cost function under attack strategy 1 is maximal. This experimental result is consistent with Theorem 4.1. Figure 4.10 shows the variation of states and control commands under attack strategy 1. We also study the effectiveness of optimal attack schedules with different *n* in Figure 4.11. It can be seen that a greater attack energy *n* will result in a larger cost *J*.

4.6 Conclusion

In this chapter, we studied the optimal DoS attack policy with the energy constraint to maximize the LQG cost function in wireless industrial control systems. We first formulated an optimization problem from the perspective of a DoS attacker, in which the attacker can jam the transmission channel with limited actions in any active period. Then, we analyzed the properties of the LQG cost function under any given feasible attack schedule. The optimal attack schedules and corresponding expected cost are obtained, which demonstrate that grouping the limited attacks together in every active period is optimal. We also studied the effectiveness of the proposed optimal jamming attack policy by software simulation and semiphysical simulation.

References

1. S. Amin, A. Cardenas, and S. Sastry. Safe and secure networked control systems under denial-of-service attacks. *Hybrid Systems: Computation and Control*, Springer, Berlin, pp. 31–45, 2009.

2. J. Chen, X. Cao, P. Cheng, Y. Xiao, and Y. Sun. Distributed collaborative control for industrial automation with wireless sensor and actuator networks. *IEEE Transactions on Industrial Electronics*, 57(12):4219–4230, 2010.

3. P. Cheng, Y. Qi, K. Xin, J. Chen, and L. Xie. Energy-efficient data forwarding for state estimation in multi-hop wireless sensor networks. *IEEE Transaction on Automatic Control*, [Epub ahead of print], 2015.

4. Y. Cho and P. Gyugyi. Control of rapid thermal processing: A system theoretic approach. *IEEE Transactions on Control Systems Technology*, 5(6):644–653, 1997.

5. H. S. Foroush and S. Martinez. On event-triggered control of linear systems under periodic denial-of-service jamming attacks. In *Proceedings of IEEE Conference on Decision and Control*, Maui, HI, pp. 2551–2556, 2012.

6. A. Gupta, C. Langbort, and T. Başar. Optimal control in the presence of an intelligent jammer with limited actions. In *Proceedings of IEEE Conference on Decision and Control*, Atlanta, GA, pp. 1096–1101, 2010.

7. V. Gupta, B. Hassibi, and R. Murray. Optimal LQG control across packet-dropping links. *Systems and Control Letters*, 56(6):439–446, 2007.

8. V. Gupta, R. Murray, L. Shi, and B. Sinopoli. Networked sensing, estimation and control systems. California Institute of Technology, Pasadena, 2009.

9. A. Kashyap, T. Başar, and R. Srikant. Correlated jamming on MIMO Gaussian fading channels. *IEEE Transactions on Information Theory*, 50(9):2119–2123, 2004.

10. P. Lancaster and L. Rodman. *Algebraic Riccati Equations*. Oxford University Press, Oxford, 1995.

11. M. Li, I. Koutsopoulos, and R. Poovendran. Optimal jamming attack strategies and network defense policies in wireless sensor networks. *IEEE Transactions on Mobile Computing*, 9(8):1119–1133, 2010.

12. Y. Mo, R. Chabukswar, and B. Sinopoli. Detecting integrity attacks on SCADA systems. *IEEE Transactions on Control Systems Technology*, 22(4):1396–1407, 2014.

13. R. Poisel. *Modern Communications Jamming: Principles and Techniques*. Artech House, Norwood, MA, 2011.

14. C. Queiroz, A. Mahmood, J. Hu, Z. Tari, and X. Yu. Building a SCADA security testbed. In *Proceedings of IEEE International Conference on Network and System Security*, Gold Coast, Queensland, pp. 357–364, 2009.

15. L. Schenato, B. Sinopoli, M. Franceschetti, K. Poolla, and S. Sastry. Foundations of control and estimation over lossy networks. *Proceedings of the IEEE*, 95(1):163–187, 2007.

16. L. Shi, Y. Yuan, and H. Zhang. Sensor data scheduling for linear quadratic Gaussian control with full state feedback. In *Proceedings of American Control Conference*, Montréal, Quebec, pp. 2030–2035, 2012.

17. B. Sinopoli, L. Schenato, M. Franceschetti, K. Poolla, and S. Sastry. Optimal control with unreliable communication: The TCP case. In *Proceedings of American Control Conference*, Portland, OR, pp. 3354–3359, 2005.

18. M. Wei, K. Kim, P. Wang, and J. Choe. Research and implementation on the security scheme of industrial wireless network. In *Proceedings of International Conference on Information Networking*, Kuala Lumpur, pp. 37–42, 2011.

19. K. Xin, X. Cao, P. Cheng, and J. Chen. Optimal controller location in wireless sensor and actuator networks. In *Proceedings of International Conference on Control Automation Robotics and Vision*, Guangzhou, Guangdong, pp. 560–565, 2012.

20. Y. Yuan, Q. Zhu, F. Sun, Q. Wang, and B. Tamer. Resilient control of cyberphysical systems against denial-of-service attacks. In *Proceedings of International Symposium on Resilient Control Systems*, San Francisco, CA, 2013.

21. H. Zhang, P. Cheng, L. Shi, and J. Chen. Optimal DoS attack policy against remote state estimation. In *Proceedings of IEEE Conference on Decision and Control*, Firenze, pp. 5444–5449, 2013.

22. H. Zhang, P. Cheng, L. Shi, and J. Chen. Optimal denial-of-service attack scheduling with energy constraint. *IEEE Transactions on Automatic Control*, 60(11), 3203–3208, 2015.

23. M. Zhu and S. Martinez. On the performance analysis of resilient networked control systems under replay attacks. *IEEE Transactions on Automatic Control*, 59(3):804–808, 2014.

24. M. Zuba, Z. Shi, Z. Peng, and J. H. Cui. Launching denial-of-service jamming attacks in underwater sensor networks. In *Proceedings of ACM International Workshop on Underwater Networks*, Seattle, WA, p. 12, 2011.

Chapter 5

Behavior Rule Specification-Based False Data Injection Detection Technique for Smart Grid

Beibei Li

School of Electrical and Electronic Engineering, Nanyang Technological University

Rongxing Lu

School of Electrical and Electronic Engineering, Nanyang Technological University

Haiyong Bao

School of Electrical and Electronic Engineering, Nanyang Technological University

CONTENTS

Cyber-physical systems (CPSs) are tightly coupled cyber and physical intelligent systems of collaborating computational units that monitor and control physical elements. Smart grid (SG), medical cyber-physical system (MCPS), unmanned aerial vehicle (UAV), and intelligent transportation system (ITS) are all representatives of CPSs, among which smart grid is the most typical form. The smart grid CPS takes advantages of information and communication technologies (ICTs) to provide reliable, efficient, and accurate power generation, transmission, and distribution services. Nevertheless, as a result of the advancement of ICTs, an increasing number of cyber attacks have targeted the smart grid due to the fact that thousands of electronic devices are interconnected via widely deployed communication networks. This makes the whole power system more vulnerable to attacks from cyber space. Therefore, cyber security is an emergent critical issue in the smart grid. In addition, since the attacks come from not only the outsider world, but also insider systems, it is quite a challenging task to guarantee security in the smart grid. Even though great efforts have been made to resist outsider attacks, less attention has been paid to insider ones. In fact, according to the 2013 U.S. State Cybercrime Survey, insider attacks constitute 34% of all surveyed attacks (external constitute 31%, and the remaining 35% have unknown/uncertain sources), which surprisingly shows that insider attacks have already become one of the main sources of threats to cyber and cyber-physical systems. Among various insider attacks, false data injection (FDI) attacks are the most substantial and fatal ones. FDI attackers report falsified measurement data to the system control center (CC); therefore, energy generation, transmission, and distribution could be erroneous due to responses to the false commands from CC, resulting in unnecessary costs, system outages, or fatal consequences. To mitigate FDI threats, a number of signature-based and anomaly-based methods have been proposed over the past few years. However, they suffer from some drawbacks, such as their inability to detect some source-unknown threats and high false alarm probability. To cope with such challenges, in this chapter, we introduce a behavior rule specification-based detection technique for insider threats (e.g., FDI attacks) in smart grid CPSs, which can improve the accuracy of detection and decrease false alarms.

5.1 Introduction

The rapid evolution of the smart grid is bound to pose new cyber and physical security challenges for at least a few years. An increasing number of outsider and insider attacks emphasize the significance of secure smart grids [9, 24, 32]. Compared to outsider attackers, insider ones are hard to detect and identify, as they often hide themselves in the systems' primary elements and launch attacks whenever they find any appropriate opportunity or on a specified date. Besides, successful insider attackers always cause tremendous loss to the system and often result in dire consequences, for example, "Stuxnet" attacks on an Iranian nuclear power plant in June 2010 at Natanz [10]. Therefore, in this chapter, we mainly discuss the detection technique for insider attackers, such as false data injection attackers, in smart grid cyber-physical systems (CPSs) [16].

Sensors, actuators, phasor measurement units (PMUs), line meter devices, smart meters, and communication devices are primary devices of smart grids, together maintaining the normal operation and stability of the whole power system. They are therefore prone to be selected as objectives by insider attackers. Since smart grids are characterized by an intelligent closed control loop formed by the integration of these devices, once they are compromised by insider attackers, secret information stored inside them (e.g., private key for message encryption and decryption) is leaked to these attackers. They obtain opportunities to eavesdrop on other devices to get access to data resources, falsify real-time data from sensors to mislead the control center into giving inappropriate commands, or directly execute false commands to the physical systems by compromising actuator devices. Although insider attackers are really hard to detect by the power system due to their invisibility, many intrusion detection techniques have been developed to overcome such challenges [1, 3–5, 13, 14, 17–20, 23, 27–30]. In general, they can be classified into three categories: signature-based, anomaly-based, and specification-based methods. In this chapter, we design a behavior rule specification-based insider attacker detection scheme to deal with hidden FDI attackers.

Compared to other related work, our contributions are as follows. First, we focus more on the behaviors of the primary devices in smart grid CPSs. Different from other communication systems, these devices in CPSs are mainly tightly connected to the physical world. Hence, behavior analysis is relatively more reasonable than communication protocols or routing path analysis. Second, we introduce an approach of transforming device behavior rules to a state machine, so that the monitored device's behaviors are easier to be identified whether or not they violate the normal rules. Third, an ElGamal encryption algorithm is proposed in this chapter to prevent data eavesdropping. Moreover, our proposed detection technique is applicable to various attack scenarios, including a single attacker, double attackers, and even multiple attackers. Furthermore, it can be demonstrated that our proposed behavior rule specification-based insider attack detection can effectively trade a higher false positive for a lower false negative to guarantee a high detection rate. In addition, we exploit spatial-temporal correlations among grid components to identify the behaviors of meter devices. Notice that our detection technique for FDI attacks is based on line meter devices' measurement data rather than PMUs'. Thus, our technique can be applied over a wide area measurement system (WAMS) network deployed with widespread line meters rather than PMUs to save a lot of fund costs.

The remainder of the chapter is organized as follows. In Section 5.2, we provide some preliminaries related to smart grids and our proposed scheme. Section 5.3 presents the smart grid power system model and threat models. We propose our behavior rule specification-based insider attacker detection technique in Section 5.4, followed by numerical results and performance analysis in Section 5.5. Then, we show the comparative analysis of our proposed technique under various threat scenarios in Section 5.6. We also discuss the related work in Section 5.7. Finally, we draw our conclusions in Section 5.8.

5.2 Preliminaries

5.2.1 Smart Grids and WAMS

The electric grid is the fundamental support to modern society, industry, and economy. With the pervasiveness of electronic devices, electric vehicles, and smart systems, a higher-quality, more reliable, more efficient, and more secure power grid is required to support all the electrical applications. Moreover, due to the requirements of substantial world energy demand, lower global CO_2 emissions, and intermittent renewable energy capacity, an evolution of the electric grid is expected to emerge to handle all these challenges. Smart grid, the envisioned future power system, is an electricity network for which information and communication technologies are leveraged to automatically monitor and control the energy generation, transmission, distribution, and consumption. Figure 5.1 shows the convergence of smart grid.

Compared to a one-way flow legacy power system providing no information about the cost of electricity and no idea on whether any area is overloaded, smart grid is characterized by two-way flow of information and electricity that is able to monitor and control everything from power generation to consumption to ensure the efficiency, reliability, and security of the whole power system [7, 15], as shown in Figure 5.2.

To maintain real-time monitoring and control of the whole smart grid system and guarantee system stability and security, the wide area measurement system (WAMS) is deployed to collect real-time measurement data, which provide much better observability and controllability of the power system. WAMS is composed

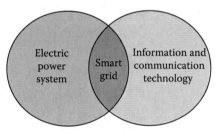

Figure 5.1: Convergence of smart grid.

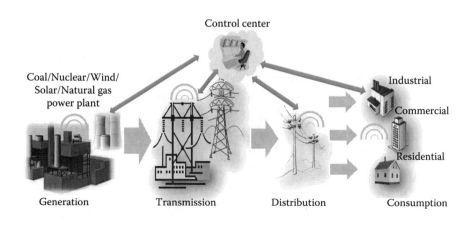

Figure 5.2: Typical architecture of smart grid.

of widespread meter devices, like phasor measurement units (PMUs) on buses, line meters on transmission lines, phasor data concentrators (PDCs), local area networks (LANs), a backbone communication network, and the system control center [12]. The hierarchical structure of the WAMS is shown in Figure 5.3.

5.2.2 Power Loss on Transmission Lines

Electric power generated at various kinds of power plants is transmitted to a distribution center or consumers through transmission lines between them. Due to the physical properties of the transmission medium, part of the transmitted power is lost during the transmission process. The relationship between available current and voltage at various points and the length of the transmission line, from 10 km to a maximum of 300 km, is presented in Table 5.1 [2].

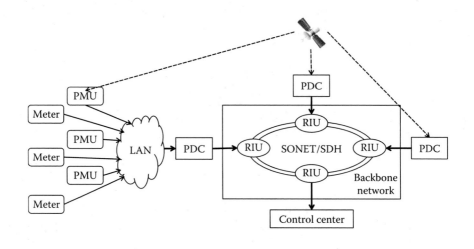

Figure 5.3: Hierarchical structure of WAMS.

Table 5.1   Available Current and Voltage along a 330 kV Single Circuit of a Typical High-Voltage Transmission Network

Length of Line	Current (A)	Voltage (kV)
10	19.14	329.5
20	19.09	329.1
50	19.00	327.6
100	19.87	325.3
200	19.60	320.7
300	19.33	316.1

The power loss, a portion of electricity that is unused and often causes additional costs, comprises [11]

- Joule loss—caused by the active resistance of the conductors to passing power flows

- Corona loss—brought by inefficient electrical insulation between conductors

- Leakage loss—due to inefficient insulation against the potential of the earth

The value of the Joule loss is given as [25]

$$P_{JL} = I^2 * R \text{ kW/km},\tag{5.1}$$

where I represents the current along the conductor and R denotes the resistivity of the conductor.

The Corona power loss for a fair weather situation is given in [2] and [31] as the value

$$P_{CL} = K * \frac{f+25}{\delta} * \sqrt[4]{\frac{r}{d}}(V - V_c)^2 * 10^{-5} \text{ kW/km},\tag{5.2}$$

where K is a constant number equal to 243, f represents the transmission frequency, δ denotes the air density factor, r is the radius of the conductor, d represents the transmission lines' interspace, V is the operating voltage, and V_c denotes the disruptive critical voltage of the transmission line. The leakage loss P_{LL} is usually negligible in practical applications. Therefore, we derive the power loss on transmission line:

$$P_{loss} = L * (P_{JL} + P_{CL}) \text{ kW/km},\tag{5.3}$$

where L is the length of the transmission line. In real power systems, to improve the energy efficiency, the operating voltage V is always pulled up approaching V_c to reduce unnecessary power loss. Thus, P_{CL} is nearly negligible as well. And we rule out the variability of the resistance and consider R a constant value. Therefore, in this chapter, we only consider

$$P_{loss} = L * P_{JL} = L * I^2 * R \text{ kW/km}\tag{5.4}$$

to estimate P_{loss}.

Referring to Figure 5.4, theoretically, we have

$$|P_i - P_j| = P_{loss}.\tag{5.5}$$

However, due to the measurement error of the meters, in reality,

$$\epsilon = |P_i - P_j| - P_{loss},\tag{5.6}$$

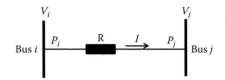

Figure 5.4: Simplified equivalent circuit of a transmission line between two buses.

where ϵ follows a normal distribution $\mathcal{N}(\mu, \sigma^2)$ with probability density function

$$f(x) = \frac{1}{\sigma\sqrt{2\pi}} e^{-\frac{(x-\mu)^2}{2\sigma^2}}. \tag{5.7}$$

5.2.3 State Estimation of Power Systems

Power system state estimation (SE) has already taken the primary place to detect bad data reports to ensure the system stability and reliability. It was first introduced by Schweppe and Rom [26]. The system control center utilizes the reported readings from meter devices to estimate the real system states. The measured state variable vector is $z = \{z_1, z_2, z_3, ..., z_m\}^T$, while the true state variable vector is presented as $x = \{x_1, x_2, ..., x_n\}^T$, where m and n are positive integers, and $z_i, x_j \in \mathbb{R}$ for $i = 1, 2, ..., m$ and $j = 1, 2, ..., n$. Then the real-time meter readings are modeled in the form of

$$z = h(x) + \eta. \tag{5.8}$$

Namely,

$$\begin{bmatrix} z_1 \\ z_2 \\ \cdot \\ \cdot \\ \cdot \\ z_m \end{bmatrix} = \begin{bmatrix} h_1(x_1, x_2, ..., x_n) \\ h_2(x_1, x_2, ..., x_n) \\ \cdot \\ \cdot \\ \cdot \\ h_m(x_1, x_2, ..., x_n) \end{bmatrix} + \begin{bmatrix} \eta_1 \\ \eta_2 \\ \cdot \\ \cdot \\ \cdot \\ \eta_m \end{bmatrix}, \tag{5.9}$$

where $h(x) = \{h_1(x), h_2(x), ..., h_m(x)\}^T$ and $h_i(x)$ is a nonlinear function of x, and $\eta = \{\eta_1, \eta_2, ..., \eta_m\}$ is the measurement error vector with zero mean.

Defining equation

$$J(x) = (z - h(x))^T * \theta * (z - h(x)), \tag{5.10}$$

θ is the covariance matrix of measurement error η. Here we derive the Jacobian matrix:

$$H(x) = \begin{bmatrix} \frac{\partial h_1(\mathbf{x})}{\partial x_1} & \frac{\partial h_1(\mathbf{x})}{\partial x_2} & \cdots & \frac{\partial h_1(\mathbf{x})}{\partial x_n} \\ \frac{\partial h_2(\mathbf{x})}{\partial x_1} & \frac{\partial h_2(\mathbf{x})}{\partial x_2} & \cdots & \frac{\partial h_2(\mathbf{x})}{\partial x_n} \\ \frac{\partial h_3(\mathbf{x})}{\partial x_1} & \frac{\partial h_3(\mathbf{x})}{\partial x_2} & \cdots & \frac{\partial h_3(\mathbf{x})}{\partial x_n} \\ \cdot & \cdot & \cdot & \cdot \\ \cdot & \cdot & \cdot & \cdot \\ \cdot & \cdot & \cdot & \cdot \\ \frac{\partial h_m(\mathbf{x})}{\partial x_1} & \frac{\partial h_m(\mathbf{x})}{\partial x_2} & \cdots & \frac{\partial h_m(\mathbf{x})}{\partial x_n} \end{bmatrix}. \tag{5.11}$$

In most cases, we use the direct current (DC) power flow model to estimate the alternating current (AC) model. Then, Equation 5.9 can be written as

$$z = Hx + \eta. \tag{5.12}$$

Normally, m is larger than n to ensure the accuracy of the estimation. Based on this assumption, maximum likelihood estimation is used to derive the estimated real system state variables:

$$\hat{\mathbf{x}} = [[\mathbf{H}]^T [\theta^{-1}][\mathbf{H}]]^{-1}][\mathbf{H}]^T [\theta^{-1}]\mathbf{z}. \tag{5.13}$$

Then, the measurement residual, like the difference between the system state measurement vector and the estimated state vector, $\mathbf{z} - \mathbf{H}\hat{\mathbf{x}}$, is calculated, and its $\mathcal{L}_2 - norm \parallel \mathbf{z} - \mathbf{H}\hat{\mathbf{x}} \parallel$ is utilized to detect the presence of bad data.

5.2.4 ElGamal Encryption

In this chapter, we propose an ElGamal encryption algorithm to ensure the data integrity to prevent eavesdropping and data falsification in communication networks. ElGamal encryption is an asymmetric key encryption algorithm for public key cryptography, which is based on the Diffie–Hellman key exchange algorithm [8]. The ElGamal cryptosystem is composed of three parts: key generation, encryption, and decryption.

5.2.4.1 Key Generation

The key generation works as follows:

■ Generate a cyclic group $\mathbb{Z}$ of order p, a large prime number, with generator g.

■ Choose a random x from $\{1, 2, ..., p-1\}$.

■ Define $h = g^x \bmod p$ and publish it.

Then we have

$$\begin{cases} \text{Public key:} & \{\mathbb{Z}, p, g, h\}, \\ \text{Private key:} & x. \end{cases}$$

5.2.4.2 Encryption

At the transmitters, we encrypt two messages $m_1 \in [0, \delta_1]$ and $m_2 \in [0, \delta_2]$ under the public key $\{\mathbb{Z}, p, g, h\}$:

■ Choose a random number r from $\{1, 2, ..., p-1\}$, and then calculate $c_1 = g^r \bmod p$.

■ Calculate the shared secret $y = h^r = g^{xr} \bmod p$.

■ Select a number s satisfying $s > \delta_1$ and $\delta_1 + s * \delta_2 < p$.

■ Then calculate $c_2 = (m_1 + s * m_2) * h^r \bmod p$.

Then we have

$$\begin{cases} c_1 = g^r \bmod p, \\ c_2 = (m_1 + s * m_2) * h^r \bmod p. \end{cases}$$

5.2.4.3 Decryption

At the receiver, we decrypt the ciphertexts (c_1, c_2, s) with the private key x:

■ Calculate the shared secret $y = c_1^x = g^{xr}$.

■ Define

$$f = \frac{c_2}{c_1^x} = \frac{(m_1 + s * m_2) * h^r}{y} \bmod p$$

$$= \frac{(m_1 + s * m_2) * g^{xr}}{g^{xr}} \bmod p = m_1 + s * m_2 \bmod p. \tag{5.14}$$

■ Calculate $m_1 = f \bmod s$.

■ Calculate $m_2 = (f - m_1)/s$.

Finally, we have the plaintexts (m_1, m_2). In smart grids, we can use this ElGamal encryption algorithm to encrypt the measured line current and power values I and P from meter devices, and the control center processes the data after decryption to prevent eavesdropping and data falsification along the communication networks.

5.3   System Model

5.3.1   System Reference Model

In this chapter, we illustrate our proposed detection scheme based on the Institute of Electrical and Electronics Engineers (IEEE) 14-bus standard system. As Figure 5.5 shows, line meter devices are deployed to each edge of a transmission line between two buses to measure and report the current and power values $\{I, P\}$ to the system control center periodically. Then, the control center processes all the data from every meter device to realize real-time monitoring and control, including (1) estimating the whole power system states and individual meter device states to acknowledge whether or not any device is in an abnormal state; (2) giving appropriate commands to improve or reduce the power generation, transmission, or distribution on specific areas; and (3) sending out orders to open or close any breaker due to system state changes.

5.3.2   Attack Model

As one primary type of insider attack, false data injection attacks may no longer be what they used to be. Normal FDI attacks are usually simple and seem silly because they are very easily detected by only power system state estimation. Nevertheless, some smart enough FDI insider attackers hide themselves in the power system for a long time to observe and derive the **H** matrix. Then

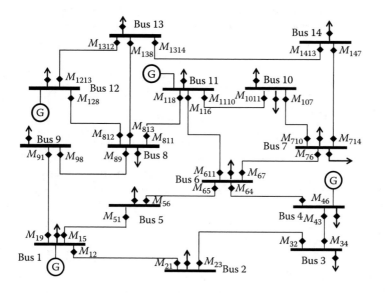

Figure 5.5: IEEE 14-bus standard system.

they acquire the capability to pass the state estimation with bad data, shown as follows:

$$\mathbf{z_{bad}} = \mathbf{Hx_{bad}} + \mathbf{\eta}. \tag{5.15}$$

Such FDI attackers are hard enough for the power system control center to detect. However, in this chapter, we propose a behavior rule specification-based FDI attack detection technique that can detect not only silly FDI attacks, but also smart ones.

According to the attacker behaviors, these attackers can be classified into three types of attack: random, reckless, and opportunistic [18]. A random attacker is insidious, performing attack behaviors at a random probability p_{att}; thus, to some extent, it can avoid system detection. Reckless attackers are quite imprudent to launch an attack whenever they have an opportunity. These attackers aim to disrupt the power system as soon as possible. As for opportunistic attackers, they make use of system noise, which is modeled by p_{err}, to conduct attacks with probability $p_{att} = C * p_{err}^{\epsilon}$.

5.3.3 Performance Analysis Metrics

Performance metrics are very important to evaluate the performance of the detection technique. Here we take p_{fn} and p_{fp} into account to measure our method. p_{fn} is the false negative probability, that is, the probability that the power control system misses an attacker's behavior, while p_{fp} denotes the false positive probability, where the system misidentifies good behavior as malicious.

In this chapter, the detection rate $R_D = 1 - p_{fn}$ and receiver operating characteristic (ROC) curve are also presented to depict the performance.

5.4 Our Proposed Scheme

5.4.1 Behavior Rules

Different from anomaly-based and signature-based attack detection methods, our scheme is a novel insider attack detection technique based on behavior rule specification. Also distinct from most other behavior rule–based methodology, both spatial and temporal rules are taken into consideration in our proposed scheme.

In this chapter, we only consider three articles of rules to present our technique. In normal conditions, line power and current real readings reported from meters to the control center should follow the rules shown in Table 5.2.

Rule 1 shows that at one certain time point, the absolute value of the difference between the transmission line power loss and the change of power amount between two edges of this transmission line should be less than a certain value τ_1, because ϵ follows a normal distribution. Rule 2 presents that at

Table 5.2   Normal Behavior Rules

Rule Index	Behavior Rules
1	$\|\epsilon\| = \|\|P_{i,t} - P_{j,t}\| - P_{loss,t}\| < \tau_1$
2	$\|\delta_1\| = \|I_{i,t} - I_{i,t-1}\| < \tau_2$
3	$\|\delta_2\| = \|P_{i,t} - \hat{P}_{i,t}\| < \tau_3$

one specific measurement point on the transmission line, the current difference between two consecutive time instants ought to be within a certain value τ_2, based on the assumption that the current on transmission lines among buses during the micro time interval cannot change dramatically due to the steady-state load flow. Rule 3 reveals that at one certain time point, the difference between the measured power and estimated value must not be larger than a certain value τ_3 according to the system state estimation introduced in Section 5.2.3.

Any kind of violation of the above three rules may be regarded as bad data, which might expose false data injection intent. Therefore, we have the abnormal behavior rules presented in Table 5.3.

5.4.2   Transform Rules to State Machines

According to the above-mentioned rules, the power system control center will calculate the values of ϵ, δ_1, and δ_2 for each meter device to see whether its measurements follow the normal behavior rules. Obedience of the behavior rule shows that the normal good state is observed, while violation of the behavior rule denotes the bad state is observed at one meter device.

5.4.2.1   Identify State Component and Component Ranges

For each state variable parameter, for example, line power $P_{i,t}$ and current $I_{i,t}$, its value varies during its own permissible range; thus, the reference parameters ϵ, δ_1, and δ_2 should also be limited to specific ranges when a system is under normal state conditions. For example, assume that $P_{i,t}$, $I_{i,t}$, and $P_{loss,t}$ are in the ranges shown in Table 5.4.

We quantized each continuous parameter at integer scale in a permissible range. Based on the three rules in Table 5.2, we may have $(1001 + 2) * 50 *$

Table 5.3   Abnormal Behavior Rules

Rule Index	Behavior Rules
1	$\|\epsilon\| = \|\|P_{i,t} - P_{j,t}\| - P_{loss,t}\| \geqslant \tau_1$
2	$\|\delta_1\| = \|I_{i,t} - I_{i,t-1}\| \geqslant \tau_2$
3	$\|\delta_2\| = \|P_{i,t} - \hat{P}_{i,t}\| \geqslant \tau_3$

Table 5.4 Power System State Parameters and Parameter Ranges

Parameter	States
Line power: P	$[0, 1000 \text{ MW}]$
Line current: I	$[0, 50 \text{ A}]$
Line power loss: P_{loss}	$[0, 2 \text{ MW}]$

$1001 = 5.02 * 10^7$ states. The automation is too large. It will cost tremendous computation resources of the system control center to estimate each device's state. Therefore, reduction of the state space is of great importance.

5.4.2.2 State Space Reduction

The number of states of each rule's reference parameter is partitioned into three levels: safe states, warning states, and malicious states, as Table 5.5 shows.

Then we use digital numbers to express different levels of state: 0 is safe, 1 is warning, and 2 is malicious. This state space reduction treatment yields a small state space with only $3 * 3 * 3 = 27$ states, out of which, we define, 1 state (000) is a safe state, 7 states (e.g., 100, 001) are warning states, and the rest, 19 states (e.g., 200, 222), are malicious states, as shown in Table 5.6.

Table 5.5 State-Level Partition of the Behavior Rules

State Type	Rule 1: ϵ	Rule 2: δ_1	Rule 3: δ_2
Safe state	$0 < |\epsilon| < \alpha_1$	$0 < |\delta_1| < \gamma_1$	$0 < |\delta_2| < \beta_1$
Warning state	$\alpha_1 < |\epsilon| < \alpha_2$	$\gamma_1 < |\delta_1| < \gamma_2$	$\beta_1 < |\delta_2| < \beta_2$
Malicious state	$\alpha_2 < |\epsilon|$	$\gamma_2 < |\delta_1|$	$\beta_2 < |\delta_2|$

Table 5.6 State Classification of All 27 States

State Type	Assumed Ranges
Safe state	000
Warning state	100, 010, 001, 110, 101, 011, 111
Malicious state	200, 020, 002, 210, 201, 021, 120, 102, 012, 211, 121, 112, 220, 202, 022, 221, 122, 212, 222

5.4.3 Collect Compliance Degree Data

To measure whether a meter device works normally, compliance degree is used as a figure of metric in our proposed method. Compliance degree c defines the extent to which a device works normally, which also shows the probability that the device is a safe one.

We model the behaviors of a meter device by a stochastic process such that it may stay in state $1, 2, ..., n$ due to normal operation, systematic noise, or even attack events, with the transition probability $P_{i,j}$ from state i to j. Then, the probability that the stochastic process stays in state i follows

$$\pi = P\pi. \tag{5.16}$$

Namely,

$$\begin{bmatrix} \pi_1 \\ \pi_2 \\ \cdot \\ \cdot \\ \cdot \\ \pi_n \end{bmatrix} = \begin{bmatrix} p_{11} & p_{21} & \cdots & p_{n1} \\ p_{12} & p_{22} & \cdots & p_{n2} \\ \cdot & \cdot & & \cdot \\ \cdot & \cdot & & \cdot \\ \cdot & \cdot & & \cdot \\ p_{1n} & p_{2n} & \cdots & p_{nn} \end{bmatrix} \begin{bmatrix} \pi_1 \\ \pi_2 \\ \cdot \\ \cdot \\ \cdot \\ \pi_n \end{bmatrix}, \tag{5.17}$$

where $\pi_1, \pi_2, ..., \pi_n$ satisfy the following relationship:

$$\sum_{i=1}^{n} \pi_i = 1. \tag{5.18}$$

We assign each state with a grade g_i showing the significance level that each observed state i contributes to the compliance degree of the meter devices. The significance level can be evaluated by the similarity between the observed behavior (in state i) and the specified normal safe behavior (state 000). With the above formulations, we derive the compliance degree of a meter device by summing up all the products of each state's grade and its probability:

$$c = \sum_{i=1}^{n} g_i * \pi_i. \tag{5.19}$$

In order to calculate the defined similarity, Euclidean distance is leveraged here [6]. If state i is $\mathbf{p_i} = (p_{i1} p_{i2} p_{i3})$, then the Euclidean distance between state i and the specified safe state $\mathbf{p_0} = (000)$ is

$$d_{i0} = \sqrt{\sum_{j=1}^{3} (p_{ij} - p_{0j})^2}$$

$$= \sqrt{(p_{i1} - p_{01})^2 + (p_{i2} - p_{02})^2 + (p_{i3} - p_{i3})^2}. \tag{5.20}$$

From the above equation, we know that the largest Euclidean distance between any state i to safe state $\mathbf{p_0} = (000)$ d_{max} is $\sqrt{(2-0)^2 + (2-0)^2 + (2-0)^2} = 2\sqrt{3}$. Then, g_i is given by

$$g_i = 1 - \frac{d_{i0}}{d_{max}}. \tag{5.21}$$

Specifically, for the malicious state $\mathbf{p_i} = (222)$, $g_i = 1 - d_{max}/d_{max} = 0$, which means that the behaviors' similarity is 0, showing full violation of the rules, thus pulling down the compliance degree; for safe state $\mathbf{p_0} = (000)$, $g_0 = 1 - 0/d_{max} = 1$, meaning that the behaviors' similarity is 1, showing full obedience to the specified rules, thus pulling up the compliance degree.

Apart from Euclidean distance, Mitchell and Chen [18] have introduced several other distance-based schemes, for example, Hamming distance, Manhattan distance, and Levenshtein distance.

5.4.4 Compliance Degree Distribution

Due to the noise and communication errors, the estimation of compliance degree is sometimes imperfect. Thus, we model the compliance degree by a random variable $X \sim Beta(\alpha, \beta)$ with probability density function (PDF)

$$f(x; \alpha, \beta) = \frac{\Gamma(\alpha + \beta)}{\Gamma(\alpha)\Gamma(\beta)} x^{\alpha-1}(1-x)^{\beta-1}, \tag{5.22}$$

where $\Gamma(t) = \int_0^\infty x^{t-1}e^{-x}\,dx$.

As we know, beta distribution is a family of continuous probability distributions defined on the interval $[0, 1]$, with 0 revealing the zero compliance and 1 indicating the perfect compliance. Thus, it is appropriate for us to model the compliance degree by beta distribution. In addition, beta distribution can specify a wide range of distributions by adjusting the parameters α and β. We can take advantage of it to obtain the compliance degree distribution by estimating the values of α and β. The cumulative density function (CDF) of beta distribution is shown as below:

$$F(x; \alpha, \beta) = \int_0^x \frac{\Gamma(\alpha + \beta)}{\Gamma(\alpha)\Gamma(\beta)} t^{\alpha-1}(1-t)^{\beta-1}\,dt, \tag{5.23}$$

and the mean value of X is given by

$$E[X] = \int_0^1 xf(x; \alpha, \beta)\,dx = \int_0^1 x\frac{\Gamma(\alpha + \beta)}{\Gamma(\alpha)\Gamma(\beta)} x^{\alpha-1}(1-x)^{\beta-1}dx = \frac{\alpha}{\alpha + \beta}. \tag{5.24}$$

To obtain the exact distribution of compliance degree, we need to estimate the values of α and β. Maximum likelihood estimation (MLE) is leveraged here with the compliance degree history $c_1, c_2, ..., c_n$.

5.4.5 *Estimation of α and β by Maximum Likelihood*

The basic principle of maximum likelihood estimation is to offer an approach to select an asymptotically efficient and best approximate estimator for a parameter or a set of parameters of a distribution [21, 22]. Here, we have a set of conditionally independent and identically beta-distributed samples $c_1, c_2, \ldots, c_n$ with parameters α and β. First, we specify the joint density function for all observations as

$$f(c_1, c_2, \ldots, c_n \mid \alpha, \beta) = f(c_1 \mid \alpha, \beta) \times f(c_2 \mid \alpha, \beta) \times \cdots \times f(c_n \mid \alpha, \beta). \qquad (5.25)$$

Now we look at this equation from a different perspective by considering the observed samples $c_1, c_2, \ldots, c_n$ to be fixed parameters of this function; then α, β will be the variables of the function, which we call the likelihood:

$$\mathcal{L}(\alpha, \beta; c_1, c_2, \ldots, c_n) = f(c_1, c_2, \ldots, c_n \mid \alpha, \beta) = \prod_{i=1}^{n} f(c_i \mid \alpha, \beta). \qquad (5.26)$$

In practice, it is usually simpler to work with the logarithm of the likelihood function:

$$\begin{aligned}
\ln \mathcal{L}(\alpha, \beta; c_1, c_2 \ldots, c_n) &= \sum_{i=1}^{n} \ln f(c_i \mid \alpha, \beta) \\
&= \sum_{i=1}^{n} \ln \left\{ \frac{\Gamma(\alpha+\beta)}{\Gamma(\alpha)\Gamma(\beta)} \, c_i^{\alpha-1}(1-c_i)^{\beta-1} \right\} \\
&= \sum_{i=1}^{n} \left\{ \ln \Gamma(\alpha+\beta) - [\ln \Gamma(\alpha) + \ln \Gamma(\beta)] \right. \qquad (5.27) \\
&\qquad \left. + (\alpha-1)\ln c_i + (\beta-1)\ln(1-c_i) \right\} \\
&= n \ln \Gamma(\alpha+\beta) - n[\ln \Gamma(\alpha) + \ln \Gamma(\beta)] \\
&\qquad + (\alpha-1) \sum_{i=1}^{n} \ln c_i + (\beta-1) \sum_{i=1}^{n} \ln(1-c_i).
\end{aligned}$$

Then, we have to find out the value that maximizes $\ln \mathcal{L}(\alpha, \beta; c_1, c_2, \ldots, c_n)$. Since logarithm is a strictly monotonically increasing function, the maximum value, if one exists, could be calculated by

$$\begin{cases} \frac{\partial \ln \mathcal{L}}{\partial \alpha} = 0, \\ \frac{\partial \ln \mathcal{L}}{\partial \beta} = 0. \end{cases} \qquad (5.28)$$

That is,

$$n \left[\frac{\frac{\partial \Gamma(\alpha+\beta)}{\partial \alpha}}{\Gamma(\alpha+\beta)} - \frac{\frac{\partial \Gamma(\alpha)}{\partial \alpha}}{\Gamma(\alpha)} \right] + \sum_{i=1}^{n} \ln c_i = 0 \qquad (5.29)$$

and

$$n\left[\frac{\frac{\partial \Gamma(\alpha+\beta)}{\partial \beta}}{\Gamma(\alpha+\beta)} - \frac{\frac{\partial \Gamma(\beta)}{\partial \beta}}{\Gamma(\beta)}\right] + \sum_{i=1}^{n} \ln(1-c_i) = 0, \tag{5.30}$$

where

$$\frac{\partial \Gamma(\alpha+\beta)}{\partial \alpha} = \int_0^\infty (\ln x) x^{\alpha+\beta-1} e^{-x} dx. \tag{5.31}$$

The above model is too complicated to estimate α and β, so a less general, though simpler model is utilized here by only considering a single variable parameter $Beta(1, \beta)$ distribution with α equal to 1. Then, the probability density function is reduced to

$$f(x; 1, \beta) = \beta(1-x)^{\beta-1}. \tag{5.32}$$

The maximum likelihood estimate of β is

$$\beta = \frac{n}{\sum_{i=1}^{n} \ln(1/(1-c_i))}. \tag{5.33}$$

5.4.6   False Negative and False Positive Probabilities

False negative and false positive probabilities, denoted by p_{fn} and p_{fp}, respectively, are two critical metrics to evaluate the performance of an intrusion detection technique. A false negative happens when a malicious meter device is misdiagnosed as a normal safe one, while a false positive occurs when a normal safe device is misidentified as a malicious one. However, neither is desirable when developing an insider attack detection technique.

For a normal safe meter device, p_{fp} is the probability that its compliance degree X_i is less than the system threshold c_{th}, which is given by

$$p_{fp} = P\{X_i < c_{th} | X_i \text{ is safe}\} = F(c_{th}). \tag{5.34}$$

Similarly, as for a malicious meter equipment, p_{fn} is the probability that its compliance degree X_i is larger than the system threshold c_{th}, which is given by

$$p_{fn} = P\{X_i > c_{th} | X_i \text{ is malicious}\} = 1 - F(c_{th}). \tag{5.35}$$

5.5   Performance and Evaluation

5.5.1   Simulation Design

In this section, we first collect compliance degree $c_1, c_2, ..., c_n$, with $n = 1000$, by means of Monte Carlo simulation, which could generate repeated random samples following the stochastic process of a device's state machines. Table 5.7 shows the design parameters that we use in the simulation.

For a normal safe device, we simulate p_{ji} as $1 - p_{err}$ when i is the single

Table 5.7   Designed Parameters of Monte Carlo Simulation

Parameters	Value (Ranges)
Number of state transitions in one run: m	200
Number of c_i: n	1000
Error probability: p_{err}	0.01, 0.02, 0.03, 0.04, 0.05
Attack probability: p_{att}	0.1, 0.2, 0.3, 0.4, 0.5, 0.6, 0.7, 0.8, 0.9, 1.0

safe state; otherwise, we simulate it as p_{err} when i is one of the 26 unsafe states (regarding both 7 warning and 19 malicious states as unsafe). And for a malicious device with attack probability p_{att}, we simulate p_{ji} as $(1 - p_{att}) * (1 - p_{err}) + p_{att} * p_{err}$ when i is the single good state, and as $p_{err} * (1 - p_{att}) + p_{att} * (1 - p_{err})$ when i is one of the 26 unsafe states. Then we use Monte Carlo simulation to run it 1000 times to collect $c_1, c_2, ..., c_{1000}$. In each run test, we start from safe state (000) and then traverse from one state to another following the stochastic process of a meter device. After m transitions, sufficient enough, we calculate the limiting probability π_i by using the ratio of the number of transitions to state i to the total number of transitions m. Then we collect one instance of c_i based on Equations 5.17 and 5.19. With 1000 repeated runs, the distribution of a compliance degree of a meter device is determined by maximum likelihood estimation with $c_1, c_2, ..., c_{1000}$.

5.5.2   Numerical Results and Performance Analysis

5.5.2.1   For Normal Safe Meter Devices

Figure 5.6 presents the raw data of compliance degree of a normal meter device with 1000 runs to different levels of p_{err}. We can see from this figure that the compliance degrees decrease when p_{err} increases, which means a larger probability of error provides a higher probability for the normal meter device to traverse to warning or unsafe states, thus decreasing its compliance degree.

With the compliance degree history values of a normal safe meter device $c_1, c_2, , ..., c_{1000}$ in hand, we use maximum likelihood estimation to estimate the β parameter value of the $Beta(\alpha, \beta)$ distribution of compliance degree of a normal safe meter device (here we consider $\alpha = 1$). Then, according to Equation 5.34, given c_{th}, we calculate p_{fp}, of which the results are shown in Figure 5.7.

Figure 5.7 shows the relationship between the probability of a false positive p_{fp} and the compliance degree threshold c_{th} to varying p_{err} values. p_{fp} increases obviously when c_{th} is approaching 1, which means the growth of c_{th} increases the chance of a normal meter device staying in unsafe states under the specific value of p_{err}. Although the rise of p_{err} does not make a big difference to p_{fp}, a higher p_{err} still has greater impact on p_{fp}.

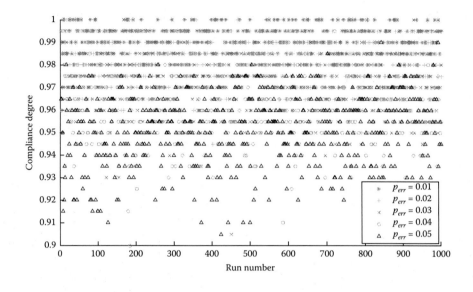

Figure 5.6: Compliance degree of a normal safe meter device with 1000 runs to various values of p_{err}.

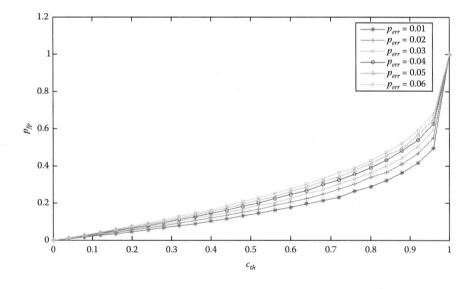

Figure 5.7: Probability of false positive vs. compliance degree with different levels of p_{err} for a normal meter device.

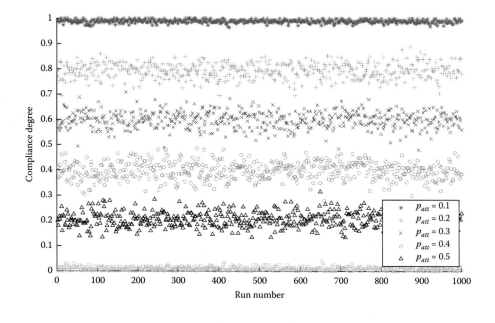

Figure 5.8: Compliance degree of a malicious meter device to p_{att} for random attackers.

5.5.2.2 For Malicious Meter Devices

Figure 5.8 presents the correlation between the compliance degree of random attackers and various attack probabilities p_{att}. It is apparent that the higher attack probability, causing the meter device to perform insecurely more often, results in a lower compliance degree. What should be emphasized here is that in the special situations where $p_{att} = 1.0$ and $p_{att} = 0.0$, the compliance degrees are nearly zero and perfect, respectively, which corresponds to a reckless attack situation and a normal safe situation.

Figure 5.9 introduces the compliance degree of reckless attackers to varying p_{err}. Reckless attackers launch attacks whenever they have a chance, leading to meter devices always staying in unsafe states, so their compliance degrees are quite low, approaching zero. Although, the error probability is higher, the increasing ambient noise will improve the compliance degree to a minor extent.

Figure 5.10 shows the compliance degree of malicious meter devices with opportunistic attack behaviors ($\epsilon = 0.9$) to different levels of p_{err}. Like Figure 5.9 shows, the compliance degree of the malicious meter devices declines while the error probability rises, but the compliance degree of devices with opportunistic attackers is more sensitive to reckless attackers whose compliance degrees are nearly bounded to below 0.1.

As such, with a compliance degree history $c_1, c_2, ..., c_{1000}$ of malicious meter devices with random attackers, reckless attackers, and opportunistic attackers,

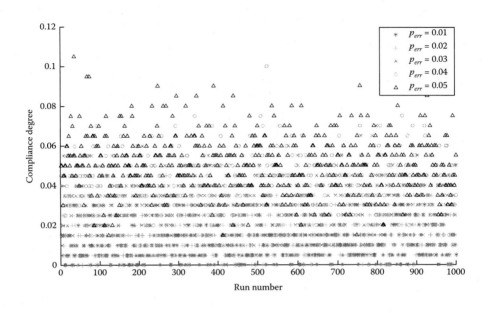

Figure 5.9: Compliance degree of a malicious meter device to p_{err} for reckless attackers.

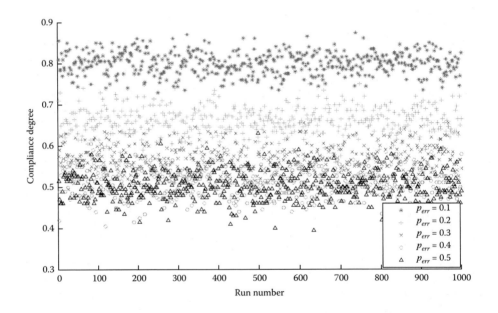

Figure 5.10: Compliance degree of a malicious meter device to p_{err} for opportunistic attackers.

respectively, in hand, we estimate the β parameter values first by Equation 5.33. Then p_{fn} for the three types of attackers could be calculated via Equation 5.35, shown in Figure 5.11.

Figure 5.11 reveals the connection between p_{fn} and c_{th} with varying p_{att} values for random attackers. As we can see, p_{fn} drops while c_{th} is rising, which reveals that a higher compliance threshold reduces the chance that a malicious meter device is misidentified as a normal safe one. Besides, the higher the attack probability, the more easily a malicious device is successfully detected.

Figure 5.12 exhibits the relationship between p_{fn} and c_{th} with various p_{err} values for reckless attackers. Unlike attackers with a lower attack probability, reckless attackers seem aggressive enough, resulting in very low compliance degrees. Thus, the probability of a false negative is quite low even under a very small compliance threshold. Apart from that, a lower error probability results in a lower p_{fn} as well.

Figure 5.13 presents the correlation between p_{fn} and c_{th} with various p_{err} values for opportunistic attackers. The p_{fn} descends as p_{err} ascends because the attack probability is higher while the error probability is higher, leading to its being detected more easily.

Figure 5.14 shows the receiver operating characteristic graph for detecting random attackers, where the detection rate is $R_D = 1 - p_{fn}$. By adjusting the compliance degree threshold c_{th}, the false negative probability p_{fn} can be reduced at

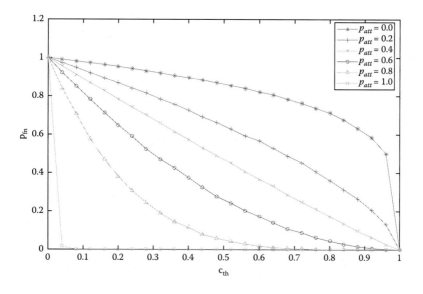

Figure 5.11: Probability of false negative vs. compliance threshold with random attackers.

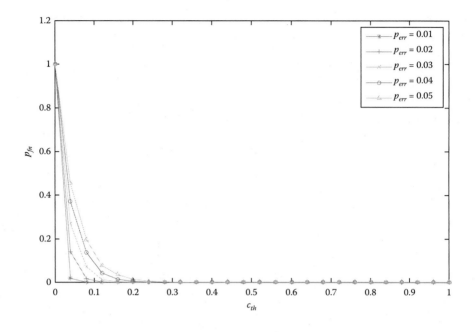

Figure 5.12: Probability of false negative vs. compliance threshold and p_{err} with reckless attackers.

the cost of the false positive probability p_{fp}, which means that our proposed method can effectively improve the detection rate with an acceptable false positive probability.

5.6 Comparative Analysis

In this chapter, we take into account four types of scenarios of the smart grid systems' operation situation:

- Case 1: The system is under normal operation condition.

- Case 2: Only one meter device on a transmission line is compromised.

- Case 3: Both meter devices on a transmission line are compromised.

- Case 4: The system is under a physical outage condition without attacks.

For Case 1, the power system operates well without any attack or physical outage, under which situation the reported data pass smoothly to all the specified rules unless a system error occurs.

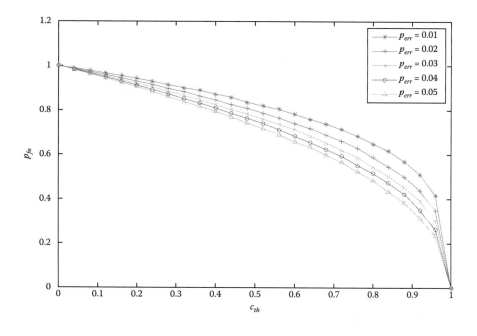

Figure 5.13: Probability of false negative vs. compliance threshold and p_{err} with opportunistic attackers.

As for Case 2, as shown in Figure 5.15, one meter device (e.g., Meter M_{65} between Bus 5 and Bus 6) on a specific transmission line is compromised as an insider attacker. Once false data injections occur in the form of current value falsification, a big change happens to the absolute difference of current values between two adjacent time points $|\delta_1| = |I_{i,t} - I_{i,t-1}|$ that is larger than the specified threshold τ_2, thus violating Rule 2. Should the line power value be forged, $|\epsilon|$ will exceed the specified value τ_1, disobeying Rule 1, and the estimated value of power will change a lot as well, going against Rule 3. Given that both the current and power values are falsified, the meter device's behaviors breach all three rules.

As for Case 3, as Figure 5.16 shows, both meters (e.g., M_{89} and M_{98} between Bus 8 and Bus 9) on a transmission line are compromised by attackers. Supposing that false current data are injected by the two meters, both meter devices will disobey Rule 2; if both false power data of the two meters are reported to the system control center, specifically if the insider attackers are insidious enough to change the two power values $p_{i,t}$ and $p_{j,t}$ at the same time, they will be lucky to pass Rule 1. However, Rule 3 will refuse their passage, showing both power values are bad; once both current and power measurement values are forged, at least two rules are violated.

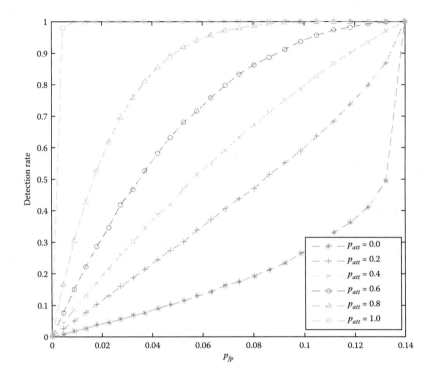

Figure 5.14: Receiver operating characteristic graph for detecting random attackers.

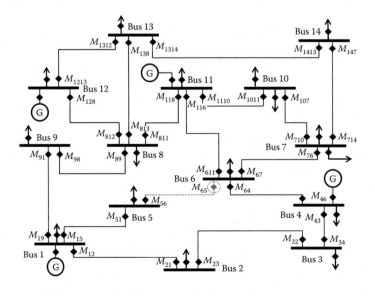

Figure 5.15: Only one meter on a transmission line is compromised.

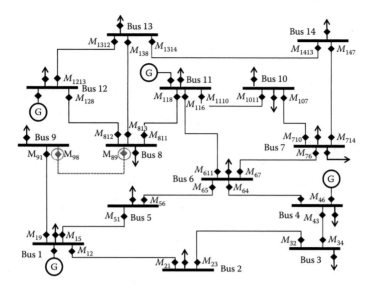

Figure 5.16: Both two meters on a transmission line are compromised.

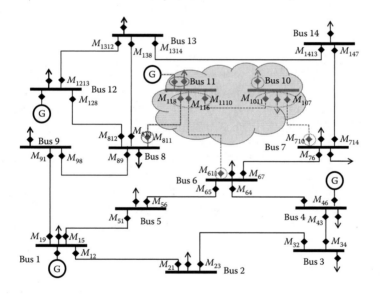

Figure 5.17: Power system outage in a specific area.

As shown in Figure 5.17, with respect to Case 4, a system physical outage happens from time to time due to either component aging failure or system routine maintenance and replacement. Provided that the system suffers a short-circuit fault, both the power and current values will change a lot, going against

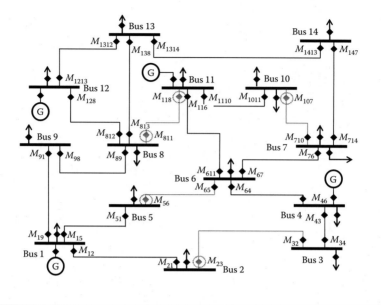

Figure 5.18: Multiple bad meters occur at the same time.

all three rules. However, in most cases, system state estimation will reveal that almost all meters in the specific fault area also suffer the same trouble, which is quite discrepant from Case 3's specific situation. In the case of an open-circuit fault, all the current and power data in a certain fault area ought to be zero, which we regard as invalid measurement values that, like a short-circuit fault, do not contribute any to meter devices' compliance degrees.

Apart from the previous cases, another, more complex case is that multiple bad meters emerge at the same time, but in different locations, as Figure 5.18 shows. Since the research objectives of our proposed detection technique are meters on a single transmission line, and bad meters on different transmission lines are independent from each other, the control center can individually estimate each meter device at the same time. Our proposed behavior rule specification-based insider attack detection is still applicable to such a complex scenario.

5.7   Related Work

The growing popularity of smart grid CPSs requires increasing attention to the security issues. For the sake of military strategic demand or individual profits, smart grid systems have been attracting more and more attackers' attention. Either successful outsider attacks or insider attacks will cause a great amount of cost to the smart grids or even to the whole nation. Outsider attacks are observable and sensible, while insider attackers always hide themselves in any area of the power system to prevent detection and removal. An increasing number of related works focus more on insider attacks, and a number of detection techniques have been developed [1, 3–5, 13, 14, 17–20, 23, 27–30]. They can be classified into three types: signature-based, anomaly-based, and behavior-based schemes.

As [23, 28, 30] introduce, signature-based detection methodologies rely drastically on the attack database of known attack patterns, and these techniques detect attacks through data mining methods, analyzing the new pattern and comparing it with the patterns in the database to see to what extent it matches any one of the known patterns. Signature-based techniques perform well when detecting already existing attacks, but badly when faced with new pattern attacks [14].

Nevertheless, anomaly-based methods, as proposed in [17, 27, 29], solve these challenges well with statistical-, model-, profile-, and distance-based means, trying to differentiate normal behaviors and abnormal behaviors. Unfortunately, legacy anomaly-based methods often consume substantial computation resources, and performance evaluation always shows a high probability of false alarms.

Similarly, a series of research is interested in rule specification-based schemes [1, 19, 20]. Although they perform well enough according to the existing work, these techniques are only applied in the context of communication networks; attention is seldom focused on smart grid CPSs. Therefore, specification-based detection techniques, especially insider attackers' detection, are still in their infancy period, and much more research work is required in the future.

5.8 Conclusion

Considerable progress has been made in being able to successfully detect and identify hidden insider attackers in intrusion defense and detection of CPSs. In this chapter, we propose a behavior rule specification-based detection technique for false data injection attacks in safety-critical smart grid systems. First, we define a set of behavior rules that normal safe meter devices are required to follow. Then, meter devices' behaviors, also referred to as system states, are classified into three levels, safe, warning, and unsafe, according to the specified rules. Third, the set of rules are transformed into state machines so that the shift of behaviors of a meter device can be regarded as state transitions. Next, compliance degree history values are collected to estimate the parameters of the beta distribution. Finally, meter devices' system thresholds of compliance degree are designed to obtain the lowest false negative probability with acceptable false positive probability. We also demonstrated that our proposed detection technique can acquire a nearly 100% detection rate (that means nearly zero false negative probability) with acceptable false positive probability.

Future work will focus more on two issues. One is our scheme's communication overheads and the computation costs of the power system's control units. Lower overhead of the system will improve the response rate and efficiency of the detection technique. The other one is how to evict an individual compromised meter device once it is detected and how to repair the system to normal conditions. Insider attackers always have dire consequences to the system's security and stability once they are launched, and they cost a great amount of time and money to repair as well. Therefore, the eviction of the insider attackers without affecting the whole system and the repair of the system to normal conditions at the lowest costs are also critical issues in the near future.

References

1. M. Anand, E. Cronin, M. Sherr, M. Blaze, Z. Ives, and I. Lee. Security challenges in next generation cyber physical systems. *Beyond SCADA: Networked Embedded Control for Cyber Physical Systems*, Pittsburgh, PA, pages 1–4, 2006.

2. O. Bamigbola, M. Ali, and M. Oke. Mathematical modeling of electric power flow and the minimization of power losses on transmission lines. *Applied Mathematics and Computation*, 241:214–221, 2014.

3. A. Cardenas, S. Amin, B. Sinopoli, A. Giani, A. Perrig, and S. Sastry. Challenges for securing cyber physical systems. In *Workshop on Future Directions in Cyber-Physical Systems Security*, Newark, NJ, 2009.

4. S. Cheung, B. Dutertre, M. Fong, U. Lindqvist, K. Skinner, and A. Valdes. Using model-based intrusion detection for scada networks. In *Proceedings of the SCADA Security Scientific Symposium*, volume 46, pages 1–12, Miami Beach, FL, 2007.

5. A. da Silva, M. Martins, B. Rocha, A. Loureiro, L. Ruiz, and H. Wong. Decentralized intrusion detection in wireless sensor networks. In *Proceedings of ACM International Workshop on Quality of Service & Security in Wireless and Mobile Networks*, pages 16–23, Montreal, 2005.

6. M. Deza and E. Deza. *Encyclopedia of Distances*. Springer, Berlin, 2009.

7. USDOE (U.S. Department of Energy) The smart grid: An introduction. USDOE, Washington, DC, 2008.

8. T. ElGamal. A public key cryptosystem and a signature scheme based on discrete logarithms. In *Advances in Cryptology*, pages 10–18. Springer, Berlin, 1985.

9. M. Farag, M. Azab, and B. Mokhtar. Cross-layer security framework for smart grid: Physical security layer. In *Proceedings of Innovative Smart Grid Technologies Conference Europe*, pages 1–7, Istanbul, 2014.

10. J. Farwell and R. Rohozinski. Stuxnet and the future of cyber war. *Survival*, 53(1):23–40, 2011.

11. V. Král, S. Rusek, and L. Rudolf. Calculation and estimation of technical losses in transmission networks. *Electrical Review*, 88(8):88–91, 2012.

12. W. Li. *Risk Assessment of Power Systems: Models, Methods, and Applications*. John Wiley & Sons, Hoboken, NJ, 2014.

13. T. Liu, Y. Gu, D. Wang, Y. Gui, and X. Guan. A novel method to detect bad data injection attack in smart grid. In *Proceedings of IEEE INFOCOM*, pages 3423–3428, Turin, 2013.

14. P. Louvieris, N. Clewley, and X. Liu. Effects-based feature identification for network intrusion detection. *Neurocomputing*, 121:265–273, 2013.

15. R. Lu, X. Liang, X. Li, X. Lin, and X. Shen. EPPA: An efficient and privacy-preserving aggregation scheme for secure smart grid communications. *IEEE Transactions on Parallel and Distributed Systems*, 23(9):1621–1631, 2012.

16. R. Lu, X. Lin, H. Zhu, X. Liang, and X. Shen. BECAN: A bandwidth-efficient cooperative authentication scheme for filtering injected false data in wireless sensor networks. *IEEE Transactions on Parallel and Distributed Systems*, 23(1):32–43, 2012.

17. M. Mahoney. Network traffic anomaly detection based on packet bytes. In *Proceedings of ACM Symposium on Applied Computing*, pages 346–350, Melbourne, FL, 2003.

18. R. Mitchell and I. Chen. Behavior rule specification-based intrusion detection for safety critical medical cyber physical systems. *IEEE Transactions on Dependable and Secure Computing*, 12(1):16–30, 2014.

19. R. Mitchell and R. Chen. Behavior-rule based intrusion detection systems for safety critical smart grid applications. *IEEE Transactions on Smart Grid*, 4(3):1254–1263, 2013.

20. R. Mitchell and R. Chen. Adaptive intrusion detection of malicious unmanned air vehicles using behavior rule specifications. *IEEE Transactions on Systems, Man, and Cybernetics: Systems*, 44(5):593–604, 2014.

21. I. Myung. Tutorial on maximum likelihood estimation. *Journal of Mathematical Psychology*, 47(1):90–100, 2003.

22. P. Paolino. Maximum likelihood estimation of models with beta-distributed dependent variables. *Political Analysis*, 9(4):325–346, 2001.

23. S. Patton, W. Yurcik, and D. Doss. An Achilles' heel in signature-based IDs: Squealing false positives in SNORT. In *Proceedings of RAID*, pages 95–114, Davis, CA, 2001.

24. P. Ray, R. Harnoor, and M. Hentea. Smart power grid security: A unified risk management approach. In *Proceedings of IEEE International Carnahan Conference on Security Technology*, pages 276–285, San Jose, CA, 2010.

25. I. Sarajcev, M. Majstrovic, and I. Medic. Calculation of losses in electric power cables as the base for cable temperature analysis. *Journal of Advanced Computational Methods in Heat Transfer*, 4:529–537, 2003.

26. F. Schweppe and D. Rom. Power system static-state estimation, Part I: Exact model, Part II: Approximate mode, Part III: Implementationl. *IEEE Transactions on Power Apparatus and Systems*, (1):120–135, 1970.

27. C. Taylor and J. Alves-Foss. NATE: Network analysis of anomalous traffic events, a low-cost approach. In *Proceedings of Workshop on New Security Paradigms*, pages 89–96, Cloudcroft, NM, 2001.

28. G. Vigna, W. Robertson, and D. Balzarotti. Testing network-based intrusion detection signatures using mutant exploits. In *Proceedings of ACM Conference on Computer and Communications Security*, pages 21–30, Washington, DC, 2004.

29. K. Wang and S. Stolfo. Anomalous payload-based network intrusion detection. In *Recent Advances in Intrusion Detection*, pages 203–222, Sophia Antipolis, 2004.

30. P. Wheeler. Techniques for improving the performance of signature-based network intrusion detection systems. PhD thesis, University of California Davis, 2006.

31. E. Yahaya, T. Jacob, M. Nwohu, and A. Abubakar. Power loss due to corona on high voltage transmission line. *IOSR Journal of Electrical and Electronics Engineering*, 3331:14–19, 2013.

32. Z. Dong. Smart grid cyber security. In *Proceedings of International Conference on Control Automation Robotics & Vision*, pages 1–2, Marina Bay Sands, Singapore, 2014.

Chapter 6

Hierarchical Architectures of Resilient Control Systems: Concepts, Metrics, and Design Principles

Quanyan Zhu

Department of Electrical and Computer Engineering, New York University

Dong Wei

Corporate Technology, Siemens Corporation

Kun Ji

Corporate Technology, Siemens Corporation

CONTENTS

6.1 Introduction

System resilience is the intrinsic ability of a system to adjust its functioning prior to, during, or following changes and disturbances, so that it can sustain required operations under conditions created by both expected and unexpected causes, such as human errors, system faults, adversarial attacks, and natural disasters. The concept of resilience is related to but also distinct from other system attributes, such as reliability, robustness, security, and fault tolerance. Resilience engineering can be understood as the science of managing multiple system properties in an integrative multilayer and multiagent fashion. Resilience has been studied in many fields, such as psychology, ecology, and organizational behavior.

Critical infrastructure refers to those systems and assets that are essential for the functioning of a society and economy. The incapacity or destruction of such systems and assets would have a debilitating impact on security, national economic security, national public health or safety, or any combination of those matters. Resilience is a desirable property of critical infrastructure. As it is mentioned in [6],

> resilience has become an important dimension of the critical infrastructure protection mission, and a key element of the value proposition for partnership with the government because it recognizes both the need for security and the reliability of business operations.

Control systems are typically used in industries such as electric generation, transmission and distribution, gas and oil plants, water and wastewater treatment, and manufacturing for monitoring and controlling physical and chemical processes in critical infrastructures [5, 8]. Therefore, it is important to ask the following questions: What does resilience mean to control systems? How do we measure or estimate it? What applications can deploy resilient control technologies? To answer the above-mentioned questions, this chapter describes a big picture of fundamental principles and applications of resilient control systems by

- Defining the scope of resilient control systems

- Developing a hierarchical viewpoint toward resilient control systems

- Defining the resilience of control systems quantitatively

- Discussing the relationship between resilience properties of a control system and other properties, such as robustness, fault tolerance, flexibility, survivability, and adaptiveness

- Proposing metrics to measure or estimate control system resiliency

- Developing design and operation principles

- Discussing applications and guidelines in the smart grid that deal with extreme events

- Disclosing related research fields, such as prognostic technologies and data fusion technologies

This chapter is organized as follows: Section 6.2 discusses the related research works conducted in the area of system resilience. In Section 6.3, we present a hierarchical viewpoint toward resilient control systems. Section 6.4 presents major performance indices that quantify the resilience of control systems. Design and operation principles are discussed in Section 6.5. An example of the power grid is used to illustrated the principles of resilient control systems in Section 6.6. We conclude the chapter with Section 6.7, discussing potential applications of resilient control systems and further research topics.

6.2   Related Work

Originally, the term *resilience* was studied in the fields of ecology and psychology. The concept of resilience in ecological systems was first described by the Canadian ecologist C. S. Holling [9] to draw attention to trade-offs between efficiency, on the one hand, and persistence, on the other, or between constancy and change, or between predictability and unpredictability. Emmy Werner was one of the first scientists to use the term *resilience* in psychology, which refers to the ability to recover from trauma or crisis. She studied children who grew up with alcoholic or mentally ill parents in the 1970s [30].

In recent years, the term *resilience* has been used to describe a movement among entities such as businesses, communities, and governments to improve their ability to respond to and quickly recover from catastrophic events such as natural disasters and terrorist attacks. The concept is gaining credence among public and private sector leaders who argue that resilience should be given equal weight to preventing terrorist attacks in U.S. homeland security policy. The study of resilience from the perspective of organizations includes business organizations [3, 16, 21, 25, 31] and government organizations [6, 7, 11].

Resilience has been studied in the field of communication networks in the past few years. Resilient communication networks [2, 4, 18] aim to provide and maintain acceptable service to the following applications in the face of faults and challenges to normal operation:

■ Enable users and applications to access information when needed, such as web browsing, distributed database access, and sensor monitoring

■ Maintain end-to-end communication association, such as computer-supported cooperative work, videoconference, and teleconference

■ Provide distributed processing and networked storage

The major topics include resilient network structure [10, 14], intrusion resilient network [26, 27], denial-of-service (DoS) resilient network [1, 24], and resilient network coding [12].

Literature on resilience study can also be found in the fields of economics, aviation industry, disaster response, nuclear power plants, oil and gas industry, emergency health care, and transportation engineering.

The area of resilient control systems (RCSs) is a new paradigm that encompasses control design for cyber security, physical security, process efficiency and stability, and process compliance in large-scale, complex systems. In [23], RCS is defined as a control system that maintains state awareness and an accepted level of operational normalcy in response to disturbances, including threats of an unexpected and malicious nature. While one might say that resilient design is primarily dependable computing coupled with fault-tolerant control, it has been argued in [23] that dependable computing views malicious faults as a source of failure, but does not consider the effect of these faults on the underlying physical processes in a large-scale complex system.

Recent literature on RCS has studied many aspects of resilience in control systems. In [22], notional examples are used to discuss fundamental aspects of resilient control systems. It has been pointed out that current research philosophies lack the depth or the focus on the control system application to satisfy many aspects of requirements of resilience, including graceful degradation of hierarchical control while under cyber attack. In [37], a hierarchical viewpoint is used to address security concerns at each level of complex systems. The paper emphasizes a holistic cross-layer philosophy for developing security solutions and provides a game-theoretical approach to model cross-layer security problems in cyber-physical systems. In [36], the authors have proposed a game-theoretic framework to analyze and design, in a quantitative and holistic way, robust and resilient control systems that can be subject to different types of disturbances at different layers of the system. In [34], a hybrid system model is used to address physical layer control design and cyber-level security policy making for cyber-physical systems that are subject to cascading effects from cyber attacks and physical disturbances. In [28], metrics for resilient control systems are developed for analyzing and designing resilient systems. In this work, we extend the concepts developed in [28] for hierarchical cyber-physical systems.

Cyber security is an essential component of the resilience of control systems. Few works have provided quantitative methods of modeling of device configurations and evaluating trade-offs among defense options. In [15], the authors have made a comprehensive survey on game-theoretic methods for different problems of network security and privacy. It has been pointed out that the quantitive methods discussed in the survey can be integrated with cyber-physical systems for analyzing and designing resilient control systems. The literature on device configurations can be found in [33, 35, 38]. In [33], a cooperative game approach has been used to address the static configuration of security devices, such as intrusion detection systems (IDSs) and intrusion prevention systems (IPSs), in the face of adversarial attacks. In [35], the authors address the dynamic counterpart of the configuration problem. The equilibrium cyber policy can be obtained from a game-theoretic analysis of a dynamic zero-sum Markov game, which has taken into account the trade-offs between different defense mechanisms. In [38], a network-level configuration of security devices has been addressed by considering the interdependence of devices in the network.

6.3 Hierarchical Resilient Control Systems

An industrial control system (ICS) is one electronic device or a set of electronic devices to monitor, manage, control, and regulate the behavior of other devices

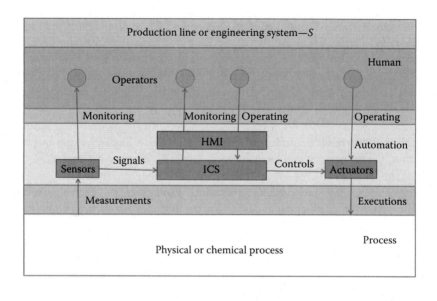

Figure 6.1: System model for a pipeline engineering system.

or systems. ICS includes supervisory control and data acquisition (SCADA) system, distributed control system (DCS), and programmable logic controllers (PLCs). A production line or engineering system S is shown in Figure 6.1: (1) the human layer sits at the top of the architecture, where operators monitor process data via sensors directly or via human machine interface (HMI) and control the process by operating actuators directly or by inputing command to HMI; (2) the physical process layer sits at the bottom of the architecture, where a physical or chemical process is monitored via sensors and controlled by actuators; and (3) the control and automation layer sits in the middle, where ICS is at the center to collect real-time data of the controlled process via sensors, provide status and diagnostic data to operators via HMI, receive commands and settings from operators via HMI, and control the controlled process via actuators. In a power grid, ICS includes the energy management system (EMS), remote terminal unit (RTU), and PLC. Meters are sensors, and circuit breakers, transformers, and tap changers are actuators; the controlled physical process is the electric power transmission.

In this section, we describe a layered architecture perspective toward resilient industrial control systems (RICSs), which helps us to identify research problems and challenges at each layer and build models for designing security measures for control systems in critical infrastructures. We also emphasize a cross-layer viewpoint toward the security issues in ICSs in that each layer can have security dependence on the other layers. We need to understand the trade-off between the information assurance and the physical layer system performance before designing defense strategies against potential cyber threats and attacks.

We hierarchically separate ICSs into six layers, namely, physical layer, control layer, communication layer, network layer, supervisory layer, and management layer. The physical and control layers constitute the physical component of the system, while the communication and network layers constitute the cyber component of the system. The supervisory and management layers constitute the human component of an ICS. This hierarchical structure is depicted in Figure 6.2. The power plant is at the physical level, and the communication network and security devices are at the network and communication layers. The controller interacts with the communication layer and the physical layer. An administrator is at the supervisory layer to monitor and control the network and the system. Security management is at the highest layer, where security policies are made against potential threats from attackers. SCADA is the fundamental monitoring and control architecture at the control area level. The control centers of all major U.S. utilities have implemented a supporting SCADA for processing data and coordinating commands to manage power generation and delivery within the electric hybrid vehicle (EHV) and HV (bulk) portions of their own electric power system [32].

In subsequent subsections, we identify problems and challenges at each layer and propose problems whose resolution requires a cross-layer viewpoint.

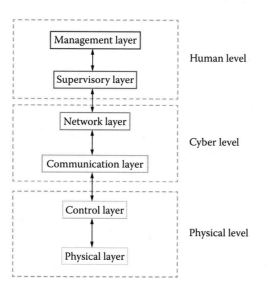

Figure 6.2: The hierarchical structure of ICSs composed of six layers. The physical layer resides at the bottom level and the management layer at the highest echelon.

6.3.1 Physical Layer

The physical layer comprises the physical plant to be controlled. It is often described by an ordinary differential equation (ODE) model from physical or chemical laws. It can also be described by difference equations, Markov models, or model-free statistics. We have the following challenges that pertain to the security and reliability of the physical infrastructure. First, it is important to find appropriate measures to protect the physical infrastructure against vandalism, environmental change, unexpected events, and so forth. Such measures often need a cost–benefit analysis involving the value assessment of a particular infrastructure. Second, it is also essential for engineers to build the physical systems with more dependable components and more reliable architecture. It brings up the concern of the physical maintenance of the control system infrastructures that demand cross-layer decision making between the management and physical levels.

6.3.2 Control Layer

The control layer consists of multiple control components, including observers and sensors, intrusion detection systems (IDSs), actuators, and other intelligent control components. An observer has the sensing capability that collects data from the physical layer and may estimate the physical state of the current system. Sensors may need to have redundancies to ensure correct reading of the states. The sensor data can be fused locally or sent to the supervisor level for global fusion. A reliable architecture of sensor data fusion will be a critical concern. An IDS protects the physical layer as well as the communication layer by performing anomaly-based or signature-based intrusion detection. An anomaly-based ID is more common for the physical layer, whereas a signature-based ID is more common for the packets or traffic at the communication layer. If an intrusion or an anomaly occurs, an IDS raises an alert to the supervisor or works hand in hand with built-in intrusion prevention systems (related to emergency responses, e.g., control reconfiguration) to take immediate action. There lies a fundamental trade-off between local decisions and a centralized decision when intrusions are detected. A local decision, for example, made by a prevention system, can react in time to unanticipated events; however, it may incur a high packet drop rate if the local decision suffers high false negative rates due to incomplete information. Hence, it is an important architectural concern on whether the diagnosis and control module needs to operate locally with IDS or globally with a supervisor.

6.3.3 Communication Layer

The communication layer is where we have a communication channel between control layer components or network layer routers. The communication channel can take multiple forms: wireless, physical cable, Bluetooth, and so forth. The communication layer handles the data communication between devices or layers. It is an important vehicle that runs between different layers and devices. It can often be vulnerable to attacks such as jamming and eavesdropping. There are also privacy concerns of the data at this layer. Such problems have been studied within the context of wireless communication networks. However, the goal of a critical infrastructure may distinguish themselves from the conventional studies of these issues.

6.3.4 Network Layer

The network layer concerns the topology of the architecture. We can see it is comprised of two major components: network formation and routing. We can randomize our routes to disguise or confuse the attacks so as to achieve certain security or secrecy or minimum delay. Moreover, once a route is chosen, how much data should be sent on that route has been long a concern for researchers in communications. In control systems, many specifics of the data form and rates may allow us to reconsider this problem in a control domain.

6.3.5 Supervisory Layer

The supervisory layer coordinates all layers by designing and sending appropriate commands. It can be viewed as the brain of the system. Its main function is to perform critical data analysis or fusion to provide immediate and precise assessment of the situation. It is also a holistic policy maker that distributes resources in an efficient way. The resources include communication resources, maintenance budget, and control efforts. In centralized control, we have one supervisory module that collects and stores all historical data and serves as a powerful data fusion and signal processing center.

6.3.6 Management Layer

The management layer is a higher-level decision-making engine where the decision makers take an economic perspective toward the resource allocation problems in control systems. At this layer, we deal with problems such as (1) how to budget resources to different systems to accomplish a goal and (2) how to manage patches for control systems, for example, disclosure of vulnerabilities to vendors and development and release of patches.

6.4 Metrics for Resilient Control Systems

Although some literature discusses the definition of resilient control systems [13, 19,23], each report covers only some parts of the resilience concept. There is no literature that defines a resilient control system quantitatively and its scope. In this chapter, we propose to define a resilient control system as follows: a *resilient control system* is designed and operated in a way that

- The incidence of undesirable incidents can be minimized.

- Most of the undesirable incidents can be mitigated or partially mitigated.

- The adverse impacts of undesirable incidents can be minimized, if these incidents cannot be mitigated completely.

- It can recover to normal operation in a short time.

Note that the undesirable incidents are not limited to those occurring on the control system itself. They can happen at the human layer, such as operators sending wrong commands or settings to ICS; they can happen at the physical process layer, such as broken cable in a power grid; they can also happen at the control layer, such as sensor damage and solenoid malfunction.

6.4.1 Properties of Resilient Industrial Control Systems

By definition, there are four desirable properties in a resilient industrial control system when it is designed and operated. Figure 6.3 illustrates those properties by using the example of cyber attacks on a power grid automation system.

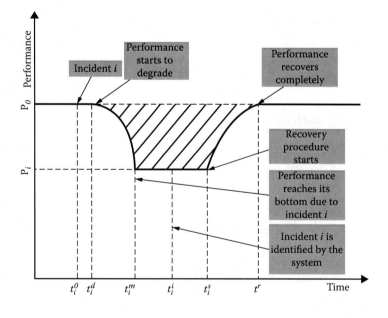

Figure 6.3: Resilience curve.

As shown in Figure 6.3, the performance axis shows the performance of the entire engineering system S (not the ICS), which can be defined as a function of production and quality, as in Equation 6.1. Let $\mathcal{P}$ denote the performance of the engineering system. p and q represent production and quality, respectively.

$$\mathcal{P}(t) = f(p(t), q(t)) \tag{6.1}$$

For instance, in a power grid, production can be measured by how much power is being delivered by the power grid, and quality is a function of voltage, frequency, harmonics, and so forth.

Property 6.1   A resilient control system is engineered and operated in a way that the incidence of undesirable incidents can be minimized.

For instance, to minimize the incidence of cyber attacks on a power grid automation, a control system can be designed and operated by using dedicated communication lines, isolating the automation network from the enterprise network, using cryptography, and hence reducing its exposure to potential hackers.

Property 6.2   A resilient control system is engineered and operated in a way that most of the undesirable incidents can be mitigated or partially mitigated.

For instance, assume that the only exposure point to the power grid automation network is the gateway to the enterprise network. In order to mitigate most cyber attacks, one firewall, which can detect intrusion and filter adverse data packets, can be implemented behind the gateway. And hence, with appropriate security policy, such as access control or an intrusion detection profile, it can mitigate most of the cyber attacks.

Property 6.3   A resilient control system is engineered and operated in a way that the adverse impacts of undesirable incidents can be minimized.

Assume incident i is a cyber attack on an automation system in one substation. In order to minimize the performance degradation, the power grid automation can be designed to be able to detect this attack and redirect some power flow to other substations before this substation automation system is compromised completely. Then the performance degradation is $P_0 - P_i$ (see notations in Table 6.1), provided that not all power flow can be shed to other substations.

Property 6.4   A resilient control system is engineered and operated in a way that it can recover from the adverse impacts of undesirable incidents to normal operation in a short time.

Assume, again, incident *i* is a cyber attack on an automation system in one substation, in Figure 6.3. In order to recover from the performance degradation, the power grid automation can be designed to be able to detect and locate this attack and isolate this attack or redirect the data path in a short time. Then the performance can come back to normal.

These properties of resilience can be regarded as four sequential steps to deal with undesirable incidents during both stages of engineering and operation, as shown in Figure 6.4.

A control system can be called *i* resilient if the engineering system is not adversely impacted by undesirable incident *i*. For instance, a power grid automation system can be called *cyber attack resilient* if (1) the control system has no exposure to hackers since the system is completely isolated; (2) if the system has exposure points to hackers, a firewall works efficiently to detect and block malicious data packets at the exposure points; or (3) the automation system possesses redundant devices and data paths and reroutes data packets to another path or uses other devices to avoid any adverse impact when it detects cyber attacks.

6.4.2   Distinguishing Terms

Terms such as *resilience, robustness, adaptiveness, survivability,* and *fault tolerance* are used interchangeably. However, they do not have the exact same meaning, although they may have some things in common. To focus on the resilience properties precisely, it is important to distinguish them.

The robustness of industrial control systems aims to function properly as long as modeling errors, in terms of uncertain parameters and disturbances at the physical process layer, as in Figure 6.1, are bounded; the adaptiveness of industrial control systems aims to function properly by adjusting their control algorithms according to uncertain parameters at the physical layer. Note that the uncertain parameters and disturbances can be regarded as undesirable incidents at the process layer. Survivability is the quantified ability of an industrial control system to continue to function during and after a nature or man-made disturbance, which may be occur at any layer in Figure 6.1. Fault-tolerant industrial control systems are focused on overcoming failures that may happen at any layer in Figure 6.1.

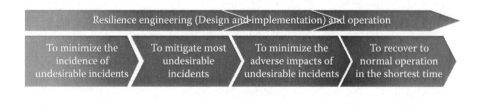

Figure 6.4: Four-step process to improve system resilience.

They try to identify failure possibilities and take precautions in order to avoid them by any means without causing a significant damage in the system; however, they do not consider the presence of intelligent adversaries, such as cyber attacks. Unlike resilience, robustness, adaptiveness, survivability, and fault tolerance do not address how quickly the industrial control system recovers to normal operation after the undesirable incident. All those properties are part of the characteristics of resilience. Therefore, resilience is a superset of all the above properties in the proposed concept.

6.4.3 Quantifying Resilience of a Control System

There is no literature report so far to measure *how resilient a control system is*, although there are reports on how to measure system resilience. For instance, it is proposed to measure resilience performance [17] by buffering capacity, margin, tolerance, and so forth. However, these metrics do not show how fast the system can recover from the undesirable incidents. Since it is objective to "measure" an object, and resilience is hard to measure, the metrics in this chapter are proposed to *estimate* rather than measure resilience of an industrial control system. Notations used in this chapter are described in Tables 6.1 and 6.2.

For an undesirable incident i, the following metrics, as shown in Table 6.2, are proposed to estimate the resilience of the control system:

■ **Protection time** T_i^p: The time that the system can withstand incident i without performance degradation:

$$T_i^p = t_i^d - t_i^0. \tag{6.2}$$

■ **Degrading time** T_i^d: The time that the system reaches its performance bottom due to incident i:

$$T_i^d = t_i^m - t_i^0. \tag{6.3}$$

■ **Identification time** T_i^i: The time that the system identifies incident i. Note that T_i^i is not necessarily greater than T_i^d—a well-designed and operated system is able to identify the incident before it reaches its performance bottom:

$$T_i^i = t_i^i - t_i^0. \tag{6.4}$$

■ **Recovery time** T_i^r: The time that the system needs to recover to normal operation from incident i:

$$T_i^r = t_i^r - t_i^s. \tag{6.5}$$

■ **Performance degradation** P_i^d: Maximal performance degradation due to incident i:

$$P_i^d = P_0 - P_i. \tag{6.6}$$

Table 6.1   Summary of Notations

Notation	Description
S	3-layered engineering system with an ICS in the center
p	Production
q	Quality
P	Performance
t_i^0	Moment that incident i occurs
t_i^d	Moment that system performance starts to degrade
t_i^m	Moment that system performance reaches the bottom due to incident i
t_i^l	Moment that incident i is identified by the ICS or operators
t_i^s	Moment that the system starts to recover; either manually initiated by operators or automatically by the ICS
t_i^r	Moment that the system completely recovers from incident i occurring
P_0	Original system performance when incident i occurs
P_i	Minimum performance due to incident i
μ_i	Number of incidents i that may occur per year
I	Set of all possible undesirable incidents
I'	Set of all possible critical undesirable incidents, where $I' \subseteq I$
M, N	Subsets of possible undesirable incidents
$\mu_{M,N}$	Probability that incidents of M occur and incidents of N do not
$L_{M,N}$	Overall potential loss that incidents of M occur and incidents of N do not

■ **Performance loss** P_i^l: Total loss of performance due to incident i:

$$P_i^l = P_0 \times (t_i^r - t_i^0) - \int_{t_i^0}^{t_i^r} P(t)dt. \qquad (6.7)$$

■ **Total loss** L_i: Total financial loss due to incident i, which includes performance loss, equipment damage, and recovery cost R_i^c:

$$L_i = f(P_i^l, R_i^c). \qquad (6.8)$$

■ **Overall potential critical loss** L': Overall loss due to all potential critical undesirable incidents I' per year:

$$L' = \sum_{\forall M,N \subseteq I'} L_{M,N} \times \mu_{M,N}, \qquad (6.9)$$

where $M, N \subseteq I'$, $M \cap N = \phi$, and $M \cup N = I'$.

Table 6.2   Resiliency Metrics for ICS

Term	Notation	Description
Protection time	T_i^p	Time that system S can withstand incident i without performance degradation
Degrading time	T_i^d	Time that system S reaches its performance bottom due to incident i
Identification time	T_i^i	Time that system S identifies incident i
Recovery time	T_i^r	Time that system S needs to recover to normal operation from incident i
Performance degradation	P_i^d	Maximal performance degradation of system S due to incident i
Performance loss	P_i^l	Total loss of performance of system S due to incident i
Total loss	L_i	Total financial loss due to incident i, which includes performance loss, equipment damage, and recovery cost
Overall potential loss	L	Overall loss due to all potential undesirable incidents
Overall potential critical loss	L'	Overall loss due to all potential critical undesirable incidents I'

Since it is not possible to enumerate all potential incidents, it is hard to compute the total loss, L_i. Thus, the overall potential critical loss, L', is proposed to calculate the overall resilience of a control system. For engineering system S, assume that there are two choices of control systems: System A and System B. System A is said to be more i-resilient than System B, or System A is more resilient than System B with regard to incident i if performance loss P_i^l and L_i of System A due to incident i are less than those of System B; System A is said to be more resilient than System B if the overall potential loss L of ICS A is less than that of ICS B.

6.5   Building, Operating, and Improving a RICS

In practice, it is not easy, if not impossible, to enumerate all potential undesirable incidents. Therefore, a reduced incident set I' is proposed to be used to analyze probability and adverse impacts of critical incidents, where $I', \overline{I'} \subseteq I, I' \cup \overline{I'} = I$, $I' \cap \overline{I'} = \phi$, and $\overline{I'}$ represent the set of undesirable incidents that can be ignored due to their insignificance of probability or adverse impacts.

Since it is also not easy to obtain precise information of all possible incidents i and their incidences μ_i, a cyclic process is proposed here, as shown in Figure 6.5, to improve the resilience of an ICS:

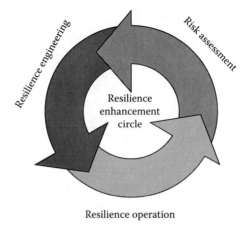

Resilience operation

Figure 6.5: Process of improving resilience of an ICS.

1. **Risk assessment:** Enumerate all possible undesirable incidents i, their incidences μ_i, and losses L_i.

2. **Resilience engineering:** Design and implement the ICS to minimize the overall possible loss L' with cost constraints.

3. **Resilience operation:** Operate system S in a way that the overall possible loss L' is minimized.

4. **Resilience enhancement:** Update information of I' and L'; repeat Steps 2 and 3.

6.5.1 Risk Assessment

The resilience properties discussed in Section 6.2 indicate that adding resilience elements to an industrial control system is focused on dealing with undesirable incidents. This requirement necessitates a control design strategy shift from reactive methods to proactive methods with consideration of assessing potential threats and taking necessary protection measures against them. In order to minimize the incidence and adverse impacts of all possible undesirable incidents, it is very important to understand them first:

■ Enumerate all possible undesirable incidents that can be imagined, I'.

■ Analyze how frequently they occur or the probability they happen, $\mu_{M,N}$.

■ Analyze the adverse impacts they can have on engineering system S, L'.

Undesirable incidents should be considered at all three levels:

- Improper commands and invalid settings from operators at the human level; wrong messages from ICS to operators could lead to wrong operation by the operators.

- Malfunctions and failures, at the control layer, of sensors and actuators; communication failure between controllers, HMIs, sensors, and actuators.

- Nonprecise and even wrong process model at the physical level.

Since it is difficult to enumerate all undesirable incidents and estimate the probabilities they happen, risk assessment cannot be performed once. After a period of operation time, this information should be updated iteratively, and risk assessment should be performed again based on the updated information.

6.5.2 Resilience Engineering

Based on the risk assessment presented in Section 6.5.1, resilience engineering is focused on dealing with some undesirable incidents, and it is a two-stage process: design and implementation. The design of a resilient industrial control system necessitates novel interactions between two engineering disciplines: computer and control engineering. From the control engineering point of view, the control of a complex dynamic industrial system is a well-studied area, such as advanced control technologies including robust control, adaptive control, and so forth, but much less is known about how to improve control system tolerance to cyber attack. As mentioned in Section 6.2, resilience is a superset of all the other properties; resilient decision and control laws should be synthesized as augmentation of existing control decisions such as robustness or adaptiveness, with the additional objective of reliable and fast recovery from undesirable incidents. The proactive control design strategy should be considered all the way from design through the implementation stages.

The following areas should be studied to improve system resilience during the stage of resilience engineering:

- To minimize the probability of undesirable incidents $\mu_{M,N}$ (1) a well-designed ICS can validate the inputs from HMI by operator authentication and authorization and input limits of data, thus identifying invalid commands from the operator; (2) it can also validate input data from sensors and pass correct data to operators; and (3) a well-designed system monitoring and prognosis tool monitors and predicts failures of some key components and enables operators to prevent them from happening.

- To mitigate undesirable incidents or minimize the adverse impacts of them $L_{M,N}$, (1) as a general paradigm, the most accepted and used implementation principle for resilience is redundancy; systems make use of redundant components along with primary components and switch to them in a failure condition; (2) deploy distributed control system, which can still work if one single controller fails; and (3) enable the control system to be aware of its states and keep distance from its operation boundaries.

■ To recover in a short time $T_{M,N}^r$, (1) enable the control system to identify the undesirable incidents accurately, and pass the corresponding information to operators, if they are in the control loop; (2) provide a functionality that can generate backup recovery plans online and automatically for some critical undesirable incidents; and (3) enable the system to start the corresponding recovery plan when the undesirable incident is identified.

6.5.3 Resilience Operation

Based on the risk assessment and resilience engineering, resilience operation should be of state awareness, cyber-attack awareness, and risk awareness. With all real-time information, a resilient industrial control system should be operated to minimize the potential loss of System S:

■ To minimize the probability of undesirable incidents $\mu_{M,N}$, (1) a well-designed and well-operated ICS can monitor System S and intelligently analyze real-time data, and identify boundary conditions and operation margins, (2) it can also pass the analysis results to operators and provide operation suggestions to operators.

■ To mitigate undesirable incidents or minimize the adverse impacts of them $L_{M,N}$, (1) a well-designed and well-operated ICS can generate and adjust control strategies online according to detected undesirable incidents or potential incidents; (2) it can be aware of its state, cyber attacks, and risks, and keep distance from boundaries; and (3) it can interpret, reduce, and prioritize undesirable incidents based on the information from the state awareness, thus providing adaptive capacity to perform corresponding responses, such as prioritized response to focus on mitigating the most critical incidents if parallel responses are limited.

■ To recover in a short time $T_{M,N}^r$, (1) a well-operated ICS can identify the undesirable incidents online accurately and pass the corresponding information to operators; (2) it can also generate backup recovery plans online and automatically for detected undesirable incidents; and (3) it can start the corresponding recovery plan once the undesirable incident is identified.

6.5.4 Resilience Enhancement

Due to the uncertainty and complexity in control system applications, control system redesign becomes inevitable to meet challenges in real applications that might have been ignored at the beginning. This is also true for resilience. Since it is hard to list all critical undesirable incidents and estimate their corresponding losses, it is important to update this information with more complete and accurate data after engineering and operation. On the other hand, with this updated information, new control strategies can be developed during engineering and executed during operation accordingly, hence further enhancing system performance in terms of resilience.

6.6 Cyber-Attack-Resilient Power Grid Automation System: An Example of Resilient ICS

This section further discusses resilience by using an example—cyber-attack-resilient power grid automation system. The approaches to improve power grid automation system resilience with respect to cyber attacks are presented. A cyber risk assessment model and a framework to protect the power grid from cyber attacks are disclosed. Some ideas in this section are part of the author's previous works in [20] and [29], and some parts are their extension.

The emerging smart grid requires the conventional power grid to operate in a way that it was not originally designed to. Bringing more participants to the smart grid will open the originally isolated automation networks to more people, if not to the public. This brings big concerns of cyber-security issues of automation systems. To improve cyber-attack resilience of the power grid automation system, a security solution framework with three major elements is proposed:

1. **Dynamic and evolutionary risk assessment model:** The model assesses the critical assets of the power grid. It uses dynamic quasi-real-time simulations to reveal potential vulnerabilities. It detects security events and activities both previously known and currently unidentified. Using the existing topology of the power grid, a risk assessment graph is created. And this graph dynamically evolves through design and real-world operation. The graph is translated into a Bayesian network where edges are weighted according to predefined economical measures and business priorities. The model provides a list of assets with utility functions that reflect the associated risks and economic loss.

2. **Integrated and distributed security system:** It overlays across the intelligent power grid network in a hierarchical and distributed manner. The system includes (a) security agents, which reside next to or are integrated to field devices and controllers, providing end-to-end security; (b) managed switches in control networks, providing quality of service (QoS)

in terms of delay and bandwidth; and (c) security managers, which are distributed across control centers and substations, managing cyber-security-related engineering, monitoring, analysis, and operation. The proposed security system enables power grid operators to monitor, analyze, and manage cyber security of the power grid.

3. **Security network topology optimization model:** It optimizes the topology on the security system without compromising the performance of control functionalities.

6.6.1 Risk Assessment Model

To construct a general model for risk assessment, an integration of physical features of power grids and substations with cyber-related and security characteristics of such systems is needed. To make the model practical as well, a level of aggregation in cyber-security analysis should be considered to avoid complexity and dimensionality that cannot be implemented with existing calculation capacities. Therefore, the proposed framework is decomposed as follows:

1. First-pass model runs at the grid level to identify the most critical substations to the power grid operation

2. Second-pass model runs at the substation level to identify the most critical components to the substation operation

This risk assessment model can be run both offline and online. When running offline (risk assessment stage), it receives inputs including power grid topology, substation primary circuit diagrams, statistical power flows, and automation system topology. It calculates and outputs all potential loss due to cyber attacks to critical components in substations. The result can help power grid operators to find critical cyber-security assets and understand the potential loss, L', related to cyber attacks on these assets. When running online (resilience operation stage), the inputs of this model replace statistical power flow data with real-time power flow data. The outputs L' are the same as those of the offline model. The results can help the operator identify critical security assets and understand the potential loss due to cyber attacks according to real-time information, and further improve its resilience during both resilience operation enhancement stages.

6.6.2 Integrated Security System

To address the security issues in power grid automation systems, a paradigm to build a distributed and scalable security framework is proposed, as shown in Figure 6.6. Note that DMS is distribution management system, IED is intelligent

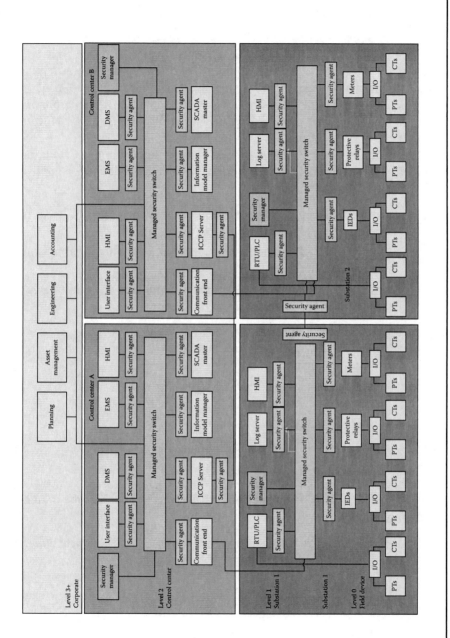

Figure 6.6: The integrated security system for power grid automation systems.

electronic device, and PT and CT are potential transmitter and current transmitter, respectively. There are three major conceptual components in the proposed integrated security framework, as follows:

■ **Security agent:** It brings security to the edges of the system by providing protection at the networked device level. These agents are firmware or software agents, depending on the layer of the control hierarchy. At the field device layer (e.g., IEDs), these agents are less intelligent—containing simple rules and decision-making capabilities—and do more of event logging and reporting. At higher control layers (e.g., RTUs), these software agents are more intelligent with more complex rules for identification and detection of intrusive events and activities within the controllers. In particular, a security agent is commissioned to accomplish the following functionalities:

  ■ Acquire and run the latest vulnerability patches from its security manager

  ■ Collect data traffic patterns and system log data and report to the security manager

  ■ Analyze traffic and access patterns with varying complexity, depending on the hierarchical layer

  ■ Run host-based intrusion detection

  ■ Detect and send alarm messages to the security manager and designated devices, such as HMI

  ■ Acquire access control policies from the security manager and enforce them

  ■ Encrypt and decrypt exchanged data

■ **Managed security switch:** To protect bandwidth and prioritize data, the managed switches are used across the automation network. These switches, working as network devices, connect controllers, RTUs, HMIs, and servers in the substation and control center. They possess the following functionalities:

  ■ Separate external and internal networks, hide the internal networks, and run network address translation (NAT) and network port address translation (NPAT)

  ■ Acquire bandwidth allocation pattern and data prioritization pattern from the security manager

  ■ Separate data according to prioritization pattern, such as operation data, log data, trace data, and engineering data

  ■ Provide QoS for important data flow, such as operation data, guaranteeing its bandwidth and delay

- Manage multiple virtual local area networks (VLANs)

- Run simple network-based intrusion detection

- **Security manager:** Security managers reside in the automation network and directly or indirectly connect to the managed switches across the automation networks. They can be protected by existing IT security solutions and are able to connect to a vendor's server, managed switches, and security agents via a virtual private network (VPN). Security managers possess the following functionalities:

    - Collect security agent information

    - Acquire vulnerability patches from a vendor's server and download them to the corresponding agents

    - Manage cryptographic keys

    - Work as an authentication, authorization, and accounting (AAA) server, which validates user identifications, authorizes user access right, and records what a user has done to controllers

    - Collect data traffic patterns and performance matrices from agents and switches

    - Collect and manage alarms and events from agents and switches

    - Generate access control policies based on collected data and download to agents

    - Run complex intrusion detection algorithms at automation network levels

    - Generate bandwidth allocation patterns and data prioritization patterns and download them to managed switches

The integrated security system monitors communication traffic, detects possible cyber attacks, and minimizes the adverse impacts of those cyber attacks.

6.6.3   *Security Optimization Model*

Based on the result of the risk assessment model (most vulnerable components such as RTUs and communication links), the security optimization model can help power grid operators place security agents and switches with constraints of cost, bandwidth, and data delay requirements, hence improving system cyber-attack resilience during the engineering stage; this model can also help operators adjust security policies to improve cyber-attack resilience during resilience operation and enhancement stages, according to online risk assessment results and detected cyber intrusions.

6.6.4 Cross-Layer Co-Design

Security solutions at the physical and cyber layers of an integrated control system need to take into account the interaction between these two layers. Figure 6.7 illustrates the concept of security interdependence between the cyber system and the physical plant. We need to adopt a holistic cross-layer viewpoint toward a hierarchical structure of industrial control systems. The physical layer is comprised of devices, controllers, and the plant, whereas the cyber layer consists of routers, protocols, and security agents and manager. The physical layer controllers are often designed to be robust, adaptive, and reliable for physical disturbances or faults. With the possibility of malicious behavior from the network, it is also essential to design controllers that take into account the disturbances and delay resulting from routing and network traffic, as well as the unexpected failure of network devices due to cyber attacks. On the other hand, the cyber-security policies are often designed without consideration of control performances. To ensure the continuous operability of the control system, it is equally important to design security policies that provide a maximum level of security enhancement but minimum level of system overhead on the networked system. The physical and cyber aspects of control systems should be viewed holistically for analysis and design.

The cyber infrastructure serves as an interface between the controller and the physical plant. The control signals are sent through a security-enhanced IT infrastructure, such as wireless networks, the Internet, and local area networks (LANs). The security architecture of the IT infrastructure is designed to enable the security practice of defense in depth for control systems. The cascading countermeasures using a multitude of security devices and agents, ranging from physical protections to firewalls and access control, can offer the administrators more opportunities for information and resource control with the advent of potential threats. However, it also creates possible issues on the latency and

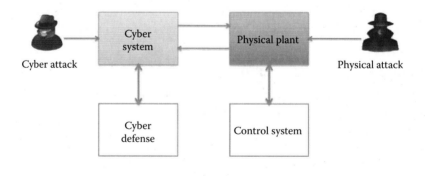

Figure 6.7: Security of cyber-physical systems: the physical and cyber layers of the control system interact with each other.

packet drop rate of communications between the controller and the plant. We propose a unifying security model for this cyber-physical scenario and investigate the mitigating strategies from the perspectives of control systems, as well as of cyber defenses. At the physical layer, we often aim to design robust or adaptive controllers that take into account the uncertainties and disturbances in the system to enhance robustness and reliability of the system. At the cyber level, we often employ IT security solutions by deploying security devices and agents in the network. The security designs at the physical and cyber layers usually follow different goals without a unifying framework. Hence, the design of optimal security policies at the cyber level requires a cross-layer approach between the interacting layers of the entire system. Hence, it is important to jointly consider the controller design of the physical layer dynamical processes together with security policy designs. The control performance of the physical process relies on the cyber-security policies that have been enforced at the communication and network layers. The time delay of data delivery, packet drop rates, and quality of service need to be taken into account when designing high-performance controllers for the underlying processes. In addition, the design of cyber policies has to consider their potential impact on the physical layer of the system.

6.6.5 *Multiagent System Design*

For large-scale complex ICSs, it is difficult to coordinate all components of the system using a centralized supervisory system. Hence, a multiagent system (MAS) architecture with communications between different components of the system is desirable. While communication and negotiation are common to all multiagent systems, data analysis and control are more specific and necessary within a control system design. The negotiation and communications aspects themselves must be tailored to the control system objectives, which are based on the desired operation of a physical process operation. Control, communications, data analysis, and negotiation comprise four technologies that characterize the functionality of an agent within a multiagent hierarchy.

6.6.5.1 *Data Analysis*

By its nature, a MAS implies a philosophy of achieving a level of global optimization based on decisions made from multiple sensors. To ensure that the degradation of sensors and resulting data does not lead to a failure in the design, mechanisms to interpret the data and ascertain status are required. Graceful degradation of sensor systems becomes a critical aspect of these designs to maintain the highest level of the MAS autonomous relationships through recognition and replacement of sensory inputs. These additional inputs can come from multiple replicant sensors, diverse technology sensors, and models coupled with sensors to provide the same sensory aspect. Salient features of this aspect of MAS should include:

■ Diagnostic detection of failure for all sensors and synthetic sources of sensory data that are used as a basis for confirming performance of the system.

■ Prognostic prediction of future degradation, specifically failures with equipment that is necessary for operation. This is especially critical for equipment with a long lead time for replacement.

■ Data fusion of inputs for data reduction, prioritization, and targeting of mixed initiative response. Characterization and measurement of blended performance, specifically where more than one measure of normal operation comes to bear. For example, the characterization of cyber health can be enhanced with the temporal data on physical security and plant stability.

6.6.5.2 Control

The type of control engineering adaptable to the multiagent design will be dependent on the application and layer of hierarchy. The type of control at the lower layers of the autonomy may often be more diverse and application dependent. At the higher layers, where more cognitive aspects and decisions are expected, a greater use of computational intelligence is expected. The following considerations should be included in the development of the control design:

■ Final implementation of the MAS design includes practitioners, and therefore the interactions should be targeted to just those aspects that require input to align operations to the particular environment.

■ The application of control to the device layer of the MAS should be based on need as determined by the designer and implementing practitioners, where advanced control is applied as needed to achieve better performance. However, while advanced control is not necessarily encouraged, the ability to implement conventional proportional-integral-derivative control with supervisory inputs is required.

■ The application of control to the coordination layer of the hierarchy is for the negotiation of resources to the MAS, as well as an interpreter of philosophy from the hierarchy superior to the subordinates. This role implies, at least in part, a hybrid control system design. That is, superior information will often come in the form of state-based direction, which must often be applied to a continuous plant. If the mechanism to dictate the philosophy to the subordinate can occur through state-based mechanisms, such as set-point adjustment, this is not an issue. However, in the case where trajectories are necessary to gracefully shift plan states, this is more difficult. Given that these trajectories can be identified as part of the device layer design, the method of dictating philosophy can still be through the variation of set points or discrete settings.

■ At the management layer of a MAS hierarchy, the decomposition of operating philosophy occurs. By operating philosophy, the intent is to define what the desired operation of the control system should be. The overall participation in the directly or indirectly human-related aspect comes from several sources. Bayesian inference, fuzzy logic, and other intelligent systems technology can provide a means of characterizing and normalizing these potentially disparate sources of philosophy. The resulting output of this layer, however, is to associate performance desires within the limitations of the system and regulations.

6.6.5.3 Communications

As the interactive framework is established, the mechanism for interacting between layers and peers is required. The types of interaction change depending on the layer and path, peer to peer or peer to superior or subordinate. The communications design will need to accommodate these interactions, which might at first glance appear to be fulfilled by current 802-based protocols. However, there are some specific MAS considerations that require accommodation:

■ Determinism and latency are important considerations and the constraints will vary, depending on the MAS layer in which the communications are occurring. Response time is slower at the higher layers of the hierarchy, and the forms of the interaction changes can be relatively simple (e.g., set points and ranges) or complex (e.g., intelligent interpreters of system and performance indices for optimal tracking).

■ Assurance of authenticity is required in the exchange of data, specifically to defend against the injection of a malicious agent. As part of the data analysis, cyber security is considered one of the performance parameters for evaluation. However, the key aspect for the communications design is to consider the unique identification of agents and mechanisms to confound a malicious attacker from determining this identification.

■ Standards such as Foundation for Intelligent Physical Agents (FIPA) and the IEEE 802 series exist to govern the fundamental aspects of communication. However, such standards are expected to evolve as techniques are codified in cyber defense, such as randomization of communications, including control system–specific defenses. These defenses will occur while still ensuring that the determinism and maximum latency are maintained for the control system application.

6.6.5.4 Negotiation

Consensus of agents within a resilient control system framework requires consideration of the dynamical attributes of the design. A MAS in the context of this chapter is intended to afford reconfiguration of agents, and ultimately control of resources to achieve an optimum performance of the overall system. The stability aspect of performance must therefore be considered during the negotiation of resources. In this regard, negotiation of a MAS system should consider the following:

■ The dynamics of the system require tracking of the optimum path or trajectory to achieve optimum stability, in addition to achieving stability. Stated another way, the performance of the system remains within its constraints for operation. This implies a performance index that considers path and endpoints.

■ The ability to share and ultimately negotiate resources is limited by the uniformity of the system. For example, an unmanned air vehicle squadron provides a highly uniform implementation of a MAS, and therefore a higher level of resource sharing is theoretically possible within the constraints of the design. The level of uniformity is defined, in this case, as the ability of an agent to provide a necessary functionality in the fulfillment of a need.

■ Decisions for shifting of resources can occur at different layers of the MAS hierarchy, with control action taken at both the middle and lower layers. However, the goal of negotiation is the same regardless of the level, which is to adjust resources to reach optimum performance. The difference lies in the sphere of influence. That is, the coordination layer has the responsibility for multiple lower-level agents and, as such, will better orchestrate shifts in operation to accommodate the philosophy of the management layer.

6.6.6 Discussion

The proposed cyber-attack-resilient power grid automation system is engineered and operated in a way that (1) the system can be aware of power grid operation states, cyber attacks, and their potential adverse impacts on power grid operation by online risk assessment and intrusion detection; (2) the system can analyze what cyber attacks are and where they occur, and pass this information to operators; (3) the system can mitigate detected cyber attacks by adjusting corresponding security policies, such as access control in security agents; (4) the system can minimize the adverse impacts by rerouting data path from the attacked communication link or redirecting power flow from the attacked substations if these cyber attacks cannot be mitigated; and (5) the system can help operators reroute the data path from the attacked communication link or redirect power flow from the compromised substations, and hence recover to normal operation quickly.

6.7 Conclusion

This chapter aimed to shed some light on resilient industrial control systems by proposing a six-layer system model and resilience curve; conceptualizing resilient industrial control systems; distinguishing the concept of resilience from other terms, such as robustness, adaptiveness, survivability, and fault tolerance; presenting desirable properties of resilient industrial control systems; and offering metrics to estimate system resiliency quantitatively. The general approaches to improve industrial control system performance in terms of resilience were discussed as well. A cyber-attack-resilient power grid automation system was presented as an example to illustrate the proposed approaches.

Acknowledgment

This material is based on work supported by the National Science Foundation under grant no. EFMA-1441140, SES 1541164, and CNS-1544782.

References

1. I. Aad, J. Hubaux, and E. Knightly. Denial of service resilience in ad hoc networks. In *Proceedings of the 10th Annual International Conference on Mobile Computing and Networking (MobiCom '04)*, New York, ACM, pp. 202–215, 2004.

2. D. Andersen, H. Balakrishnan, F. Kaashoek, and R. Morris. Resilient overlay networks. *ACM SIGOPS Operating Systems Review*, 35:131–145, 2001.

3. D. van Opstal. Transform—The resilient economy: Integrating competitiveness and security. Technical report, Compete, Council on Competitiveness, Washington, DC, July 2007.

4. da Fontoura Costa, L. Reinforcing the resilience of complex networks. *Physical Review E*, 69(6):066127, 2004.

5. Department of Homeland Security. National infrastructure protection plan—Critical manufacturing sector. Department of Homeland Security, Washington, DC, April 2008.

6. Department of Homeland Security. Critical infrastructure resilience final report and recommendations. Department of Homeland Security, Washington, DC, September 2009.

7. Department of Homeland Security. Information technology sector baseline risk assessment. Department of Homeland Security, Washington, DC, August 2009.

8. Department of Homeland Security. National infrastructure protection plan—Partnering to enhance protection and resiliency. Department of Homeland Security, Washington, DC, 2009.

9. C. S. Holling. Resilience and stability of ecological systems. *Annual Review of Ecology and Systematics*, 4:1–23, 1973.

10. M. Huynh, S. Goose, and P. Mohapatra. Resilience technologies in ethernet. *Computer Networks*, 54(1):57–78, 2010.

11. S. E. Flynn. Building a resilient nation: Enhancing security, ensuring a strong economy. Technical report; keynote address, Reform Institute, New York, October 2008.

12. S. Jaggi, M. Langberg, S. Katti, T. Ho, D. Katabi, M. Medard, and M. Effros. Resilient network coding in the presence of Byzantine adversaries. *IEEE Transactions on Information Theory*, 54(6):2596–2603, 2008.

13. A. Krings. Design for survivability: A tradeoff space. In *Proceedings of the 4th Annual Workshop on Cyber Security and Information Intelligence Research: Developing Strategies to Meet the Cyber Security and Information Intelligence Challenges Ahead* (CSIIRW '08), F. Sheldon, A. Krings, R. Abercrombie, and A. Mili (eds). ACM, New York, Article 12, 2008.

14. D. Liu. Resilient cluster formation for sensor networks. In *Distributed Computing Systems, 2007. ICDCS '07. 27th International Conference on*, pp. 40–40, Toronto, Canada, June 25–29, 2007.

15. M. H. Manshaei, Q. Zhu, T. Alpcan, T. Başar, and J. P. Hubaux. Game theory meets network security and privacy. *ACM Computing Surveys (CSUR)*, 45(3):25, 2013.

16. N. McDonald. *Organizational Resilience and Industrial Risk*, pp. 155–180. Ashgate, Surrey, 2006.

17. D. Mendonca. *Measures of Resilient Performance*, vol. 1, pp. 29–47. Ashgate, Surrey, 2008.

18. M. Menth, M. Duelli, R. Martin, and J. Milbrandt. Resilience analysis of packet-switched communication networks. In *IEEE/ACM Transactions on Networking (TON)*, 17(6):1950–1963, 2009.

19. S. M. Mitchell and M. S. Mannan. Designing resilient engineered systems. Technical Report 4, Chemical Engineering Progress, New York, April 2006.

20. Z. Mohajerani, F. Farzan, M. Jafari, Y. Lu, D. Wei, N. Kalenchits, B. Boyer, M. Muller, and P. Skare. Cyber-related risk assessment and critical asset

identification within the power grid. In *Proceedings of IEEE PES Transmission and Distribution Conference and Exposition*, New Orleans, LA, April 2010.

21. D. Nathanael and N. Marmaras. *Work Practices and Prescription: A Key Issue for Organizational Resilience*, vol. 1, pp. 101–118. Ashgate, Surrey, 2008.

22. C. G. Rieger. Notional examples and benchmark aspects of a resilient control system. In *Proceedings of International Symposium on Resilient Control Systems*, Idaho Falls, ID, pp. 64–71, 2010.

23. C. G. Rieger, D. Gertman, and M. A. McQueen. Resilient control systems: Next generation design research. In *Proceedings of IEEE Conference on Human System Interactions*, Catania, pp. 632–636, May 2009.

24. S. Roy, S. Setia, and S. Jajodia. Attack-resilient hierarchical data aggregation in sensor networks. In *Proceedings of ACM Workshop on Security of Ad Hoc and Sensor Networks*, Alexandria, VA, pp. 71–82, 2006.

25. Y. Sheffi. *The Resilient Enterprise: Overcoming Vulnerability for Competitive Enterprise*. MIT Press, Cambridge, MA, 2005.

26. P. Sousa, A. Bessani, M. Correia, N. Neves, and P. Verissimo. Resilient intrusion tolerance through proactive and reactive recovery. In *Proceedings of Pacific Rim International Symposium on Dependable Computing*, Melbourne, 17–19 December, pp. 373–380, 2007.

27. P. Sousa, N. F. Neves, and P. Verissimo. How resilient are distributed f fault/intrusion-tolerant systems? In *Proceedings of International Conference on Dependable Systems and Networks*, Yokohama, Japan, June 28–July 1, pp. 98–107, 2005.

28. D. Wei and K. Ji. Resilient industrial control system (RICS): Concepts, formulation, metrics, and insights. In *Proceedings of International Symposium on Resilient Control Systems*, Idaho Falls, ID, pp. 15–22, 2010.

29. D. Wei, Y. Lu, M. Jafari, P. Skare, and K. Rohde. An integrated security system of protecting smart grid against cyber attacks. In *Innovative Smart Grid Technologies (ISGT)*, Gothenburg, Sweden, pp. 1–7. IEEE, 2010.

30. E. E. Werner. *The Children of Kauai: A Longitudinal Study from the Prenatal Period to Age Ten*. University of Hawaii Press, Honolulu, HI, 1971.

31. J. Wreathall. *Properties of Resilient Organizations: An Initial View*, pp. 275–285. Ashgate, Surrey, 2006.

32. F. F. Wu, K. Moslehi, and A. Bose. Power system control centers: Past, present, and future. *Proceedings of the IEEE*, 93(11):1890–1908, 2005.

33. Q. Zhu and T. Başar. Indices of power in optimal IDS default configuration: Theory and examples. In *Proceedings of International Conference on Decision and Game Theory for Security*, College Park, MD, pp. 7–21, 2011.

34. Q. Zhu and T. Başar. A dynamic game-theoretic approach to resilient control system design for cascading failures. In *Proceedings of International Conference on High Confidence Networked Systems*, Beijing, pp. 41–46, 2012.

35. Q. Zhu and T. Basar. Dynamic policy-based IDS configuration. In *Proceedings of IEEE Conference on Decision and Control*, Shanghai, pp. 8600–8605, 2009.

36. Q. Zhu and T. Basar. Robust and resilient control design for cyber-physical systems with an application to power systems. In *Proceedings of IEEE Conference on Decision and Control and European Control Conference*, Orlando, FL, pp. 4066–4071, 2011.

37. Q. Zhu, C. Rieger, and T. Basar. A hierarchical security architecture for cyber-physical systems. In *Proceedings of International Symposium on Resilient Control Systems*, Boise, ID, pp. 15–20, 2011.

38. Q. Zhu, H. Tembine, and T. Basar. Network security configurations: A nonzero-sum stochastic game approach. In *Proceedings of American Control Conference*, Baltimore, MD, pp. 1059–1064, 2010.

Chapter 7

Topology Control in Secure Wireless Sensor Networks

Jun Zhao

CyLab and Department of Electrical and Computer Engineering,
Carnegie Mellon University

CONTENTS

Wireless sensor networks (WSNs) are a distributed collection of small sensor nodes that gather security-sensitive data and control security-critical operations in a wide range of industrial, home, and business applications [1]. Many applications require deploying sensor nodes in hostile environments where an adversary can eavesdrop sensor communications and can even capture a number of sensors and surreptitiously use them to compromise the network. Therefore, cryptographic protection is required to secure the sensor communication as well as to detect sensor capture and revoke the compromised keys. Given the limited communication and computational resources available at each sensor, security is expected to be a key challenge in WSNs [14, 23, 67].

In this chapter, we discuss topology control in secure WSNs; in particular, we present how to design secure WSNs such that their topologies have certain connectivity properties.

7.1   Random Key Predistribution Schemes

This section discusses random key predistribution schemes for secure wireless sensor networks.

Random key predistribution schemes have been widely recognized as appropriate solutions to secure communications in resource-constrained wireless sensor networks (WSNs) [14, 23, 41, 46, 47, 49, 67]. The idea of randomly assigning cryptographic keys to sensors before deployment has been introduced in the seminal work of Eschenauer and Gligor [23]. Following that, Chan et al. [14] proposed the *s*-composite key predistribution scheme as an extension of the Eschenauer–Gligor (EG) scheme [23] (the EG scheme is a special case of the *s*-composite scheme with $s = 1$). Over the last decade, the EG scheme has been extensively studied in the literature [16, 17, 26, 46, 47, 49, 60, 61, 64, 67, 70], and this applies to the *s*-composite scheme as well [9, 16, 17, 38, 46, 55].

7.1.1 Eschenauer–Gligor Random Key Predistribution Scheme

The EG scheme works as follows. For keying a network comprising n sensors, this scheme uses an offline *key pool* containing P_n keys, with P_n being a function of n for generality reasons. Before deployment, each sensor is independently equipped with K_n distinct keys selected uniformly at random from the key pool, where K_n is also a function of n. The K_n keys in each sensor constitute the sensor's *key ring*. After deployment, two communicating sensors can establish a secure link only if they have at least one key in common, as message secrecy and authenticity are achieved by using symmetric-key encryption modes [25, 28, 39].

7.1.2 s-Composite Random Key Predistribution Scheme

The s-composite scheme differs from the EG scheme only in that under the former scheme, a secure link between two sensors demands their sharing of at least s different keys, where $1 \leq s \leq K_n$, while for the latter scheme, only at least one overlapped key is necessary. The s-composite scheme with $s \geq 2$ outperforms the EG scheme in terms of the strength against small-scale sensor capture attacks while trading off increased vulnerability in the face of large-scale attacks. The parameter s is chosen according to the desired resilience of the network against different scales of sensor capture.

7.2 Link Constraint Models

Different link constraint models have been used to explore WSNs employing the EG scheme or the s-composite scheme. Among them, the simplest one is the following full-visibility model.

7.2.1 Full-Visibility Model

The full-visibility model assumes that any two sensors have a direct communication link in between. Under such a model, a secure link exists between two sensors if and only if they possess a certain amount of shared keys (1 for the EG scheme and s for the s-composite scheme).

In resolving the impracticality of the full-visibility model, different models of *link constraints* are considered as its alterative. To this end, the existence of a secure link between two sensors requires not only some amount of shared keys, but also that the link constraint between them is satisfied. Specifically, two different kinds of link constraints—the on/off channel model and the disk model—are as follows.

7.2.2 On/Off Channel Model

The on/off channel model consists of independent channels, each of which is either *on* with probability p_n or *off* with probability $(1 - p_n)$, where p_n is a function of n for generality. Under such a model, the channel between two sensors has to be *on* for direct communication.

7.2.3 Disk Model

In the disk model, each node's* transmission area is a disk with a transmission radius r_n, with r_n being a function of n for generality. Two nodes have to be within a distance of no greater than r_n for direct communication. For the node distribution, we consider that all n nodes are uniformly and independently deployed in a bounded area of a Euclidean plane. Such network area $\mathcal{A}$ in our analysis is either a torus $\mathcal{T}$ or a square $\mathcal{S}$, each of unit area, depending on whether the *boundary effect* exists. The boundary effect arises whenever part of the transmission area of a node may fall outside the network area $\mathcal{A}$. That is, $\mathcal{T}$ does not have the boundary effect, whereas $\mathcal{S}$ does.

Topological properties of the WSNs employing the EG scheme or the *s*-composite scheme have received much interest in the literature [3, 5, 9, 30, 31, 40, 41, 47, 49]. Most of the existing research assumes the full-visibility model [3, 5, 9, 40, 41, 49], with only a few recent works considering the on/off channel model [47] or the disk model [30, 31]. Among various topological properties, (*k*-)connectivity and node degree, detailed below, are of particular interest [3, 5, 30, 31, 40, 41, 47, 49].

7.3 System Model

7.3.1 Individual Graphs

The EG and *s*-composite key predistribution schemes are modeled by a *uniform random 1-intersection graph* [3, 4, 6, 9, 32, 35, 40, 46, 47, 49, 60–62, 67] and a *uniform random s-intersection graph* [5, 8, 9, 55], respectively, where the former is also known as a *random key graph* [3, 4, 9, 35, 40, 46, 47, 62]. The formal definitions of a uniform random 1-intersection graph and a uniform random *s*-intersection graph are as follows.

Definition 7.1 (uniform random 1-intersection graph [49]). A uniform random 1-intersection graph $G_1(n, K_n, P_n)$ is constructed on a set of n nodes as follows. Each node is independently assigned a set of K_n distinct objects, selected uniformly at random from a pool of P_n objects. There exists an edge between two nodes if and only

---

*Throughout the chapter, the terms *sensor* and *node*, as well as *network* and *graph*, are interchangeable.

if they possess at least one common object. Note that an object is a cryptographic key in the EG key predistribution scheme [23].

Definition 7.2 (uniform random s-intersection graph [55]).   A uniform random s-intersection graph denoted by $G_s(n, K_n, P_n)$ is constructed on a node set with size n as follows. Each node is independently assigned a set of K_n different objects, selected uniformly at random from a pool of P_n objects. An edge exists between two nodes if and only if they have at least s objects in common. Note that an object is a cryptographic key in the s-composite key predistribution scheme [14].

Random intersection graphs were introduced by Singer-Cohen [44]. These graphs have received considerable attention in the literature [1–11, 28–35, [55–63, 65–68]. In a general random intersection graph, each node is assigned a set of items in a *random* manner, and any two nodes establish an *undirected* edge in between if and only if they have at least a certain number of items in common.

In *a binomial random s-intersection graph* with n nodes, each item from a pool of P_n distinct items is assigned to each node *independently* with probability t_n, and any two nodes have an edge in between upon sharing at least s items, where s_n and P_n are functions of n for generality. We denote a binomial random s-intersection graph by $H_s(n, t_n, P_n)$. The term *binomial* is used since the number of items assigned to each node follows a binomial distribution with parameters P_n (the number of trials) and t_n (the success probability in each trial).

Random intersection graphs have numerous application areas, including secure wireless communication [60–63], social networks [2, 12, 13, 26], cryptanalysis [4], circuit design [44], recommender systems [35], classification [27], and clustering [8, 15]. Random intersection graphs are natural models for social networks [7], examples of which are given below as common-interest networks, researcher networks, and actor networks. In a common-interest network [67], each user has several interests following some distribution, and two users are said to have a common-interest relation if they share at least s interests. In a researcher network (an example of a collaboration network) [5, 15], each researcher publishes a number of papers, and two researchers are adjacent if coauthoring at least s papers. In an actor network [12, 13], each actor contributes to a number of films, and two actors are adjacent if acting in at least s common films. Examples can be extended to other types of social networks. For all examples, clearly the induced topologies are represented by random intersection graphs.

With $\text{network}_{\text{full-visibility}}^{\text{EG}}$ (respectively, $\text{network}_{\text{full-visibility}}^{s\text{-composite}}$) denoting a sensor network employing the EG (respectively, s-composite) scheme under the full-visibility model, we formally present Lemma 7.1 for the topologies of $\text{network}_{\text{full-visibility}}^{\text{EG}}$ and $\text{network}_{\text{full-visibility}}^{s\text{-composite}}$.

$n = 50, P = 200$

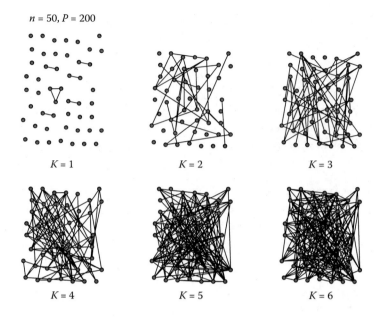

Figure 7.1: Uniform random 1-intersection graphs $G_1(n, K, P)$, with $n = 50$, $P = 200$, and $K = 1,2,3,4,5,6$.

Lemma 7.1
The following results hold:

- The graph topology of network$_{\text{full-visibility}}^{\text{EG}}$ is a uniform random 1-intersection graph $G_1(n, K_n, P_n)$.

- The graph topology of network$_{\text{full-visibility}}^{s\text{-composite}}$ is a uniform random s-intersection graph $G_s(n, K_n, P_n)$.

Figures 7.1 and 7.2 illustrate uniform random 1-intersection graphs and uniform random s-intersection graphs.

The on/off channel model induces an *Erdős–Rényi graph* $G_{ER}(n, p_n)$ [20], while the disk model results in a *random geometric graph* $G_{RGG}(n, r_n, \mathcal{A})$ [37]. The formal definitions of an Erdős–Rényi graph and a random geometric graph are as follows.

Definition 7.3 (Erdős–Rényi graph [20]).   An Erdős–Rényi graph $G_{ER}(n, p_n)$ is defined on a set of n nodes such that any two nodes establish an edge in between independently with probability p_n.

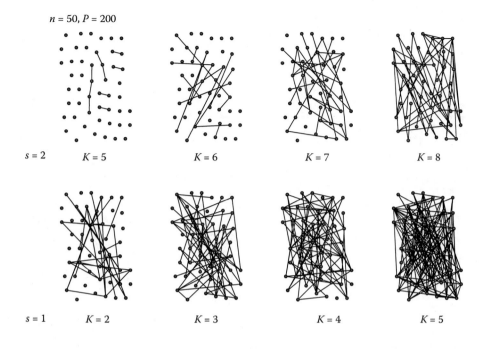

$n = 50, P = 200$

$s = 2$ $K = 5$ $K = 6$ $K = 7$ $K = 8$

$s = 1$ $K = 2$ $K = 3$ $K = 4$ $K = 5$

Figure 7.2: Uniform random s-intersection graphs $G_s(n, K, P)$, with $n = 50$, $P = 200$, and (1) $s = 2$, $K = 5, 6, 7, 8$ or (2) $s=1$, $K = 2, 3, 4, 5$.

Definition 7.4 (random geometric graph [37]). A random geometric graph $G_{RGG}(n, r_n, \mathcal{A})$ is defined on n nodes that are independently and uniformly deployed in a network area $\mathcal{A}$. An edge exists between two nodes if and only if their distance is no greater than r_n.

To explain the difference between the unit torus $\mathcal{T}$ without the boundary effect and the unit square $\mathcal{S}$ with the boundary effect, let us consider a Cartesian coordinate system. With $\mathcal{T}$ being $[0, 1]^2$, the distance between two nodes v_1 and v_2 at coordinates (x_1, y_1) and (x_2, y_2), respectively, in $\mathcal{T}$ is

$$\sqrt{(x_1 - x_2)^2 + (y_1 - y_2)^2}.$$

In $\mathcal{S}$ represented by $[0, 1)^2$, the distance between two nodes v_1' and v_2' having coordinates (x_1', y_1') and (x_2', y_2') in $\mathcal{S}$, respectively, is

$$\sqrt{(\min\{|x_1' - x_2'|, 1 - |x_1' - x_2'|\})^2 + (\min\{|y_1' - y_2'|, 1 - |y_1' - y_2'|\})^2}.$$

Figure 7.3 presents an illustration of the boundary effect. As we can see from Figure 7.3, two nodes close to opposite edges of the square may be unable to

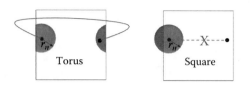

Figure 7.3: The boundary effect.

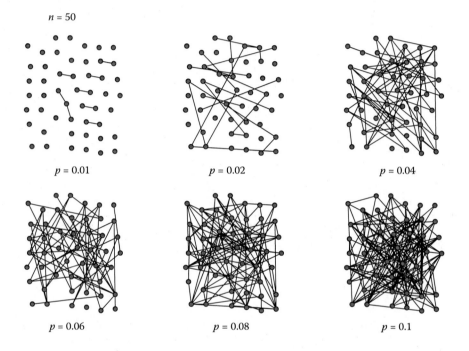

Figure 7.4: Erdős–Rényi graphs $G_{ER}(n, p)$, with $n = 50$ and $p = 0.01, 0.02, 0.04, 0.06, 0.08, 0.1$.

establish an edge (i.e., a communication link) on the square S but may have an edge in between on the torus T because of possible wraparound connections on the torus.

Figures 7.4 and 7.5 illustrate Erdős–Rényi graphs and random geometric graphs.

7.3.2   Graph Intersections

We introduce the notation for different networks modeled by graph intersections in Table 7.1.

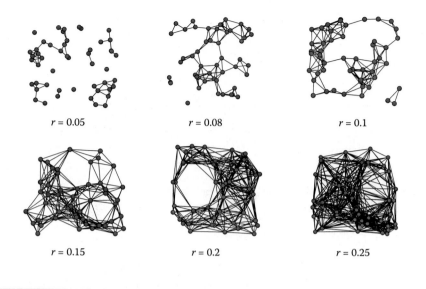

$r = 0.05$ $r = 0.08$ $r = 0.1$

$r = 0.15$ $r = 0.2$ $r = 0.25$

Figure 7.5: Random geometric graphs $G_{\text{RGG}}(n, r, \mathcal{S})$, with $n = 50$, $r = 0.05, 0.08, 0.1$, $0.15, 0.2, 0.25$, and $\mathcal{S}$ being a square of unit area.

We will explain how the topology of each network in Table 7.1 can be given by the intersection of different random graphs. With two graphs G_1 and G_2 defined on the same node set, two nodes have an edge in between in the intersection $G_1 \cap G_2$ if and only if these two nodes have an edge in G_1 and also have an edge in G_2. Figure 7.6 illustrates graph intersections.

Lemma 7.2 below formally gives graph intersections for networks in Table 7.1.

Lemma 7.2
The following results hold:

- The topology of network$^{\text{EG}}_{\text{on/off channel model}}$
  is $G_1(n, K_n, P_n) \cap G_{ER}(n, p_n)$.

- The topology of network$^{\text{EG}}_{\text{disk model on a unit torus}}$
  is $G_1(n, K_n, P_n) \cap G_{RGG}(n, r_n, \mathcal{T})$.

- The topology of network$^{\text{EG}}_{\text{disk model on a unit square}}$
  is $G_1(n, K_n, P_n) \cap G_{RGG}(n, r_n, \mathcal{S})$.

- The topology of network$^{\text{EG}}_{\substack{\text{disk model on a unit torus} \\ \text{on/off channel model}}}$
  is $G_1(n, K_n, P_n) \cap G_{RGG}(n, r_n, \mathcal{T}) \cap G_{ER}(n, p_n)$.

- The topology of network$^{\substack{\text{EG} \\ \text{disk model on a unit square} \\ \text{on/off channel model}}}$
  is $G_1(n, K_n, P_n) \cap G_{RGG}(n, r_n, \mathcal{S}) \cap G_{ER}(n, p_n)$.

Table 7.1: Notation for Different Networks Modeled by Graph Intersections

Notation	Networks
$\text{network}^{EG}_{\text{on/off channel model}}$	WSN with the EG scheme under the on/off channel model
$\text{network}^{EG}_{\text{disk model on a unit torus}}$	WSN with the EG scheme under the disk model on a unit torus
$\text{network}^{EG}_{\text{disk model on a unit square}}$	WSN with the EG scheme under the disk model on a unit square
$\text{network}^{EG, \text{disk model on a unit torus}}_{\text{on/off channel model}}$	WSN with the EG scheme under both the disk model on a unit torus and the on/off channel model
$\text{network}^{EG, \text{disk model on a unit square}}_{\text{on/off channel model}}$	WSN with the EG scheme under both the disk model on a unit square, and the on/off channel model
$\text{network}^{s\text{-composite}}_{\text{on/off channel model}}$	WSN with the s-composite scheme under the on/off channel model
$\text{network}^{s\text{-composite}}_{\text{disk model on a unit torus}}$	WSN with the s-composite scheme under the disk model on a unit torus
$\text{network}^{s\text{-composite}}_{\text{disk model on a unit square}}$	WSN with the s-composite scheme under the disk model on a unit square
$\text{network}^{s\text{-composite, disk model on a unit torus}}_{\text{on/off channel model}}$	WSN with the s-composite scheme under both the disk model on a unit torus and the on/off channel model
$\text{network}^{s\text{-composite, disk model on a unit square}}_{\text{on/off channel model}}$	WSN with the s-composite scheme under both the disk model on a unit square and the on/off channel model

■ The topology of $\text{network}^{s\text{-composite}}_{\text{on/off channel model}}$
 is $G_s(n, K_n, P_n) \cap G_{ER}(n, p_n)$.

■ The topology of $\text{network}^{s\text{-composite}}_{\text{disk model on a unit torus}}$
 is $G_s(n, K_n, P_n) \cap G_{RGG}(n, r_n, \mathcal{T})$.

■ The topology of $\text{network}^{s\text{-composite}}_{\text{disk model on a unit square}}$
 is $G_s(n, K_n, P_n) \cap G_{RGG}(n, r_n, \mathcal{S})$.

■ The topology of $\text{network}^{s\text{-composite, disk model on a unit torus}}_{\text{on/off channel model}}$
 is $G_s(n, K_n, P_n) \cap G_{RGG}(n, r_n, \mathcal{T}) \cap G_{ER}(n, p_n)$.

■ The topology of $\text{network}^{s\text{-composite, disk model on a unit square}}_{\text{on/off channel model}}$
 is $G_s(n, K_n, P_n) \cap G_{RGG}(n, r_n, \mathcal{S}) \cap G_{ER}(n, p_n)$.

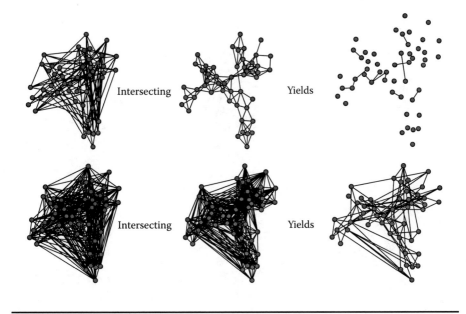

Figure 7.6: Graph intersections.

Table 7.2: Different Settings of Secure Sensor Networks and Their Corresponding Random Graph Models

Setting		Graph
EG scheme	Full visibility	$G_1(n,K_n,P_n)$
	Link unreliability	$G_1(n,K_n,P_n) \cap G_{ER}(n,p_n)$
	Transmission constraints	$G_1(n,K_n,P_n) \cap G_{RGG}(n,r_n,\mathcal{A})$
CPS scheme	Full visibility	$G_s(n,K_n,P_n)$
	Link unreliability	$G_s(n,K_n,P_n) \cap G_{ER}(n,p_n)$
	Transmission constraints	$G_s(n,K_n,P_n) \cap G_{RGG}(n,r_n,\mathcal{A})$

Table 7.2 summarizes different settings of secure sensor networks and their corresponding random graph models.

Figure 7.7 presents intersections of uniform random 1-intersection graphs $G_1(n,K,P)$ and Erdős–Rényi graphs $G_{ER}(n,t)$, and Figure 7.8 presents intersections of uniform random 1-intersection graphs $G_1(n,K,P)$ and random geometric graphs $G_{RGG}(n,r,\mathcal{S})$, with $\mathcal{S}$ being a square of unit area.

7.3.3 (k-)Connectivity and Node Degree

A network (or a graph) is said to be *k-node*-connected (respectively, *k-edge*-connected) for a positive integer k if it remains connected despite the deletion of

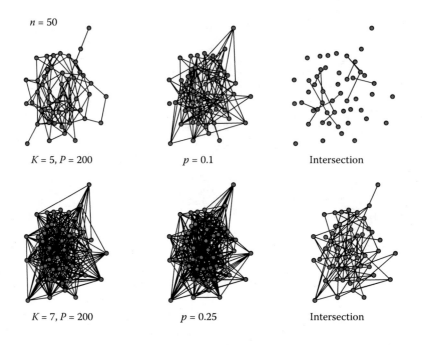

Figure 7.7: Intersections of uniform random 1-intersection graphs $G_1(n, K, P)$ and Erdős–Rényi graphs $G_{ER}(n, p)$, where $n = 50$, $P = 200$, and (i) $K = 5$, $p = 0.1$, or (ii) $K = 7$, $p = 0.25$.

at most $(k - 1)$ nodes (respectively, edges). An equivalent definition of k-node-connectivity is that for each pair of nodes there exists at least k mutually disjoint paths connecting them [21]. In the case of k being 1, 1-node-connectivity and 1-edge-connectivity simply become connectivity, meaning that each node in the network can find at least one path to any other node, either directly or with the help of other nodes acting as relays.

A weaker property related to and implied by k-connectivity is that the minimum degree of the network or the graph is at least k; namely, each node has at least k other neighboring nodes. As detailed in our work [60, 67], k-connectivity and the property that the minimum degree is no less than k are particularly significant and useful in WSN applications where sensors are deployed in hostile environments (e.g., battlefield surveillance) or are unattended for long periods of time (e.g., environmental monitoring) or are used in life-critical applications (e.g., patient monitoring). Examples of the benefits include improving resiliency against the failure of links or sensors (e.g., under battery depletion [45] or sensor capture attacks [14]), enabling flexible communication-load balancing across multiple paths [24], and achieving consensus despite adversarial nodes in the network [18, 54].

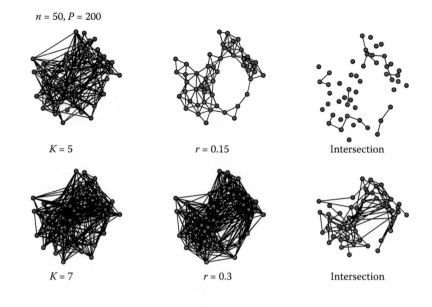

$n = 50, P = 200$

$K = 5$ $r = 0.15$ Intersection

$K = 7$ $r = 0.3$ Intersection

Figure 7.8: Intersections of uniform random 1-intersection graphs $G_1(n, K, P)$ and random geometric graphs $G_{RGG}(n, r, S)$, with S being a square of unit area, where $n = 50$, $P = 200$, and (1) $K = 5$, $r = 0.15$ or (2) $K = 7$, $r = 0.3$.

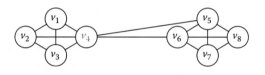

Figure 7.9: A graph that has a minimum node degree of 3, but is only 2-edge-connected and 1-node-connected.

k-e implies k-edge-connectivity, which further implies that the minimum node degree is at least k. However, the reverse directions do not hold. In this chapter, k-node-connectivity and k-edge-connectivity are often called together as k-connectivity for convenience.

Figure 7.9 presents a graph that has a minimum node degree of 3, but is only 2-edge-connected and 1-node-connected (i.e., connected). If node v_4 is removed, the graph becomes disconnected. If any single edge is deleted, the graph is still connected.

We present below the results for random intersection graphs and their compositions with other random graphs in terms of various properties, including

k-connectivity, perfect matching containment, Hamilton cycle containment, and k-robustness. These properties are defined as follows:

1. A graph is k-connected if each pair of nodes has at least k internally node-disjoint paths between them, and a graph is connected if it is 1-connected.

2. A perfect matching is a set of edges that do not have common nodes and cover all nodes with the exception of missing at most one node.

3. A Hamiltonian cycle is a closed loop that visits each node exactly once.

4. The notion of k-robustness proposed by Zhang and Sundaram [51] measures the effectiveness of local-information-based diffusion algorithms in the presence of adversarial nodes; formally, a graph with a node set V is k-robust if at least one of (a) and (b) below holds for each nonempty and strict subset T of V: (a) there exists at least a node $v_a \in T$ such that v_a has no less than k neighbors inside $V \setminus T$, and (b) there exists at least a node $v_b \in V \setminus T$ such that v_b has no less than k neighbors inside T, where two nodes are neighbors if they have an edge in between. This notion of k-robustness has received much attention [33, 34, 52, 53, 59, 62].

7.3.4 Notation and Conventions

We define several notations that will be used throughout.

Let $V = \{v_1, \ldots, v_n\}$ be the node set in the studied graphs and networks. In the EG/s-composite scheme, each node $v_i \in V$ is assigned a key ring S_i that consists of K_n distinct keys selected uniformly at random from a key pool $\mathcal{P}$ of size P_n. The uniform random 1-intersection graph $G_1(n, K_n, P_n)$ modeling the EG scheme is defined on the node set V such that two distinct nodes v_i and v_j are adjacent, denoted T_{ij}, if their key rings have at least one key in common, that is,

$$T_{ij} = [S_i \cap S_j \neq \emptyset]. \tag{7.1}$$

The uniform random s-intersection graph $G_s(n, K_n, P_n)$ modeling the s-composite scheme is defined on the node set V such that two distinct nodes v_i and v_j are adjacent, denoted $T_{ij,s}$, if their key rings have at least s keys in common, that is,

$$T_{ij,s} = [|S_i \cap S_j| \geq s], \tag{7.2}$$

where $|A|$ with A as a set means the cardinality of A. We will use S_{ij} to denote $S_i \cap S_j$ at some places.

In the on/off channel model, each channel is *on* with probability p_n or *off* with probability $1 - p_n$, independently from any other channel. Namely, with C_{ij} denoting the event that the channel between v_i and v_j is on, $\{C_{ij},\ 1 \leq i < j \leq n\}$ are mutually independent such that

$$\mathbb{P}[C_{ij}] = p_n, \quad 1 \leq i < j \leq n. \tag{7.3}$$

This on/off channel model is represented by an Erdős–Rényi (ER) graph $G_{ER}(n, p_n)$ on the node set $\mathcal{V}$ such that there exists an edge between nodes v_i and v_j if the link between them is on, that is, if the event C_{ij} takes place.

In network$_{\text{on/off channel model}}^{\text{EG}}$, two distinct nodes v_i and v_j have an edge in between, denoted E_{ij}, if the events T_{ij} and C_{ij} take place at the same time. In other words, we have

$$E_{ij} = T_{ij} \cap C_{ij}, \quad 1 \leq i < j \leq n, \tag{7.4}$$

so the topology of network$_{\text{on/off channel model}}^{\text{EG}}$ is given by $G_1(n, K_n, P_n) \cap G_{ER}(n, p_n)$, which we write as $G_{RKG-ER}(n, K_n, P_n, p_n)$ or simply G_{RKG-ER}.

In network$_{\text{on/off channel model}}^{s\text{-composite}}$, two distinct nodes v_i and v_j have an edge in between, denoted $E_{ij,s}$, if the events $T_{ij,s}$ and C_{ij} take place at the same time. In other words, we have

$$E_{ij,s} = T_{ij,s} \cap C_{ij}, \quad 1 \leq i < j \leq n, \tag{7.5}$$

so the topology of network$_{\text{on/off channel model}}^{s\text{-composite}}$ is given by $G_s(n, K_n, P_n) \cap G_{ER}(n, p_n)$, which we write as $G_{RKG-ER}(n, K_n, P_n, p_n, s)$ or simply G_{RKG-ER_s}.

The disk model induces a *random geometric graph* denoted by $G_{RGG}(n, r_n, \mathcal{A})$, in which an edge exists between two nodes if and only if their distance is no greater than r_n. We denote by R_{ij} the event that distinct nodes v_i and v_j are within a distance of r_n.

In network$_{\text{disk model}}^{\text{EG}}$, two distinct nodes v_i and v_j have an edge in between, denoted F_{ij}, if the events T_{ij} and $R_{ij}(\mathcal{A})$ happen simultaneously. In other words, we have

$$F_{ij}(\mathcal{A}) = T_{ij} \cap R_{ij}(\mathcal{A}), \quad 1 \leq i < j \leq n, \tag{7.6}$$

so the topology of network$_{\text{disk model}}^{\text{EG}}$ is given by $G_1(n, K_n, P_n) \cap G_{RGG}(n, r_n, \mathcal{A})$, which we write as $G_{RKG-RGG}(n, K_n, P_n, r_n, \mathcal{A})$ or simply $G_{RKG-RGG}$.

In network$_{\text{disk model}}^{s\text{-composite}}$, two distinct nodes v_i and v_j have an edge in between, denoted $F_{ij,s}(\mathcal{A})$, if the events $T_{ij,s}$ and $R_{ij}(\mathcal{A})$ take place at the same time. In other words, we have

$$F_{ij,s}(\mathcal{A}) = T_{ij,s} \cap R_{ij}(\mathcal{A}), \quad 1 \leq i < j \leq n, \tag{7.7}$$

so the topology of network$_{\text{disk model}}^{s\text{-composite}}$ is given by $G_s(n, K_n, P_n) \cap G_{RGG}(n, r_n, \mathcal{A})$, which we write as $G_{RKG-RGG}(n, K_n, P_n, r_n, \mathcal{A}, s)$ or simply $G_{RKG-RGG_s}$.

Note that $\mathbb{P}[C_{ij}] = p_n$ from (7.3). Now we analyze probabilities $\mathbb{P}[T_{ij}]$, $\mathbb{P}[T_{ij,s}]$, $\mathbb{P}[E_{ij}]$, $\mathbb{P}[E_{ij,s}]$, $\mathbb{P}[F_{ij}(\mathcal{A})]$, $\mathbb{P}[F_{ij,s}(\mathcal{A})]$, and $\mathbb{P}[R_{ij}(\mathcal{A})]$ below based on (7.1) through (7.7).

We define the following:

$$x(K_n, P_n) := \mathbb{P}[T_{ij}], \tag{7.8}$$

$$x_s(K_n, P_n) := \mathbb{P}[T_{ij,s}], \tag{7.9}$$

$$t(K_n, P_n, p_n) := \mathbb{P}[E_{ij}], \tag{7.10}$$

$$t_s(K_n, P_n, p_n) := \mathbb{P}[E_{ij,s}], \tag{7.11}$$

$$\varrho_n(\mathcal{A}) := \mathbb{P}[F_{ij}(\mathcal{A})]. \tag{7.12}$$

In other words, $x(K_n, P_n)$ (respectively, $x_s(K_n, P_n)$) is the probability that the key rings of two distinct nodes share at least one key (respectively, s keys).

It is easy to derive $x(K_n, P_n)$ in terms of K_n and P_n as shown in previous work [3,40,49]. In fact, we have

$$x(K_n, P_n) = \mathbb{P}[S_i \cap S_j \neq \emptyset] = \begin{cases} 1 - \frac{\binom{P_n - K_n}{K_n}}{\binom{P_n}{K_n}}, & \text{if } P_n \geq 2K_n, \\ 1, & \text{if } P_n < 2K_n. \end{cases} \tag{7.13}$$

By [67, Lemma 8], if $K_n^2/P_n = o(1)$, then

$$x(K_n, P_n) = K_n^2/P_n \cdot \left[1 \pm O(K_n^2/P_n)\right] \sim K_n^2/P_n. \tag{7.14}$$

We will frequently use (7.14) throughout the chapter.

Given (7.4), the independence of the events C_{ij} and K_{ij} gives

$$t(K_n, P_n, p_n) := \mathbb{P}[E_{ij}] = \mathbb{P}[C_{ij}] \cdot \mathbb{P}[K_{ij}] = p_n \cdot x(K_n, P_n) \tag{7.15}$$

from (7.3) and (7.8). Substituting (7.13) into (7.15), we obtain

$$t(K_n, P_n, p_n) = p_n \cdot \left[1 - \frac{\binom{P_n - K_n}{K_n}}{\binom{P_n}{K_n}}\right] \quad \text{if } P_n \geq 2K_n. \tag{7.16}$$

By definition, $x_s(K_n, P_n)$ is determined through

$$x_s(K_n, P_n) = \mathbb{P}[T_{ij,s}] = \sum_{u=s}^{K_n} \mathbb{P}[|S_i \cap S_j| = u], \tag{7.17}$$

where it is shown [9,55] that

$$\mathbb{P}[|S_i \cap S_j| = u]$$
$$= \begin{cases} \frac{\binom{K_n}{u}\binom{P_n - K_n}{K_n - u}}{\binom{P_n}{K_n}} & \text{for } \max\{0, 2K_n - P_n\} \leq u \leq K_n, \\ 0 & \text{otherwise.} \end{cases} \tag{7.18}$$

From $E_{ij} = T_{ij,s} \cap C_{ij}$ and the independence of C_{ij} and $T_{ij,s}$, we obtain

$$t_s(K_n, P_n, p_n) = \mathbb{P}[E_{ij}] = \mathbb{P}[C_{ij}] \cdot \mathbb{P}[T_{ij,s}] = p_n \cdot x_s(K_n, P_n). \qquad (7.19)$$

From (7.12), $\varrho_n(\mathcal{A})$ is the probability that a secure link exists between two sensors in the WSN modeled by graph $G_{RKG-RGG}(n, K_n, P_n, r_n, \mathcal{A})$; that is, $\varrho_n(\mathcal{A})$ is the edge probability in graph $G_{RKG-RGG}(n, K_n, P_n, r_n, \mathcal{A})$. When $\mathcal{A}$ is the torus $\mathcal{T}$, clearly $\varrho_n(\mathcal{T})$ equals $\pi r_n^2 \cdot s_n$, and if $K_n^2/P_n = o(1)$, then $\varrho_n(\mathcal{T}) \sim \pi r_n^2 \cdot K_n^2/P_n$ by* (7.14). When $\mathcal{A}$ is the square $\mathcal{S}$, it is a simple matter to show $\varrho_n(\mathcal{S}) \geq (1 - 2r_n)^2 \cdot \pi r_n^2 \cdot s_n$ (with $r_n \leq \frac{1}{2}$) and $\varrho_n(\mathcal{S}) \leq \pi r_n^2 \cdot s_n$. The reason is that for the position of v_x satisfying the condition that v_x's distances to all four edges of $\mathcal{S}$ are at least $2r_n$, with $r_n \leq \frac{1}{2}$, given the position of v_x, the probability that v_y falls in v_x's transmission area is πr_n^2; for the position of v_x not satisfying the above condition, the probability that v_y falls in v_x's transmission area is upper bounded by πr_n^2. Then on $\mathcal{S}$ it holds that $\varrho_n(\mathcal{S}) \sim \pi r_n^2 \cdot s_n$ if $r_n = o(1)$ (note that $r_n = o(1)$ implies $r_n \leq \frac{1}{2}$ for all n sufficiently large). Therefore, on $\mathcal{S}$, if $r_n = o(1)$ and $K_n^2/P_n = o(1)$, we further obtain $\varrho_n(\mathcal{S}) \sim \pi r_n^2 \cdot K_n^2/P_n$ in view of (7.14).

We define the edge probability of graph $H_1(n, t_n, P_n)$ (respectively, $H_s(n, t_n, P_n)$) as $y(t_n, P_n)$ (respectively, $y_s(t_n, P_n)$).

Throughout the chapter, both k and s are positive constant integers and do not scale with n. All asymptotic statements are understood with $n \to \infty$. Also, $\mathbb{P}[\mathcal{E}]$ denotes the probability that event $\mathcal{E}$ occurs. An event happens *almost surely* if its probability converges to 1 as $n \to \infty$.

7.4 Results of Random Intersection Graphs

7.4.1 Results of Uniform Random 1-Intersection Graphs

Theorem 7.1
(k-connectivity in uniform random 1-intersection graphs by our work [62]).

---

*We use the standard asymptotic notation $o(\cdot), \omega(\cdot), O(\cdot), \Omega(\cdot), \Theta(\cdot)$, and $\sim$. Specifically, given two positive functions $f(n)$ and $g(n)$,

1. $f(n) = o(g(n))$ signifies $\lim_{n \to \infty} [f(n)/g(n)] = 0$.

2. $f(n) = \omega(g(n))$ means $\lim_{n \to \infty} [f(n)/g(n)] = \infty$.

3. $f(n) = O(g(n))$ signifies that there exists a positive constant c_1 such that $f(n) \leq c_1 g(n)$ for all n sufficiently large.

4. $f(n) = \Omega(g(n))$ means that there exists a positive constant c_2 such that $f(n) \geq c_2 g(n)$ for all n sufficiently large.

5. $f(n) = \Theta(g(n))$ means that there exist positive constants c_3 and c_4 with $c_3 \leq c_4$ such that $c_3 g(n) \leq f(n) \leq c_4 g(n)$ for all n sufficiently large.

6. $f(n) \sim g(n)$ stands for $\lim_{n \to \infty} [f(n)/g(n)] = 1$.

1. For a uniform random 1-intersection graph $G_1(n, K_n, P_n)$, if there is a sequence α_n with $\lim_{n\to\infty} \alpha_n \in [-\infty, \infty]$ such that

$$x(K_n, P_n) = \frac{\ln n + (k-1)\ln\ln n + \alpha_n}{n},$$

then under $P_n = \Omega(n)$, it holds that

$$\lim_{n\to\infty} \mathbb{P}[\, G_1(n, K_n, P_n) \text{ is } k\text{-connected.}\,]$$

$$= \lim_{n\to\infty} \mathbb{P}[\, G_1(n, K_n, P_n) \text{ has a minimum degree at least } k.\,]$$

$$= e^{-\frac{e^{-\lim_{n\to\infty}\alpha_n}}{(k-1)!}} = \begin{cases} 0 & \text{if } \lim_{n\to\infty}\alpha_n = -\infty, \\ 1 & \text{if } \lim_{n\to\infty}\alpha_n = \infty, \\ e^{-\frac{e^{-\alpha^*}}{(k-1)!}} & \text{if } \lim_{n\to\infty}\alpha_n = \alpha^* \in (-\infty, \infty). \end{cases}$$

2. For a uniform random 1-intersection graph $G_1(n, K_n, P_n)$, if there is a sequence β_n with $\lim_{n\to\infty} \beta_n \in [-\infty, \infty]$ such that

$$\frac{K_n^2}{P_n} = \frac{\ln n + (k-1)\ln\ln n + \beta_n}{n},$$

then under $P_n = \Omega(n)$, it holds that

$$\lim_{n\to\infty} \mathbb{P}[\, G_1(n, K_n, P_n) \text{ is } k\text{-connected.}\,]$$

$$= \lim_{n\to\infty} \mathbb{P}[\, G_1(n, K_n, P_n) \text{ has a minimum degree at least } k.\,]$$

$$= e^{-\frac{e^{-\lim_{n\to\infty}\beta_n}}{(k-1)!}} = \begin{cases} 0 & \text{if } \lim_{n\to\infty}\beta_n = -\infty, \\ 1 & \text{if } \lim_{n\to\infty}\beta_n = \infty, \\ e^{-\frac{e^{-\beta^*}}{(k-1)!}} & \text{if } \lim_{n\to\infty}\beta_n = \beta^* \in (-\infty, \infty). \end{cases}$$

Remark 7.1. Theorem 7.1 presents the asymptotically exact probability of k-connectivity in a uniform random 1-intersection graph, while a zero–one law is implicitly obtained by Rybarczyk [41] and explicitly given by us as a side result [60, 67]. For connectivity (i.e., k-connectivity in the case of $k = 1$), Blackburn and Gerke [3] and Yağan and Makowski [49] show different granularities of zero–one laws, while Rybarczyk [40] derives the asymptotically exact probability.

Theorem 7.2
(perfect matching containment in uniform random 1-intersection graphs by our work [58]).

1. For a uniform random 1-intersection graph $G_1(n, K_n, P_n)$, if there is a sequence α_n with $\lim_{n\to\infty} \alpha_n \in [-\infty, \infty]$ such that

$$x(K_n, P_n) = \frac{\ln n + \alpha_n}{n},$$

then under $P_n = \omega(n(\ln n)^5)$, it holds that

$$\lim_{n\to\infty} \mathbb{P}[\, G_1(n, K_n, P_n) \text{ has at least one perfect matching.}\,]$$

$$= e^{-e^{-\lim_{n\to\infty} \alpha_n}} = \begin{cases} 0 & \text{if } \lim_{n\to\infty} \alpha_n = -\infty, \\ 1 & \text{if } \lim_{n\to\infty} \alpha_n = \infty, \\ e^{-e^{-\alpha^*}} & \text{if } \lim_{n\to\infty} \alpha_n = \alpha^* \in (-\infty, \infty). \end{cases}$$

2. For a uniform random 1-intersection graph $G_1(n, K_n, P_n)$, if there is a sequence β_n with $\lim_{n\to\infty} \beta_n \in [-\infty, \infty]$ such that

$$\frac{K_n^2}{P_n} = \frac{\ln n + \beta_n}{n},$$

then under $P_n = \omega(n(\ln n)^5)$, it holds that

$$\lim_{n\to\infty} \mathbb{P}[\, G_1(n, K_n, P_n) \text{ has at least one perfect matching.}\,]$$

$$= e^{-e^{-\lim_{n\to\infty} \beta_n}} = \begin{cases} 0 & \text{if } \lim_{n\to\infty} \beta_n = -\infty, \\ 1 & \text{if } \lim_{n\to\infty} \beta_n = \infty, \\ e^{-e^{-\beta^*}} & \text{if } \lim_{n\to\infty} \beta_n = \beta^* \in (-\infty, \infty). \end{cases}$$

Remark 7.2. Theorem 7.2 presents the asymptotically exact probability of perfect matching containment in a uniform random 1-intersection graph. A similar result is given by setting s as 1 in the work of Bloznelis and Łuczak [10] studying $G_s(n, K_n, P_n)$. However, they use conditions $K_n = O((\ln n)^{1/5})$ and

$$\frac{K_n^2}{P_n} = O(\frac{\ln n}{n}).$$

Furthermore, for the one law (i.e., the case where $G_1(n, K_n, P_n)$ contains a perfect matching almost surely), their result relies on $P_n = o(n(\ln n)^{-3/5})$, whereas our result uses $P_n = \omega(n(\ln n)^5)$. We note that P_n is expected to be at least on the order of n in the sensor network applications of uniform random 1-intersection graphs [23]. In addition, Blackburn et al. [4] derive a result that is weaker than Theorem 7.2 to analyze cryptographic hash functions. Specifically, they show that for a uniform random 1-intersection graph $G_1(n, K_n, P_n)$ under $P_n = \Omega(n^c)$ with a constant $c > 1$, $G_1(n, K_n, P_n)$ contains (respectively, does not contain) a perfect matching almost surely if

$$\lim_{n\to\infty} (\frac{K_n^2}{P_n} / \frac{\ln n}{n}) > 1 \text{ (respectively, } < 1).$$

Theorem 7.3

(Hamilton cycle containment in uniform random 1-intersection graphs by our work [58]).

1. For a uniform random 1-intersection graph $G_1(n, K_n, P_n)$, if there is a sequence α_n with $\lim_{n\to\infty} \alpha_n \in [-\infty, \infty]$ such that

$$x(K_n, P_n) = \frac{\ln n + \ln \ln n + \alpha_n}{n},$$

then under $P_n = \omega(n(\ln n)^5)$, it holds that

$$\lim_{n\to\infty} \mathbb{P}[\, G_1(n, K_n, P_n) \text{ has at least one Hamilton cycle.}\,]$$

$$= e^{-e^{-\lim_{n\to\infty}\alpha_n}} = \begin{cases} 0 & \text{if } \lim_{n\to\infty} \alpha_n = -\infty, \\ 1 & \text{if } \lim_{n\to\infty} \alpha_n = \infty, \\ e^{-e^{-\alpha^*}} & \text{if } \lim_{n\to\infty} \alpha_n = \alpha^* \in (-\infty, \infty). \end{cases}$$

2. For a uniform random 1-intersection graph $G_1(n, K_n, P_n)$, if there is a sequence β_n with $\lim_{n\to\infty} \beta_n \in [-\infty, \infty]$ such that

$$\frac{K_n^2}{P_n} = \frac{\ln n + \ln \ln n + \beta_n}{n},$$

then under $P_n = \omega(n(\ln n)^5)$, it holds that

$$\lim_{n\to\infty} \mathbb{P}[\, G_1(n, K_n, P_n) \text{ has at least one Hamilton cycle.}\,]$$

$$= e^{-e^{-\lim_{n\to\infty}\beta_n}} = \begin{cases} 0 & \text{if } \lim_{n\to\infty} \beta_n = -\infty, \\ 1 & \text{if } \lim_{n\to\infty} \beta_n = \infty, \\ e^{-e^{-\beta^*}} & \text{if } \lim_{n\to\infty} \beta_n = \beta^* \in (-\infty, \infty). \end{cases}$$

Remark 7.3. Nikoletseas et al. [36] proves that $G_1(n, K_n, P_n)$ under $K_n \geq 2$ has a Hamilton cycle with high probability if it holds for some constant $\delta > 0$ that $n \geq (1+\delta)\left(\frac{P_n}{K_n}\right)\ln\left(\frac{P_n}{K_n}\right)$, which implies that P_n is much smaller than n ($P_n = O(\sqrt{n})$ given $K_n \geq 2$, $P_n = O(\sqrt[3]{n})$ if $K_n \geq 3$, $P_n = O(\sqrt[4]{n})$ if $K_n \geq 4$, etc.). Different from the result of Nikoletseas et al. [36], our Theorem 7.3 is for $P_n = \omega(n(\ln n)^5)$. Furthermore, Theorem 7.3 presents the asymptotically exact probability, whereas Nikoletseas et al. [36] only derive conditions for $G_1(n, K_n, P_n)$ to have a Hamilton cycle almost surely. They do not provide conditions for $G_1(n, K_n, P_n)$ to have no Hamilton cycle with high probability, or to have a Hamilton cycle with an asymptotic probability in $(0, 1)$.

Theorem 7.4

(k-robustness in uniform random 1-intersection graphs by our work [62]).

1. For a uniform random 1-intersection graph $G_1(n, K_n, P_n)$, with a sequence α_n defined by

$$x(K_n, P_n) = \frac{\ln n + (k-1)\ln\ln n + \alpha_n}{n},$$

under $P_n = \Omega\left(n(\ln n)^5\right)$, it holds that

$$\lim_{n\to\infty} \mathbb{P}\left[\, G_1(n, K_n, P_n) \text{ is } k\text{-robust.}\,\right]$$

$$= \begin{cases} 0 & \text{if } \lim_{n\to\infty}\alpha_n = -\infty, \\ 1 & \text{if } \lim_{n\to\infty}\alpha_n = \infty. \end{cases}$$

2. For a uniform random 1-intersection graph $G_1(n, K_n, P_n)$, with a sequence β_n defined by

$$\frac{K_n^{\,2}}{P_n} = \frac{\ln n + (k-1)\ln\ln n + \beta_n}{n},$$

under $P_n = \Omega\left(n(\ln n)^5\right)$, it holds that

$$\lim_{n\to\infty} \mathbb{P}\left[\, G_1(n, K_n, P_n) \text{ is } k\text{-robust.}\,\right]$$

$$= \begin{cases} 0 & \text{if } \lim_{n\to\infty}\beta_n = -\infty, \\ 1 & \text{if } \lim_{n\to\infty}\beta_n = \infty. \end{cases}$$

Remark 7.4. As mentioned earlier, we use the definition of k-robustness proposed by Zhang and Sundaram [51]. They present results on k-robustness in Erdős–Rényi graphs and one-dimensional random geometric graphs, whereas we study their notion of k-robustness in random intersection graphs [59, 62].

7.4.2 Results of Binomial Random 1-Intersection Graphs

Theorem 7.5
(k-connectivity in binomial random 1-intersection graphs by our work [62]).

1. For a binomial random 1-intersection graph $H_1(n, t_n, P_n)$, if there is a sequence α_n with $\lim_{n\to\infty}\alpha_n \in [-\infty, \infty]$ such that

$$y(t_n, P_n) = \frac{\ln n + (k-1)\ln\ln n + \alpha_n}{n},$$

then under $P_n = \omega(n(\ln n)^5)$, it holds that

$$\lim_{n\to\infty} \mathbb{P}[\,H_1(n,t_n,P_n) \text{ is } k\text{-connected.}\,]$$

$$= \lim_{n\to\infty} \mathbb{P}[\,H_1(n,t_n,P_n) \text{ has a minimum degree at least } k.\,]$$

$$= e^{-\frac{e^{-\lim_{n\to\infty}\alpha_n}}{(k-1)!}} = \begin{cases} 0 & \text{if } \lim_{n\to\infty}\alpha_n = -\infty, \\ 1 & \text{if } \lim_{n\to\infty}\alpha_n = \infty, \\ e^{-\frac{e^{-\alpha^*}}{(k-1)!}} & \text{if } \lim_{n\to\infty}\alpha_n = \alpha^* \in (-\infty,\infty). \end{cases}$$

2. For a binomial random 1-intersection graph $H_1(n,t_n,P_n)$, if there is a sequence β_n with $\lim_{n\to\infty}\beta_n \in [-\infty,\infty]$ such that

$$t_n^2 P_n = \frac{\ln n + (k-1)\ln\ln n + \beta_n}{n},$$

then under $P_n = \omega(n(\ln n)^5)$, it holds that

$$\lim_{n\to\infty} \mathbb{P}[\,H_1(n,t_n,P_n) \text{ is } k\text{-connected.}\,]$$

$$= \lim_{n\to\infty} \mathbb{P}[\,H_1(n,t_n,P_n) \text{ has a minimum degree at least } k.\,]$$

$$= e^{-\frac{e^{-\lim_{n\to\infty}\beta_n}}{(k-1)!}} = \begin{cases} 0 & \text{if } \lim_{n\to\infty}\beta_n = -\infty, \\ 1 & \text{if } \lim_{n\to\infty}\beta_n = \infty, \\ e^{-\frac{e^{-\beta^*}}{(k-1)!}} & \text{if } \lim_{n\to\infty}\beta_n = \beta^* \in (-\infty,\infty). \end{cases}$$

Remark 7.5. Theorem 7.5 presents the asymptotically exact probability of k-connectivity in a binomial random 1-intersection graph, while zero–one laws are obtained by Rybarczyk [41,42]. Connectivity (i.e., k-connectivity in the case of $k=1$) results are presented by Rybarczyk [41, 42], Shang [43], and Singer-Cohen [44].

Theorem 7.6
(perfect matching containment in binomial random 1-intersection graphs by Rybarczyk [41, 42]).

1. For a binomial random 1-intersection graph $H_1(n,t_n,P_n)$, if there is a sequence α_n with $\lim_{n\to\infty}\alpha_n \in [-\infty,\infty]$ such that

$$y(t_n,P_n) = \frac{\ln n + \alpha_n}{n}, \tag{7.20}$$

then under $P_n = \Omega(n^c)$ for a constant $c > 1$, it holds that

$$\lim_{n\to\infty} \mathbb{P}[\,H_1(n,t_n,P_n) \text{ has at least one perfect matching.}\,]$$

$$= e^{-e^{-\lim_{n\to\infty}\alpha_n}} = \begin{cases} 0 & \text{if } \lim_{n\to\infty}\alpha_n = -\infty, \\ 1 & \text{if } \lim_{n\to\infty}\alpha_n = \infty, \\ e^{-e^{-\alpha^*}} & \text{if } \lim_{n\to\infty}\alpha_n = \alpha^* \in (-\infty,\infty). \end{cases}$$

2. For a binomial random 1-intersection graph $H_1(n, t_n, P_n)$, if there is a sequence β_n with $\lim_{n \to \infty} \beta_n \in [-\infty, \infty]$ such that

$$t_n^2 P_n = \frac{\ln n + \beta_n}{n}, \qquad (7.21)$$

then under $P_n = \Omega(n^c)$ for a constant $c > 1$, it holds that

$$\lim_{n \to \infty} \mathbb{P}[\, H_1(n, t_n, P_n) \text{ has at least one perfect matching.}\,]$$

$$= e^{-e^{-\lim_{n \to \infty} \beta_n}} = \begin{cases} 0 & \text{if } \lim_{n \to \infty} \beta_n = -\infty, \\ 1 & \text{if } \lim_{n \to \infty} \beta_n = \infty, \\ e^{-e^{-\beta^*}} & \text{if } \lim_{n \to \infty} \beta_n = \beta^* \in (-\infty, \infty). \end{cases}$$

Remark 7.6.   For perfect matching containment in a binomial random 1-intersection graph, in addition to Theorem 7.6 above under $P_n = \Omega(n^c)$ for a constant $c > 1$, Rybarczyk [41,42] also derives results under $P_n = \Omega(n^c)$ for a constant $c < 1$, with a scaling condition different from (7.21).

Theorem 7.7
(Hamilton cycle containment in binomial random 1-intersection graphs by our work [56]).

1. For a binomial random 1-intersection graph $H_1(n, t_n, P_n)$, if there is a sequence α_n with $\lim_{n \to \infty} \alpha_n \in [-\infty, \infty]$ such that

$$y(t_n, P_n) = \frac{\ln n + \ln \ln n + \alpha_n}{n},$$

then under $P_n = \omega(n(\ln n)^5)$, it holds that

$$\lim_{n \to \infty} \mathbb{P}[\, H_1(n, t_n, P_n) \text{ has at least one Hamilton cycle.}\,]$$

$$= e^{-e^{-\lim_{n \to \infty} \alpha_n}} = \begin{cases} 0 & \text{if } \lim_{n \to \infty} \alpha_n = -\infty, \\ 1 & \text{if } \lim_{n \to \infty} \alpha_n = \infty, \\ e^{-e^{-\alpha^*}} & \text{if } \lim_{n \to \infty} \alpha_n = \alpha^* \in (-\infty, \infty). \end{cases}$$

2. For a binomial random 1-intersection graph $H_1(n, t_n, P_n)$, if there is a sequence β_n with $\lim_{n \to \infty} \beta_n \in [-\infty, \infty]$ such that

$$t_n^2 P_n = \frac{\ln n + \ln \ln n + \beta_n}{n},$$

then under $P_n = \omega\left(n(\ln n)^5\right)$, it holds that

$$\lim_{n\to\infty} \mathbb{P}[\,H_1(n,t_n,P_n) \text{ has at least one Hamilton cycle.}\,]$$

$$= e^{-e^{-\lim_{n\to\infty}\beta_n}} = \begin{cases} 0 & \text{if } \lim_{n\to\infty}\beta_n = -\infty, \\ 1 & \text{if } \lim_{n\to\infty}\beta_n = \infty, \\ e^{-e^{-\beta^*}} & \text{if } \lim_{n\to\infty}\beta_n = \beta^* \in (-\infty,\infty). \end{cases}$$

Remark 7.7. Theorem 7.7 presents the asymptotically exact probability of Hamilton cycle containment in a binomial random 1-intersection graph, while zero–one laws are obtained by Efthymioua and Spirakis [19] and Rybarczyk [41, 42].

Theorem 7.8
(*k*-robustness in binomial random 1-intersection graphs by our work [62]).

1. For a binomial random 1-intersection graph $H_1(n,t_n,P_n)$, with a sequence α_n defined by

$$y(t_n, P_n) = \frac{\ln n + (k-1)\ln\ln n + \alpha_n}{n},$$

under $P_n = \Omega\left(n(\ln n)^5\right)$, it holds that

$$\lim_{n\to\infty} \mathbb{P}[\,H_1(n,t_n,P_n) \text{ is } k\text{-robust.}\,]$$

$$= \begin{cases} 0 & \text{if } \lim_{n\to\infty}\alpha_n = -\infty, \\ 1 & \text{if } \lim_{n\to\infty}\alpha_n = \infty. \end{cases}$$

2. For a binomial random 1-intersection graph $H_1(n,t_n,P_n)$, with a sequence β_n defined by

$$t_n^2 P_n = \frac{\ln n + (k-1)\ln\ln n + \beta_n}{n},$$

under $P_n = \Omega\left(n(\ln n)^5\right)$, it holds that

$$\lim_{n\to\infty} \mathbb{P}[\,H_1(n,t_n,P_n) \text{ is } k\text{-robust.}\,]$$

$$= \begin{cases} 0 & \text{if } \lim_{n\to\infty}\beta_n = -\infty, \\ 1 & \text{if } \lim_{n\to\infty}\beta_n = \infty. \end{cases}$$

7.4.3   Results of Uniform Random *s*-Intersection Graphs

Theorem 7.9
(*k*-connectivity in uniform random *s*-intersection graphs by our work [68]).

1. For a uniform random s-intersection graph $G_s(n, K_n, P_n)$, if there is a sequence α_n with $\lim_{n \to \infty} \alpha_n \in [-\infty, \infty]$ such that

$$x_s(K_n, P_n) = \frac{\ln n + (k-1)\ln\ln n + \alpha_n}{n},$$

   then under $P_n = \Omega(n^c)$ for a constant $c > 2 - \frac{1}{s}$, it holds that

$$\lim_{n \to \infty} \mathbb{P}\left[\, G_s(n, K_n, P_n) \text{ is } k\text{-connected.}\,\right]$$

$$= \lim_{n \to \infty} \mathbb{P}\left[\, G_s(n, K_n, P_n) \text{ has a minimum degree at least } k.\,\right]$$

$$= e^{-\frac{e^{-\lim_{n\to\infty}\alpha_n}}{(k-1)!}} = \begin{cases} 0 & \text{if } \lim_{n\to\infty}\alpha_n = -\infty, \\ 1 & \text{if } \lim_{n\to\infty}\alpha_n = \infty, \\ e^{-\frac{e^{-\alpha^*}}{(k-1)!}} & \text{if } \lim_{n\to\infty}\alpha_n = \alpha^* \in (-\infty, \infty). \end{cases}$$

2. For a uniform random s-intersection graph $G_s(n, K_n, P_n)$, if there is a sequence β_n with $\lim_{n \to \infty} \beta_n \in [-\infty, \infty]$ such that

$$\frac{1}{s!} \cdot \frac{K_n^{2s}}{P_n^s} = \frac{\ln n + (k-1)\ln\ln n + \beta_n}{n},$$

   then under $P_n = \Omega(n^c)$ for a constant $c > 2 - \frac{1}{s}$, it holds that

$$\lim_{n \to \infty} \mathbb{P}\left[\, G_s(n, K_n, P_n) \text{ is } k\text{-connected.}\,\right]$$

$$= \lim_{n \to \infty} \mathbb{P}\left[\, G_s(n, K_n, P_n) \text{ has a minimum degree at least } k.\,\right]$$

$$= e^{-\frac{e^{-\lim_{n\to\infty}\beta_n}}{(k-1)!}} = \begin{cases} 0 & \text{if } \lim_{n\to\infty}\beta_n = -\infty, \\ 1 & \text{if } \lim_{n\to\infty}\beta_n = \infty, \\ e^{-\frac{e^{-\beta^*}}{(k-1)!}} & \text{if } \lim_{n\to\infty}\beta_n = \beta^* \in (-\infty, \infty). \end{cases}$$

Remark 7.8.   Theorem 7.9 presents the asymptotically exact probability of k-connectivity in a uniform random s-intersection graph, while a similar result for k-connectivity is given by Bloznelis and Rybarczyk [11], and a similar result for connectivity (i.e., k-connectivity in the case of $k = 1$) is shown by Bloznelis and Łuczak [10], but both results [10, 11] assume $K_n = O\left((\ln n)^{\frac{1}{5s}}\right)$, which limits their applications to secure sensor networks [14].

Theorem 7.10
(perfect matching containment in uniform random s-intersection graphs by our work [59]).

1. For a uniform random s-intersection graph $G_s(n, K_n, P_n)$, if there is a sequence α_n with $\lim_{n\to\infty} \alpha_n \in [-\infty, \infty]$ such that

$$x_s(K_n, P_n) = \frac{\ln n + \alpha_n}{n},$$

then under $P_n = \Omega(n^c)$ for a constant $c > 2 - \frac{1}{s}$, it holds that

$$\lim_{n\to\infty} \mathbb{P}[\, G_s(n, K_n, P_n) \text{ has at least one perfect matching.}\,]$$

$$= e^{-e^{-\lim_{n\to\infty} \alpha_n}} = \begin{cases} 0 & \text{if } \lim_{n\to\infty} \alpha_n = -\infty, \\ 1 & \text{if } \lim_{n\to\infty} \alpha_n = \infty, \\ e^{-e^{-\alpha^*}} & \text{if } \lim_{n\to\infty} \alpha_n = \alpha^* \in (-\infty, \infty). \end{cases}$$

2. For a uniform random s-intersection graph $G_s(n, K_n, P_n)$, if there is a sequence β_n with $\lim_{n\to\infty} \beta_n \in [-\infty, \infty]$ such that

$$\frac{1}{s!} \cdot \frac{K_n^{2s}}{P_n^s} = \frac{\ln n + \beta_n}{n},$$

then under $P_n = \Omega(n^c)$ for a constant $c > 2 - \frac{1}{s}$, it holds that

$$\lim_{n\to\infty} \mathbb{P}[\, G_s(n, K_n, P_n) \text{ has at least one perfect matching.}\,]$$

$$= e^{-e^{-\lim_{n\to\infty} \beta_n}} = \begin{cases} 0 & \text{if } \lim_{n\to\infty} \beta_n = -\infty, \\ 1 & \text{if } \lim_{n\to\infty} \beta_n = \infty, \\ e^{-e^{-\beta^*}} & \text{if } \lim_{n\to\infty} \beta_n = \beta^* \in (-\infty, \infty). \end{cases}$$

Remark 7.9. Theorem 7.10 presents the asymptotically exact probability of perfect matching containment in a uniform random s-intersection graph, while a similar result is given by Bloznelis and Łuczak [10] under $K_n = O\big((\ln n)^{\frac{1}{5s}}\big)$.

Theorem 7.11

(Hamilton cycle containment in uniform random s-intersection graphs by our work [59]).

1. For a uniform random s-intersection graph $G_s(n, K_n, P_n)$, if there is a sequence α_n with $\lim_{n\to\infty} \alpha_n \in [-\infty, \infty]$ such that

$$x_s(K_n, P_n) = \frac{\ln n + \ln \ln n + \alpha_n}{n},$$

then under $P_n = \Omega(n^c)$ for a constant $c > 2 - \frac{1}{s}$, it holds that

$$\lim_{n\to\infty} \mathbb{P}[\, G_s(n, K_n, P_n) \text{ has at least one Hamilton cycle.}\,]$$

$$= e^{-e^{-\lim_{n\to\infty} \alpha_n}} = \begin{cases} 0 & \text{if } \lim_{n\to\infty} \alpha_n = -\infty, \\ 1 & \text{if } \lim_{n\to\infty} \alpha_n = \infty, \\ e^{-e^{-\alpha^*}} & \text{if } \lim_{n\to\infty} \alpha_n = \alpha^* \in (-\infty, \infty). \end{cases}$$

2. For a uniform random s-intersection graph $G_s(n, K_n, P_n)$, if there is a sequence β_n with $\lim_{n\to\infty} \beta_n \in [-\infty, \infty]$ such that

$$\frac{1}{s!} \cdot \frac{K_n^{2s}}{P_n^s} = \frac{\ln n + \ln\ln n + \beta_n}{n},$$

then under $P_n = \Omega(n^c)$ for a constant $c > 2 - \frac{1}{s}$, it holds that

$$\lim_{n\to\infty} \mathbb{P}[\, G_s(n, K_n, P_n) \text{ has at least one Hamilton cycle.}\,]$$

$$= e^{-e^{-\lim_{n\to\infty} \beta_n}} = \begin{cases} 0 & \text{if } \lim_{n\to\infty} \beta_n = -\infty, \\ 1 & \text{if } \lim_{n\to\infty} \beta_n = \infty, \\ e^{-e^{-\beta^*}} & \text{if } \lim_{n\to\infty} \beta_n = \beta^* \in (-\infty, \infty). \end{cases}$$

Theorem 7.12
(k-robustness in uniform random s-intersection graphs by our work [59]).

1. For a uniform random s-intersection graph $G_s(n, K_n, P_n)$, with a sequence α_n defined by

$$x_s(K_n, P_n) = \frac{\ln n + (k-1)\ln\ln n + \alpha_n}{n},$$

under $P_n = \Omega(n^c)$ for a constant $c > 2 - \frac{1}{s}$, it holds that

$$\lim_{n\to\infty} \mathbb{P}[\, G_s(n, K_n, P_n) \text{ is } k\text{-robust.}\,]$$

$$= \begin{cases} 0 & \text{if } \lim_{n\to\infty} \alpha_n = -\infty, \\ 1 & \text{if } \lim_{n\to\infty} \alpha_n = \infty. \end{cases}$$

2. For a uniform random s-intersection graph $G_s(n, K_n, P_n)$, with a sequence β_n defined by

$$\frac{1}{s!} \cdot \frac{K_n^{2s}}{P_n^s} = \frac{\ln n + (k-1)\ln\ln n + \beta_n}{n},$$

$P_n = \Omega(n^c)$ for a constant $c > 2 - \frac{1}{s}$, it holds that

$$\lim_{n\to\infty} \mathbb{P}[\, G_s(n, K_n, P_n) \text{ is } k\text{-robust.}\,]$$

$$= \begin{cases} 0 & \text{if } \lim_{n\to\infty} \beta_n = -\infty, \\ 1 & \text{if } \lim_{n\to\infty} \beta_n = \infty. \end{cases}$$

7.4.4 Results of Binomial Random s-Intersection Graphs

Theorem 7.13
(*k*-connectivity in binomial random *s*-intersection graphs by our work [68]).

1. For a binomial random *s*-intersection graph $H_s(n, t_n, P_n)$, if there is a sequence α_n with $\lim_{n \to \infty} \alpha_n \in [-\infty, \infty]$ such that

$$y_s(t_n, P_n) = \frac{\ln n + (k-1) \ln \ln n + \alpha_n}{n},$$

then under $P_n = \Omega(n^c)$ for a constant $c > 2 - \frac{1}{s}$, it holds that

$$\lim_{n \to \infty} \mathbb{P}\big[\,H_s(n, t_n, P_n) \text{ is } k\text{-connected.}\,\big]$$

$$= \lim_{n \to \infty} \mathbb{P}\big[\,H_s(n, t_n, P_n) \text{ has a minimum degree at least } k.\,\big]$$

$$= e^{-\frac{e^{-\lim_{n\to\infty}\alpha_n}}{(k-1)!}} = \begin{cases} 0 & \text{if } \lim_{n\to\infty}\alpha_n = -\infty, \\ 1 & \text{if } \lim_{n\to\infty}\alpha_n = \infty, \\ e^{-\frac{e^{-\alpha^*}}{(k-1)!}} & \text{if } \lim_{n\to\infty}\alpha_n = \alpha^* \in (-\infty, \infty). \end{cases}$$

2. For a binomial random *s*-intersection graph $H_s(n, t_n, P_n)$, if there is a sequence β_n with $\lim_{n \to \infty} \beta_n \in [-\infty, \infty]$ such that

$$\frac{1}{s!} \cdot t_n^{2s} P_n^{\,s} = \frac{\ln n + (k-1) \ln \ln n + \beta_n}{n},$$

then under $P_n = \Omega(n^c)$ for a constant $c > 2 - \frac{1}{s}$, it holds that

$$\lim_{n \to \infty} \mathbb{P}\big[\,H_s(n, t_n, P_n) \text{ is } k\text{-connected.}\,\big]$$

$$= \lim_{n \to \infty} \mathbb{P}\big[\,H_s(n, t_n, P_n) \text{ has a minimum degree at least } k.\,\big]$$

$$= e^{-\frac{e^{-\lim_{n\to\infty}\beta_n}}{(k-1)!}} = \begin{cases} 0 & \text{if } \lim_{n\to\infty}\beta_n = -\infty, \\ 1 & \text{if } \lim_{n\to\infty}\beta_n = \infty, \\ e^{-\frac{e^{-\beta^*}}{(k-1)!}} & \text{if } \lim_{n\to\infty}\beta_n = \beta^* \in (-\infty, \infty). \end{cases}$$

Theorem 7.14
(perfect matching containment in binomial random *s*-intersection graphs [59]).

1. For a binomial random *s*-intersection graph $H_s(n, t_n, P_n)$, if there is a sequence α_n with $\lim_{n \to \infty} \alpha_n \in [-\infty, \infty]$ such that

$$y_s(t_n, P_n) = \frac{\ln n + \alpha_n}{n},$$

then under $P_n = \Omega(n^c)$ for a constant $c > 2 - \frac{1}{s}$, it holds that

$$\lim_{n \to \infty} \mathbb{P}[\, H_s(n, t_n, P_n) \text{ has at least one perfect matching.}\,]$$

$$= e^{-e^{-\lim_{n \to \infty} \alpha_n}} = \begin{cases} 0 & \text{if } \lim_{n \to \infty} \alpha_n = -\infty, \\ 1 & \text{if } \lim_{n \to \infty} \alpha_n = \infty, \\ e^{-e^{-\alpha^*}} & \text{if } \lim_{n \to \infty} \alpha_n = \alpha^* \in (-\infty, \infty). \end{cases}$$

2. For a binomial random s-intersection graph $H_s(n, t_n, P_n)$, if there is a sequence β_n with $\lim_{n \to \infty} \beta_n \in [-\infty, \infty]$ such that

$$\frac{1}{s!} \cdot t_n{}^{2s} P_n{}^s = \frac{\ln n + \beta_n}{n},$$

then under $P_n = \Omega(n^c)$ for a constant $c > 2 - \frac{1}{s}$, it holds that

$$\lim_{n \to \infty} \mathbb{P}[\, H_s(n, t_n, P_n) \text{ has at least one perfect matching.}\,]$$

$$= e^{-e^{-\lim_{n \to \infty} \beta_n}} = \begin{cases} 0 & \text{if } \lim_{n \to \infty} \beta_n = -\infty, \\ 1 & \text{if } \lim_{n \to \infty} \beta_n = \infty, \\ e^{-e^{-\beta^*}} & \text{if } \lim_{n \to \infty} \beta_n = \beta^* \in (-\infty, \infty). \end{cases}$$

Theorem 7.15
(Hamilton cycle containment in binomial random s-intersection graphs [59]).

1. For a binomial random s-intersection graph $H_s(n, t_n, P_n)$, if there is a sequence α_n with $\lim_{n \to \infty} \alpha_n \in [-\infty, \infty]$ such that

$$y_s(t_n, P_n) = \frac{\ln n + \ln \ln n + \alpha_n}{n},$$

then under $P_n = \Omega(n^c)$ for a constant $c > 2 - \frac{1}{s}$, it holds that

$$\lim_{n \to \infty} \mathbb{P}[\, H_s(n, t_n, P_n) \text{ has at least one Hamilton cycle.}\,]$$

$$= e^{-e^{-\lim_{n \to \infty} \alpha_n}} = \begin{cases} 0 & \text{if } \lim_{n \to \infty} \alpha_n = -\infty, \\ 1 & \text{if } \lim_{n \to \infty} \alpha_n = \infty, \\ e^{-e^{-\alpha^*}} & \text{if } \lim_{n \to \infty} \alpha_n = \alpha^* \in (-\infty, \infty). \end{cases}$$

2. For a binomial random s-intersection graph $H_s(n, t_n, P_n)$, if there is a sequence β_n with $\lim_{n \to \infty} \beta_n \in [-\infty, \infty]$ such that

$$\frac{1}{s!} \cdot t_n{}^{2s} P_n{}^s = \frac{\ln n + \ln \ln n + \beta_n}{n},$$

then under $P_n = \Omega(n^c)$ for a constant $c > 2 - \frac{1}{s}$, it holds that

$$\lim_{n \to \infty} \mathbb{P}[\, H_s(n, t_n, P_n) \text{ has at least one Hamilton cycle.}\,]$$

$$= e^{-e^{-\lim_{n \to \infty} \beta_n}} = \begin{cases} 0 & \text{if } \lim_{n \to \infty} \beta_n = -\infty, \\ 1 & \text{if } \lim_{n \to \infty} \beta_n = \infty, \\ e^{-e^{-\beta^*}} & \text{if } \lim_{n \to \infty} \beta_n = \beta^* \in (-\infty, \infty). \end{cases}$$

Theorem 7.16
(*k*-robustness in binomial random *s*-intersection graphs [59]).

1. For a binomial random *s*-intersection graph $H_s(n, t_n, P_n)$, if there is a sequence α_n with $\lim_{n \to \infty} \alpha_n \in [-\infty, \infty]$ such that

$$y_s(t_n, P_n) = \frac{\ln n + (k-1)\ln \ln n + \alpha_n}{n},$$

then under $P_n = \Omega(n^c)$ for a constant $c > 2 - \frac{1}{s}$, it holds that

$$\lim_{n \to \infty} \mathbb{P}[H_s(n, t_n, P_n) \text{ is } k\text{-robust.}] = \begin{cases} 0 \text{ if } \alpha^* = -\infty, & \text{(7.22a)} \\ 1 \text{ if } \alpha^* = \infty. & \text{(7.22b)} \end{cases}$$

2. For a binomial random *s*-intersection graph $H_s(n, t_n, P_n)$, if there is a sequence β_n with $\lim_{n \to \infty} \beta_n \in [-\infty, \infty]$ such that

$$\frac{1}{s!} \cdot t_n^{2s} P_n{}^s = \frac{\ln n + (k-1)\ln \ln n + \beta_n}{n},$$

then under $P_n = \Omega(n^c)$ for a constant $c > 2 - \frac{1}{s}$, it holds that

$$\lim_{n \to \infty} \mathbb{P}[H_s(n, t_n, P_n) \text{ is } k\text{-robust.}] = \begin{cases} 0 \text{ if } \beta^* = -\infty, & \text{(7.23a)} \\ 1 \text{ if } \beta^* = \infty. & \text{(7.23b)} \end{cases}$$

7.5 Results of Random Intersection Graphs Composed with Erdős–Rényi Graphs

Theorem 7.17
(*k*-connectivity in uniform random 1-intersection graphs ∩ Erdős–Rényi graphs by our work [60, 66, 67]).

1. Consider a graph $G_1(n, K_n, P_n) \cap G_{ER}(n, p_n)$ induced by the composition of a uniform random 1-intersection graph $G_1(n, K_n, P_n)$ and an Erdős–Rényi graph $G_{ER}(n, p_n)$. With s_n denoting the edge probability of

$G_1(n,K_n,P_n) \cap G_{ER}(n,p_n)$, if there is a sequence α_n with $\lim_{n\to\infty}\alpha_n \in [-\infty,\infty]$ such that

$$x_s(K_n,P_n)\cdot p_n = \frac{\ln n + (k-1)\ln\ln n + \alpha_n}{n},$$

then under $P_n = \Omega(n)$ and $\frac{K_n}{P_n} = o(1)$, it holds that

$$\lim_{n\to\infty} \mathbb{P}\left[\, G_1(n,K_n,P_n) \cap G_{ER}(n,p_n) \text{ is } k\text{-connected.}\,\right]$$

$$= \lim_{n\to\infty} \mathbb{P}\begin{bmatrix} G_1(n,K_n,P_n) \cap G_{ER}(n,p_n) \\ \text{has a minimum degree at least } k. \end{bmatrix}$$

$$= e^{-\frac{e^{-\lim_{n\to\infty}\alpha_n}}{(k-1)!}} = \begin{cases} 0 & \text{if } \lim_{n\to\infty}\alpha_n = -\infty, \\ 1 & \text{if } \lim_{n\to\infty}\alpha_n = \infty, \\ e^{-e^{-\alpha^*}} & \text{if } \lim_{n\to\infty}\alpha_n = \alpha^* \in (-\infty,\infty). \end{cases}$$

2. Consider a graph $G_1(n,K_n,P_n) \cap G_{ER}(n,p_n)$ induced by the composition of a uniform random 1-intersection graph $G_1(n,K_n,P_n)$ and an Erdős–Rényi graph $G_{ER}(n,p_n)$. With s_n denoting the edge probability of $G_1(n,K_n,P_n) \cap G_{ER}(n,p_n)$, if there is a sequence β_n with $\lim_{n\to\infty}\beta_n \in [-\infty,\infty]$ such that

$$\frac{K_n^2}{P_n}\cdot p_n = \frac{\ln n + (k-1)\ln\ln n + \beta_n}{n},$$

then under $P_n = \Omega(n)$ and $\frac{K_n^2}{P_n} = o\left(\frac{1}{\ln n}\right)$, it holds that

$$\lim_{n\to\infty} \mathbb{P}\left[\, G_1(n,K_n,P_n) \cap G_{ER}(n,p_n) \text{ is } k\text{-connected.}\,\right]$$

$$= \lim_{n\to\infty} \mathbb{P}\begin{bmatrix} G_1(n,K_n,P_n) \cap G_{ER}(n,p_n) \\ \text{has a minimum degree at least } k. \end{bmatrix}$$

$$= e^{-\frac{e^{-\lim_{n\to\infty}\beta_n}}{(k-1)!}} = \begin{cases} 0 & \text{if } \lim_{n\to\infty}\beta_n = -\infty, \\ 1 & \text{if } \lim_{n\to\infty}\beta_n = \infty, \\ e^{-e^{-\beta^*}} & \text{if } \lim_{n\to\infty}\beta_n = \beta^* \in (-\infty,\infty). \end{cases}$$

Remark 7.10. As summarized in Theorem 7.17, for k-connectivity in a uniform random 1-intersection graph composed with an Erdős–Rényi graph, our papers [60, 67] show a zero–one law, and later another work of ours [66] derives the asymptotically exact probability. For connectivity, Yağan [47] shows a zero–one law under a weaker scaling.

Theorem 7.18
(k-connectivity in uniform random s-intersection graphs $\cap$ Erdős–Rényi graphs by our work [63]).

1. Consider a graph $G_s(n,K_n,P_n) \cap G_{ER}(n,p_n)$ induced by the composition of a uniform random s-intersection graph $G_s(n,K_n,P_n)$ and an Erdős–Rényi graph $G_{ER}(n,p_n)$. With s_n denoting the edge probability of $G_s(n,K_n,P_n) \cap G_{ER}(n,p_n)$, if there is a sequence α_n with $\lim_{n\to\infty}\alpha_n \in [-\infty,\infty]$ such that

$$x_s(K_n,P_n) \cdot p_n = \frac{\ln n + (k-1)\ln\ln n + \alpha_n}{n},$$

then under $P_n = \Omega(n)$ and $\frac{K_n}{P_n} = o(1)$, it holds that

$$\lim_{n\to\infty} \mathbb{P}\begin{bmatrix} G_s(n,K_n,P_n) \cap G_{ER}(n,p_n) \\ \text{has a minimum degree at least } k. \end{bmatrix}$$

$$= e^{-\frac{e^{-\lim_{n\to\infty}\alpha_n}}{(k-1)!}} = \begin{cases} 0 & \text{if } \lim_{n\to\infty}\alpha_n = -\infty, \\ 1 & \text{if } \lim_{n\to\infty}\alpha_n = \infty, \\ e^{-e^{-\alpha^*}} & \text{if } \lim_{n\to\infty}\alpha_n = \alpha^* \in (-\infty,\infty). \end{cases}$$

2. Consider a graph $G_s(n,K_n,P_n) \cap G_{ER}(n,p_n)$ induced by the composition of a uniform random s-intersection graph $G_s(n,K_n,P_n)$ and an Erdős–Rényi graph $G_{ER}(n,p_n)$. With s_n denoting the edge probability of $G_s(n,K_n,P_n) \cap G_{ER}(n,p_n)$, if there is a sequence β_n with $\lim_{n\to\infty}\beta_n \in [-\infty,\infty]$ such that

$$\frac{K_n^2}{P_n} \cdot p_n = \frac{\ln n + (k-1)\ln\ln n + \beta_n}{n},$$

then under $P_n = \Omega(n)$ and $\frac{K_n^2}{P_n} = o\left(\frac{1}{\ln n}\right)$, it holds that

$$\lim_{n\to\infty} \mathbb{P}\begin{bmatrix} G_s(n,K_n,P_n) \cap G_{ER}(n,p_n) \\ \text{has a minimum degree at least } k. \end{bmatrix}$$

$$= e^{-\frac{e^{-\lim_{n\to\infty}\beta_n}}{(k-1)!}} = \begin{cases} 0 & \text{if } \lim_{n\to\infty}\beta_n = -\infty, \\ 1 & \text{if } \lim_{n\to\infty}\beta_n = \infty, \\ e^{-e^{-\beta^*}} & \text{if } \lim_{n\to\infty}\beta_n = \beta^* \in (-\infty,\infty). \end{cases}$$

7.6 Results of Random Intersection Graphs Composed with Random Geometric Graphs

Theorem 7.19
(connectivity in uniform random 1-intersection graphs $\cap$ random geometric graphs without the boundary effect by our work [61]).

1. Consider a graph $G_1(n,K_n,P_n) \cap G_{RGG}(n,r_n,\mathcal{T})$ induced by the composition of a uniform random s-intersection graph $G_s(n,K_n,P_n)$ and a random geometric graph $G_{RGG}(n,r_n,\mathcal{T})$, where $\mathcal{T}$ is a torus of unit area. If

$$\pi r_n^2 \cdot x(K_n,P_n) \sim a \cdot \frac{\ln n}{n}$$

for some positive constant a, then under $K_n = \omega(\ln n)$, $\frac{K_n^2}{P_n} = O\left(\frac{1}{\ln n}\right)$, $\frac{K_n^2}{P_n} = \omega\left(\frac{\ln n}{n}\right)$, and $\frac{K_n}{P_n} = o\left(\frac{1}{n}\right)$, it holds that

$$\lim_{n\to\infty} \mathbb{P}[\, G_1(n,K_n,P_n) \cap G_{RGG}(n,r_n,\mathcal{T}) \text{ is connected.}\,]$$

$$= \begin{cases} 0 & \text{if } a < 1, \\ 1 & \text{if } a > 1. \end{cases}$$

2. Consider a graph $G_1(n,K_n,P_n) \cap G_{RGG}(n,r_n,\mathcal{T})$ induced by the composition of a uniform random s-intersection graph $G_s(n,K_n,P_n)$ and a random geometric graph $G_{RGG}(n,r_n,\mathcal{T})$, where $\mathcal{T}$ is a torus of unit area. If

$$\pi r_n^2 \cdot \frac{K_n^2}{P_n} \sim a \cdot \frac{\ln n}{n} \tag{7.24}$$

for some positive constant a, then under $K_n = \omega(\ln n)$, $\frac{K_n^2}{P_n} = O\left(\frac{1}{\ln n}\right)$, $\frac{K_n^2}{P_n} = \omega\left(\frac{\ln n}{n}\right)$, and $\frac{K_n}{P_n} = o\left(\frac{1}{n}\right)$, it holds that

$$\lim_{n\to\infty} \mathbb{P}[\, G_1(n,K_n,P_n) \cap G_{RGG}(n,r_n,\mathcal{T}) \text{ is connected.}\,]$$

$$= \begin{cases} 0 & \text{if } a < 1, \\ 1 & \text{if } a > 1. \end{cases}$$

Theorem 7.20
(connectivity in uniform random 1-intersection graphs ∩ random geometric graphs with the boundary effect by our work [61]).

1. Consider a graph $G_1(n,K_n,P_n) \cap G_{RGG}(n,r_n,\mathcal{S})$ induced by the composition of a uniform random s-intersection graph $G_s(n,K_n,P_n)$ and a random geometric graph $G_{RGG}(n,r_n,\mathcal{S})$, where $\mathcal{S}$ is a square of unit area. If

$$\pi r_n^2 \cdot x(K_n,P_n) \sim \begin{cases} b \cdot \dfrac{\ln \frac{nP_n}{K_n^2}}{n}, & \text{for } \dfrac{K_n^2}{P_n} = \omega\left(\dfrac{1}{n^{1/3}\ln n}\right), \\[4mm] b \cdot \dfrac{4\ln \frac{P_n}{K_n^2}}{n}, & \text{for } \dfrac{K_n^2}{P_n} = O\left(\dfrac{1}{n^{1/3}\ln n}\right), \end{cases}$$

for some positive constant b, then under $K_n = \omega(\ln n)$, $\frac{K_n^2}{P_n} = O\left(\frac{1}{\ln n}\right)$, $\frac{K_n^2}{P_n} = \omega\left(\frac{\ln n}{n}\right)$, and $\frac{K_n}{P_n} = o\left(\frac{1}{n}\right)$, it holds that

$$\lim_{n \to \infty} \mathbb{P}\left[\, G_1(n, K_n, P_n) \cap G_{RGG}(n, r_n, \mathcal{S}) \text{ is connected.}\,\right]$$

$$= \begin{cases} 0 & \text{if } b < 1, \\ 1 & \text{if } b > 1. \end{cases}$$

2. Consider a graph $G_1(n, K_n, P_n) \cap G_{RGG}(n, r_n, \mathcal{S})$ induced by the composition of a uniform random s-intersection graph $G_s(n, K_n, P_n)$ and a random geometric graph $G_{RGG}(n, r_n, \mathcal{S})$, where $\mathcal{S}$ is a square of unit area. If

$$\pi r_n^2 \cdot \frac{K_n^2}{P_n} \sim \begin{cases} b \cdot \dfrac{\ln \frac{nP_n}{K_n^2}}{n}, & \text{for } \dfrac{K_n^2}{P_n} = \omega\left(\dfrac{1}{n^{1/3} \ln n}\right), \\[3ex] b \cdot \dfrac{4 \ln \frac{P_n}{K_n^2}}{n}, & \text{for } \dfrac{K_n^2}{P_n} = O\left(\dfrac{1}{n^{1/3} \ln n}\right), \end{cases}$$

for some positive constant b, then under $K_n = \omega(\ln n)$, $\frac{K_n^2}{P_n} = O\left(\frac{1}{\ln n}\right)$, $\frac{K_n^2}{P_n} = \omega\left(\frac{\ln n}{n}\right)$, and $\frac{K_n}{P_n} = o\left(\frac{1}{n}\right)$, it holds that

$$\lim_{n \to \infty} \mathbb{P}\left[\, G_1(n, K_n, P_n) \cap G_{RGG}(n, r_n, \mathcal{S}) \text{ is connected.}\,\right]$$

$$= \begin{cases} 0 & \text{if } b < 1, \\ 1 & \text{if } b > 1. \end{cases}$$

Remark 7.11. For the graph $G_1(n, K_n, P_n) \cap G_{RGG}(n, r_n, \mathcal{S})$, Krzywdziński and Rybarczyk [31] and Krishnan et al. [30] also obtain connectivity results, but their results are weaker than those in Theorem 7.20 above; see [61, Section VIII] for details. Furthermore, Pishro-Nik *et al.* [38] and Yi et al. [50] investigate the absence of isolated nodes.

7.7 Comparison between Random Graphs

We have studied random intersection graphs (respectively, their intersections with other random graphs) and Erdős–Rényi graphs. To compare our studied graphs with Erdős–Rényi graphs, we summarize below the results of Erdős–Rényi graphs shown in prior work.

Lemma 7.3

(k-connectivity in Erdős–Rényi graphs by [21, Theorem 1]). For an Erdős–Rényi graph $G_{ER}(n, p_n)$, if there is a sequence α_n with $\lim_{n \to \infty} \alpha_n \in [-\infty, \infty]$ such that

$$p_n = \frac{\ln n + (k-1) \ln \ln n \alpha_n}{n},$$

then it holds that

$$\lim_{n\to\infty} \mathbb{P}[\, G_{ER}(n,p_n) \text{ is } k\text{-connected.} \,]$$

$$= \lim_{n\to\infty} \mathbb{P}[\, G_{ER}(n,p_n) \text{ has a minimum degree at least } k. \,]$$

$$= e^{-\frac{e^{-\lim_{n\to\infty}\alpha_n}}{(k-1)!}} = \begin{cases} 0 & \text{if } \lim_{n\to\infty}\alpha_n = -\infty, \\ 1 & \text{if } \lim_{n\to\infty}\alpha_n = \infty, \\ e^{-\frac{e^{-\alpha^*}}{(k-1)!}} & \text{if } \lim_{n\to\infty}\alpha_n = \alpha^* \in (-\infty,\infty). \end{cases}$$

Lemma 7.4
(perfect matching containment in Erdős–Rényi graphs by [22, Theorem 1]). For an Erdős–Rényi graph $G_{ER}(n,p_n)$, if there is a sequence β_n with $\lim_{n\to\infty}\beta_n \in [-\infty,\infty]$ such that

$$p_n = \frac{\ln n + \beta_n}{n},$$

then it holds that

$$\lim_{n\to\infty} \mathbb{P}[\, G_{ER}(n,p_n) \text{ has a perfect matching.} \,] = e^{-e^{-\lim_{n\to\infty}\beta_n}}.$$

Lemma 7.5
(Hamilton cycle containment in Erdős–Rényi graphs by [29, Theorem 1]). For an Erdős–Rényi graph $G_{ER}(n,p_n)$, if there is a sequence γ_n with $\lim_{n\to\infty}\gamma_n \in [-\infty,\infty]$ such that

$$p_n = \frac{\ln n + \ln\ln n + \gamma_n}{n},$$

then it holds that

$$\lim_{n\to\infty} \mathbb{P}[\, G_{ER}(n,p_n) \text{ has a Hamilton cycle.} \,] = e^{-e^{-\lim_{n\to\infty}\gamma_n}}.$$

Lemma 7.6
(k-robustness in Erdős–Rényi graphs by [51, Theorem 3] and [62, Lemma 1]). For an Erdős–Rényi graph $G_{ER}(n,p_n)$, with a sequence δ_n for all n through

$$p_n = \frac{\ln n + (k-1)\ln\ln n + \delta_n}{n}, \tag{7.25}$$

it holds that

$$\lim_{n\to\infty} \mathbb{P}[\, G_{ER}(n,p_n) \text{ is } k\text{-robust.} \,] = \begin{cases} 0 \text{ if } \lim_{n\to\infty}\delta_n = -\infty, \\ 1 \text{ if } \lim_{n\to\infty}\delta_n = \infty. \end{cases} \tag{7.26}$$

From Theorems 7.1 through 7.19 and Lemmas 7.4 through 7.6, random graphs $G_1(n, K_n, P_n)$, $G_s(n, K_n, P_n)$, $H_1(n, t_n, P_n)$, $H_s(n, t_n, P_n)$, $G_1(n, K_n, P_n) \cap G_{ER}(n, p_n)$, $G_s(n, K_n, P_n) \cap G_{ER}(n, p_n)$, and $G_1(n, K_n, P_n) \cap G_{RGG}(n, r_n, \mathcal{T})$ under the conditions in the respective theorems have threshold behaviors for the respective properties similar to those of the Erdős–Rényi graphs with the same edge probabilities. However, these graphs may be different from Erdős–Rényi graphs under other conditions or for other properties; for example, $G_1(n, K_n, P_n)$ is shown to be more clustered than an Erdős–Rényi graph with the same edge probability [48].

7.8 Conclusion

We have made the following contributions:

1. For a WSN with the EG scheme under the on/off channel model, we have established the asymptotically exact probabilities [66] and sharp zero–one laws [60, 67] for k-connectivity and for the property that the minimum node degree is at least k. These results [60,66,67] present the first complete analysis of k-connectivity in WSNs with key predistribution under link constraints other than the full visibility. Moreover, our results with k set as 1 are novel as well and improve prior zero–one laws for connectivity [47]. In addition, our results also imply those on k-connectivity under the full visibility, and our implied results [62] also outperform analytical findings in previous work [40].

2. For a WSN with the s-composite scheme under the on/off channel model, we have derived the asymptotically exact probabilities and sharp zero–one laws for k-connectivity [69] and for the property that the minimum node degree is at least k [63]. These results [63,69] present the first analysis of k-connectivity in WSNs with the s-composite scheme under link constraints other than the full visibility. Furthermore, our results also imply those on k-connectivity under the full visibility, and our implied results [68] also surpass theoretical findings in prior work [10, 11].

3. For a WSN with the EG scheme under the disk model, we have obtained sharp zero–one laws [61] for connectivity and the property that the minimum node degree is at least 1 (i.e., the absence of isolated node). These laws significantly improve recent results [30, 31] and specify the *critical transmission ranges* for connectivity. Further, we have shown sharp zero–one laws [61] for the property that the minimum node degree is at least k.

4. For a WSN with the s-composite scheme under the disk model, we have given sharp zero–one laws [65] for connectivity and the property that the minimum node degree is at least 1 (i.e., the absence of isolated

node). Similar to result number 3 above, these laws determine the *critical transmission ranges* for connectivity. In addition, we have proved sharp zero–one laws [61] for the property that the minimum node degree is at least k.

We note that our results 3 and 4 above under the disk model consider both of the following cases on the network field: the absence of boundary effect and the presence of boundary effect. It is worth noting that the consideration of boundary effect on the network field renders the analysis much more challenging [37].

References

1. I. F. Akyildiz, W. Su, Y. Sankarasubramaniam, and E. Cayirci. Wireless sensor networks: A survey. *Computer Networks*, 38, 2002.

2. F. G. Ball, D. J. Sirl, and P. Trapman. Epidemics on random intersection graphs. *Annals of Applied Probability*, 24(3):1081–1128, 2014.

3. S. Blackburn and S. Gerke. Connectivity of the uniform random intersection graph. *Discrete Mathematics*, 309(16), 2009.

4. S. Blackburn, D. Stinson, and J. Upadhyay. On the complexity of the herding attack and some related attacks on hash functions. *Designs, Codes and Cryptography*, 64(1–2):171–193, 2012.

5. M. Bloznelis. Degree and clustering coefficient in sparse random intersection graphs. *Annals of Applied Probability*, 23(3):1254–1289, 2013.

6. M. Bloznelis and J. Damarackas. Degree distribution of an inhomogeneous random intersection graph. *Electronic Journal of Combinatorics*, 20(3):P3, 2013.

7. M. Bloznelis, J. Jaworski, E. Godehardt, V. Kurauskas, and K. Rybarczyk. *Recent Progress in Complex Network Analysis Models of Random Intersection Graphs*. Studies in Classification, Data Analysis, and Knowledge Organization, Springer Series. Springer, Berlin, 2015.

8. M. Bloznelis, J. Jaworski, and V. Kurauskas. Assortativity and clustering of sparse random intersection graphs. *Electronic Journal of Probability*, 18(38):1–24, 2013.

9. M. Bloznelis, J. Jaworski, and K. Rybarczyk. Component evolution in a secure wireless sensor network. *Networks*, 53:19–26, 2009.

10. M. Bloznelis and T. Łuczak. Perfect matchings in random intersection graphs. *Acta Mathematica Hungarica*, 138(1–2):15–33, 2013.

11. M. Bloznelis and K. Rybarczyk. k-Connectivity of uniform s-intersection graphs. *Discrete Mathematics*, 333(0):94–100, 2014.

12. M. Bradonjić, A. Hagberg, N. Hengartner, and A. Percus. Component evolution in general random intersection graphs. In *Workshop on Algorithms and Models for the Web Graph (WAW)*, Stanford, CA, pp. 36–49, 2010.

13. M. Bradonjić, A. Hagberg, N. W. Hengartner, N. Lemons, and A. G. Percus. The phase transition in inhomogeneous random intersection graphs. Arxiv e-prints, 2013. Available at http://arxiv.org/abs/1301.7320.

14. H. Chan, A. Perrig, and D. Song. Random key predistribution schemes for sensor networks. In *IEEE Symposium on Security and Privacy*, May 2003.

15. M. Deijfen and W. Kets. Random intersection graphs with tunable degree distribution and clustering. *Probability in the Engineering and Informational Sciences*, 23:661–674, 2009.

16. R. Di Pietro, L. V. Mancini, A. Mei, A. Panconesi, and J. Radhakrishnan. Connectivity properties of secure wireless sensor networks. In *ACM Workshop on Security of Ad-Hoc and Sensor Networks*, Washington, DC, pp. 53–58, 2004.

17. R. Di Pietro, L. V. Mancini, A. Mei, A. Panconesi, and J. Radhakrishnan. Redoubtable sensor networks. *ACM Transactions on Information and System Security*, 11(3):13:1–13:22, 2008.

18. D. Dolev. The byzantine generals strike again. *Journal of Algorithms*, 3(1):14–30, 1982.

19. C. Efthymiou and P. Spirakis. Sharp thresholds for Hamiltonicity in random intersection graphs. *Theoretical Computer Science*, 411(40–42):3714–3730, 2010.

20. P. Erdős and A. Rényi. On random graphs, I. *Publicationes Mathematicae (Debrecen)*, 6:290–297, 1959.

21. P. Erdős and A. Rényi. On the strength of connectedness of random graphs. *Acta Mathematica Academiae Scientiarum Hungarica*, 261–267, 1961.

22. P. Erdős and A. Rnyi. On the existence of a factor of degree one of a connected random graph. *Acta Mathematica Academiae Scientiarum Hungarica*, 17(3–4):359–368, 1966.

23. L. Eschenauer and V. Gligor. A key-management scheme for distributed sensor networks. In *ACM Conference on Computer and Communications Security (CCS)*, 2002.

24. D. Ganesan, R. Govindan, S. Shenker, and D. Estrin. Highly-resilient, energy-efficient multipath routing in wireless sensor networks. *ACM SIGMOBILE Mobile Computing and Communications Review*, 5:11–25, 2001.

25. V. Gligor and P. Donescu. Fast encryption and authentication: XCBC encryption and XECB authentication modes. In *Fast Software Encryption*,

vol. 2355 of Lecture Notes in Computer Science, pp. 92–108. Springer, Berlin, 2001.

26. V. Gligor, A. Perrig, and J. Zhao. Brief encounters with a random key graph. In B. Christianson, J. Malcolm, V. Maty, and M. Roe, editors, *Security Protocols XVII*, vol. 7028 of Lecture Notes in Computer Science, pp. 157–161. Springer, Berlin, 2013.

27. E. Godehardt and J. Jaworski. Two models of random intersection graphs for classification. In *Exploratory Data Analysis in Empirical Research*, pp. 67–81. 2003.

28. C. S. Jutla. Encryption modes with almost free message integrity. In *International Conference on the Theory and Applications of Cryptographic Techniques (Eurocrypt)*, Innsbruck, pp. 529–544, 2001.

29. J. Komls and E. Szemerdi. Limit distribution for the existence of Hamiltonian cycles in a random graph. *Discrete Mathematics*, 43(1):55–63, 1983.

30. B. Krishnan, A. Ganesh, and D. Manjunath. On connectivity thresholds in superposition of random key graphs on random geometric graphs. In *IEEE International Symposium on Information Theory (ISIT)*, Istanbul, pp. 2389–2393, 2013.

31. K. Krzywdziński and K. Rybarczyk. Geometric graphs with randomly deleted edges—Connectivity and routing protocols. *Mathematical Foundations of Computer Science*, 6907:544–555, 2011.

32. R. La and M. Kabkab. A new graph model with random edge values: Connectivity and diameter. In *Annual Allerton Conference on Communication, Control, and Computing (Allerton)*, Monticello, IL, pp. 861–868, 2013.

33. H. LeBlanc, H. Zhang, X. Koutsoukos, and S. Sundaram. Resilient asymptotic consensus in robust networks. *IEEE Journal on Selected Areas in Communications (JSAC)*, 31(4):766–781, 2013.

34. H. LeBlanc, H. Zhang, S. Sundaram, and X. Koutsoukos. Resilient continuous-time consensus in fractional robust networks. In *American Control Conference (ACC)*, 2013.

35. P. Marbach. A lower-bound on the number of rankings required in recommended systems using collaborative filtering. In *IEEE Conference on Information Sciences and Systems (CISS)*, 2008.

36. S. Nikoletseas, C. Raptopoulos, and P. Spirakis. On the independence number and Hamiltonicity of uniform random intersection graphs. *Theoretical Computer Science*, 412(48):6750–6760, 2011.

37. M. Penrose. *Random Geometric Graphs*. Oxford University Press, Oxford, 2003.

38. H. Pishro-Nik, K. Chan, and F. Fekri. Connectivity properties of large-scale sensor networks. *Wireless Networks*, 15:945–964, 2009.

39. P. Rogaway, M. Bellare, J. Black, and T. Krovetz. OCB: A block-cipher mode of operation for efficient authenticated encryption. In *ACM Conference on Computer and Communications Security (CCS)*, Philadelphia, PA, pp. 196–205, 2001.

40. K. Rybarczyk. Diameter, connectivity and phase transition of the uniform random intersection graph. *Discrete Mathematics*, 311, 2011.

41. K. Rybarczyk. Sharp threshold functions for the random intersection graph via a coupling method. *Electronic Journal of Combinatorics*, 18:36–47, 2011.

42. K. Rybarczyk. The coupling method for inhomogeneous random intersection graphs. ArXiv e-prints, January 2013. Available at http://arxiv.org/abs/1301.0466.

43. Y. Shang. On the isolated vertices and connectivity in random intersection graphs. *International Journal of Combinatorics*, 2011, 872703, 2011.

44. K. B. Singer-Cohen. Random intersection graphs. PhD thesis, Johns Hopkins University, Baltimore, MD, 1995.

45. F. Stajano and R. J. Anderson. The resurrecting duckling: Security issues for ad-hoc wireless networks. In *International Workshop on Security Protocols*, Cambridge, pp. 172–194, 2000.

46. O. Yağan. Random graph modeling of key distribution schemes in wireless sensor networks. PhD thesis, University of Maryland, College Park, June 2011. Available at http://hdl.handle.net/1903/11910.

47. O. Yağan. Performance of the Eschenauer–Gligor key distribution scheme under an on/off channel. *IEEE Transactions on Information Theory*, 58(6): 3821–3835, 2012.

48. O. Yağan and A. M. Makowski. On the existence of triangles in random key graphs. In *47th Annual Allerton Conference on Communication, Control, and Computing (Allerton 2009)*, Monticello, IL, pp. 1567–1574, October 2009.

49. O. Yağan and A. M. Makowski. Zero–one laws for connectivity in random key graphs. *IEEE Transactions on Information Theory*, 58(5):2983–2999, 2012.

50. C.-W. Yi, P.-J. Wan, K.-W. Lin, and C.-H. Huang. Asymptotic distribution of the number of isolated nodes in wireless ad hoc networks with unreliable nodes and links. In *IEEE Global Communications Conference (GLOBECOM)*, November 2006.

51. H. Zhang and S. Sundaram. Robustness of complex networks with implications for consensus and contagion. In *IEEE Conference on Decision and Control (CDC)*, Kihei, HI, pp. 3426–3432, December 2012.

52. H. Zhang and S. Sundaram. Robustness of information diffusion algorithms to locally bounded adversaries. In *IEEE American Control Conference (ACC)*, Montreal, pp. 5855–5861, 2012.

53. H. Zhang and S. Sundaram. A simple median-based resilient consensus algorithm. In *Annual Allerton Conference on Communication, Control, and Computing (Allerton)*, Monticello, IL, pp. 1734–1741, 2012.

54. H. Zhang and S. Sundaram. Robustness of complex networks: Reaching consensus despite adversaries. Arxiv e-prints, February 2013. Available at arXiv:1203.6119 [cs.SI].

55. J. Zhao. Topological properties of wireless sensor networks under the q-composite key predistribution scheme with unreliable links. Technical report Carnegie Mellon University CyLab, Pittsburgh, PA, 2014.

56. J. Zhao. On Hamiltonicity in binomial random intersection graphs. Technical report, Carnegie Mellon University CyLab, Pittsburgh, PA, 2015.

57. J. Zhao. Parameter control in predistribution schemes of cryptographic keys. In *IEEE Global Conference on Signal and Information Processing (GlobalSIP)*, Orlando, FL, 2015.

58. J. Zhao. Sharp transitions in random key graphs. In *Allerton Conference on Communication, Control, and Computing*, Monticello, IL, 2015.

59. J. Zhao. Threshold functions in random s-intersection graphs. In *Allerton Conference on Communication, Control, and Computing*, Monticello, IL, 2015.

60. J. Zhao, O. Yağan, and V. Gligor. Secure k-connectivity in wireless sensor networks under an on/off channel model. In *IEEE International Symposium on Information Theory (ISIT)*, Istanbul, 2013.

61. J. Zhao, O. Yağan, and V. Gligor. Connectivity in secure wireless sensor networks under transmission constraints. In *Allerton Conference on Communication, Control, and Computing*, Monticello, IL, 2014.

62. J. Zhao, O. Yağan, and V. Gligor. On the strengths of connectivity and robustness in general random intersection graphs. In *IEEE Conference on Decision and Control (CDC)*, Los Angeles, CA, 2014.

63. J. Zhao, O. Yağan, and V. Gligor. On topological properties of wireless sensor networks under the q-composite key predistribution scheme with on/off channels. In *IEEE International Symposium on Information Theory (ISIT)*, Honolulu, HI, 2014.

64. J. Zhao, O. Yağan, and V. Gligor. Applications of random intersection graphs to secure sensor networks—Connectivity results. In *SIAM Workshop on Network Science (NS)*, Snowbird, UT, 2015.

65. J. Zhao, O. Yağan, and V. Gligor. Designing securely and reliably connected wireless sensor networks. ArXiv e-prints, 2015. Available at http://arxiv.org/abs/1501.01826.

66. J. Zhao, O. Yağan, and V. Gligor. Exact analysis of k-connectivity in secure sensor networks with unreliable links. In *International Symposium on Modeling and Optimization in Mobile, Ad Hoc and Wireless Networks (WiOpt)*, Mumbai, India, 2015.

67. J. Zhao, O. Yağan, and V. Gligor. k-Connectivity in random key graphs with unreliable links. *IEEE Transactions on Information Theory*, 2015.

68. J. Zhao, O. Yağan, and V. Gligor. On k-connectivity and minimum vertex degree in random s-intersection graphs. In *ACM-SIAM Meeting on Analytic Algorithmics and Combinatorics (ANALCO)*, San Diego, CA, 2015.

69. J. Zhao. On secure connectivity in wireless sensor networks under q-composite key predistribution with on/off channels. Technical report, Carnegie Mellon University CyLab, Pittsburgh, PA, 2015.

70. J. Zhao, O. Yağan, and V. Gligor. Random intersection graphs and their applications in security, wireless communication, and social networks. In *Information Theory and Applications Workshop (ITA)*, San Diego, CA, 2015.

Chapter 8

Resilient Distributed Control in Cyber-Physical Energy Systems

Wente Zeng

Department of Electrical and Computer Engineering, North Carolina State University

Mo-Yuen Chow

Department of Electrical and Computer Engineering, North Carolina State University

CONTENTS

With the rapid advancements and use of modern embedded systems, communication technologies, and novel control strategies, the legacy power system is gradually evolving into an advanced cyber-physical energy system—smart grid—with the goal of improving energy efficiency and availability. A variety of distributed control algorithms are being developed for smart grid energy management applications because of their advantages in terms of flexibility, robustness, and scalability. However, these algorithms also increase the vulnerability of the smart grid to adversaries. Thus, there is an urgent need to protect the distributed energy management systems from malicious cyber-attacks. Preliminary work to address the resilience and security issues of distributed energy management in the smart grid is reviewed in this chapter. As an example, a reputation-based resilient distributed energy management algorithm is discussed and demonstrated to guarantee accurate control computations in the presence of misbehaving generation units in the system. The characteristics of the reputation-based resilient distributed energy management algorithm under different network and adversary scenarios are illustrated through several simulation case studies.

8.1 Introduction

A cyber-physical system (CPS) is a spatially distributed system for which the communication among distributed sensors, distributed actuators, and distributed controllers is supported by a shared communication network [5]. These systems integrate computing and communication capabilities with monitoring and controlling entities in the physical world [1]. With its widespread applications in areas such as defense systems, medical devices, autonomous vehicles, and industrial control systems, CPS has been at the core of smart structure of national critical infrastructures (e.g., smart grid and intelligent transportation systems).

As a typical CPS, the legacy power grid is evolving into the smart grid gradually [6]. The goal of the smart grid is to improve the efficiency and availability of power by adding more communication and control capabilities. In the last few decades, there have been significant innovations and advances in technologies in areas such as networking technology (e.g., *ad hoc* networks) [2], embedded systems (e.g., field-programmable gate array [FPGA]), and communications technologies (e.g., the Internet and wireless communications) [4], which give us much more freedom in how we operate the power grid. Moreover, due to the significant increasing amount of distributed energy resources (DERs), such as renewable energy resources, energy storage systems, and other distributed generations in the smart grid, a variety of novel distributed control algorithms have been proposed for grid energy management applications because of their flexibility and scalability advantages when compared to conventional centralized control strategies [15, 23, 27]. However, these advancements also introduce new vulnerabilities into the smart grid [9, 20]. Although it has been safe in the past, several infamous emerging malicious cyber-attacks and malware targeting the smart grid have been exposed recently [12, 13].

Recently, researchers have realized the importance of power system security and developed techniques to mitigate the cyber threats. A new type of malicious attacks—false data injection attack—is formulated and analyzed in [11, 14]. It compromises the power system state estimator by injecting false data into sensors. An efficient algorithm to find sparse attacks and a phasor measurement units placement algorithm to prevent sparse attacks are both developed in [10]. The impact of data integrity attacks to automatic generation control (AGC) on power system frequency and electricity market operations is demonstrated in [19]. It also develops an anomaly detection algorithm using a statistical method to predict AGC operation over a given time period. Yuan et al. [24] design another type of false data injection attack—load redistribution attack. Under this attack, the load bus injection measurements are compromised to disrupt the power system operation. The economic impact of false data injection attacks on electricity market operation is examined in [3, 22], in which the undetectable and profitable attack strategies that exploit the virtual bidding mechanism and generation ramp constraints are proposed, respectively.

However, these state-of-the-art efforts of power grid protection are mainly focused on the communication security and conventional centralized control algorithms. The vulnerability of newly developed distributed control algorithms has not been addressed. Because of the lack of a centralized controller that monitors the activities of all the units in the system, distributed control algorithms are prone to attacks and component failures [25, 26]. Let us take the incremental cost consensus (ICC) algorithm developed in [27] as an example to illustrate the use of distributed control in a smart grid to solve the economic dispatch problem (EDP). In ICC, each generation unit updates its incremental cost (IC) as a weighted combination of its own value and the values received from its neighbors at each time instant [8]. Due to its distributed nature, there is no central authority to monitor the activities of all the generation units and demands in the system. The misconduct of one single unit may lead to performance deterioration or even system outage of the grid. Thus, it is important to guarantee the accurate control computation in the distributed control algorithms in the presence of failures or adversaries in a smart grid environment.

Furthermore, in the state-of-the-art security methods addressing grid resilience, all the units need to obtain the global information of the entire system in a centralized manner, which requires high computational overhead. Moreover, the anomaly detection mechanisms and the distributed control algorithms are always separated. There is no feedback to the control parameters when the compromised units are identified. Therefore, a resilient distributed control algorithm that has been set up with an inherent security design is more likely to remain in normal operation in the face of faults and malicious cyber-attacks.

Motivated by the arising security need for a smart grid, we have developed a reputation-based resilient distributed energy management algorithm to solve the EDP in an unreliable smart grid environment [28]. For comparison purposes, the ICC algorithm is used as an example of the conventional distributed energy management algorithms without resilience to adversaries. The proposed reputation-based resilient distributed energy management algorithm outperforms the conventional security mechanisms for distributed energy management systems; it has the following features:

1. It utilizes a fully distributed detection structure that is resilient to the single point of failure because no central fusion node is required, in comparison with the conventional centralized anomaly detection schemes.

2. It embeds the anomaly detection mechanism as a feedback loop inside the distributed control computation process, enabling the distributed energy management algorithm itself to be resilient to misbehaving generation units.

The rest of this chapter organized as follows. In Section 8.2, basic graph theory notation, linear consensus, and the ICC algorithm are introduced. The vulnerabilities of the ICC algorithm and two corresponding adversary scenarios are analyzed in Section 8.3. In Section 8.4, the reputation-based resilient distributed energy management algorithm is described. Several simulation case studies in the Institute of Electrical and Electronics Engineers (IEEE) 30-bus system under different scenarios are given in Section 8.5. Finally, conclusions are drawn in Section 8.6.

8.2 Distributed Energy Management Algorithm Preliminaries

8.2.1 Graph Theory Notation

The topology of a network can be modeled as a graph G. The graph G is an ordered pair $G = (V, E)$ comprising a set V of nodes and a set E of edges.

An adjacency matrix is a means of representing which nodes of a graph are adjacent to which other nodes. Specifically, the adjacency matrix A of a graph G with n nodes is an $n \times n$ matrix where the off-diagonal entry a_{ij} is the number of edges from node i to node j. It is a $(0, 1)$ matrix with zeros on its diagonal in the case of a finite simple graph. The Laplacian matrix L is another matrix representation of a graph. It is a symmetric, positive, and semidefinite matrix for an undirected graph. Let matrix $L = [l_{ij}]$ be

$$l_{ii} = \sum_{i \neq j} a_{ij} \text{ for on-diagonal elements,} \tag{8.1}$$

$$l_{ij} = -a_{ij} \text{ for off-diagonal elements.} \tag{8.2}$$

The eigenvalue of the L matrix contains information about the graph [16]. For example, the second smallest eigenvalue of Laplacian matrix λ_2 is the *algebraic connectivity* of the graph.

8.2.2 Linear Consensus Algorithm

Let χ_i denote the state variable of the ith node in an n-node network. All the nodes in the network reach a consensus if and only if $\chi_i = \chi_j$ for all i, j [18]. Assuming each node has a linear dynamic, a continuous-time first-order consensus algorithm can be formulated as

$$\dot{\chi}_i(t) = -\sum_{i=1}^{n} a_{ij}(\chi_i(t) - \chi_j(t)), \tag{8.3}$$

where a_{ij} is the (i, j) entry of the adjacency matrix A. In a matrix form, the algorithm is written as

$$\dot{X} = -L_n X, \tag{8.4}$$

where L_n is the $n \times n$ Laplacian matrix.

Since the data packet in a communications network arrives discretely, the discrete-time form of the first-order linear consensus algorithm is then formulated in [17] and described by

$$\chi_i[k+1] = \sum_{i=1}^{n} d_{ij}\chi_j[k],$$ (8.5)

where k is the discrete time-step and d_{ij} is the (i, j) entry of a row-stochastic matrix D_n that is defined as

$$d_{ij} = \frac{l_{ij}}{\sum_{h=1}^{n} l_{ih}}.$$ (8.6)

8.2.3 Incremental Cost Consensus Algorithm

As one of the basic energy management problems in power systems, the objective of EDP is to determine the most economic generation dispatch that minimizes the total power generation cost while satisfying the operational constraints, such as power balance and generation limits.

Let us assume each generation unit has a quadratic cost function:

$$C_i(P_{Gi}) = \alpha_i P_{Gi}^2 + \beta_i P_{Gi} + \gamma_i,$$ (8.7)

where α_i, β_i, and γ_i are generator coefficients, and P_{Gi} is the output power of the ith generation unit in the system.

Then, the overall EDP optimization of an n-generation unit system is formulated as

$$\min_{P_G} J = \sum_{i=1}^{n} C_i(P_{Gi})$$ (8.8)

subject to the power balance constraint

$$P_D - \sum_{i=1}^{n} P_{Gi} = 0$$ (8.9)

and the generation limit of each generation unit

$$P_{Gi,min} \leq P_{Gi} \leq P_{Gi,max},$$ (8.10)

where P_D denotes the total system power demand, and $P_{Gi,min}$ and $P_{Gi,max}$ are the lower and upper limits of the ith generation unit.

The ICC algorithm uses the first-order linear consensus algorithm as a foundation to solve the EDP in a distributed manner [27]. In ICC, the incremental cost (IC) for each generation unit is selected as the consensus variable:

$$IC_i = \frac{\partial C_i(P_{Gi})}{\partial P_{Gi}} = \lambda_i.$$ (8.11)

Following the discrete-time first-order consensus algorithm, each generation unit updates its IC in each time-step k with the following updating rule:

$$IC_i[k+1] = \sum_{i=1}^{n} d_{ij}IC_j[k], \quad (8.12)$$

where d_{ij} is the (i, j) entry of a row-stochastic matrix D_n.

To satisfy the power balance constraint, $\Delta P[k]$ is defined to measure the power mismatch between the total generation and demand in the system in the kth time-step:

$$\Delta P[k] = P_D[k] - \sum_{i=1}^{n} P_{Gi}[k], \quad (8.13)$$

One leader generation unit is chosen to monitor the power mismatch $\Delta P[k]$ of the overall system. The update rule for this leader generation unit becomes

$$IC_i[k+1] = \sum_{i=1}^{n} d_{ij}IC_j[k] + \epsilon \Delta P[k], \quad (8.14)$$

where ϵ is a positive scalar. It is defined as the convergence coefficient, and it lets the leader generation unit control the convergence rate of the ICC algorithm.

By considering the generation limits of each unit, we also have the following constraints:

$$P_{Gi}[k] = P_{Gi,min}, \text{ if } P_{Gi}[k] < P_{Gi,min}, \quad (8.15)$$

$$P_{Gi}[k] = \frac{IC_i[k] - \beta_i}{2\alpha_i}, \text{ if } P_{Gi,min} \leq P_{Gi}[k] \leq P_{Gi,max}, \quad (8.16)$$

$$P_{Gi}[k] = P_{Gi,max}, \text{ if } P_{Gi}[k] > P_{Gi,max}. \quad (8.17)$$

The mathematical formulation of the ICC algorithm is Equations 8.12 through 8.17. By following these update rules, the IC of all the generation units will converge to a common system IC asymptotically. Thus, the EDP is solved iteratively in a distributed way. The local controller of each generation unit will set its generation power output based on the converged IC value in the system.

8.3 Adversary Scenarios

In the ICC algorithm, all the generation units are assumed to cooperate and follow the IC update rule exactly. However, if one generation unit in the system is compromised by adversaries and does not follow the update rule, as shown in Figure 8.1, the consensus of the ICs cannot be guaranteed to be reached. Due to its distributed nature, the ICC algorithm is prone to the misbehaving generation units.

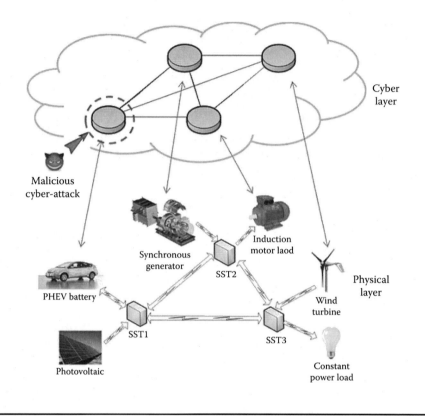

Figure 8.1: Vulnerability of distributed energy management systems in the smart grid.

For the simplicity of discussion, let us assume (1) all the adversaries are independent without colluding with each other, and (2) the leader generation unit is never compromised.

The misbehaving generation unit in the power system is modeled as one whose local distributed controller suffers from a malicious cyber-attack launched by adversaries. The cyber-attack is modeled as an exogenous input to the generation unit's IC calculation. Thus, the IC update rule of the ith generation unit being compromised at the kth time-step is formulated as

$$IC_i[k+1] = \sum_{i=1}^{n} d_{ij}IC_j[k] + b_{mi}u_{mi}[k], \qquad (8.18)$$

where u_{mi} is the adversary input and b_{mi} is the input coefficient.

By the definition above, when there are no cyber-attacks, the corresponding adversary input u_{mi} is equal to zero. When the generation unit is compromised, it is allowed to update its IC value in an arbitrary way with different adversary input

u_{mi}. Therefore, depending on the capabilities of different cyber-attacks [25], two adversary scenarios are described in this chapter.

8.3.1 Fault Attack

The first type of cyber-attack is denoted as the fault attack. It simply aims to stop the IC updating of the compromised generation unit, having it remain a constant value toward its own benefit. The compromised generation unit will act as a faulty generator in this adversary scenario. The adversary input of the fault attack is modeled as

$$u_{mi}[k] = c - \sum_{i=1}^{n} d_{ij} IC_j[k], \tag{8.19}$$

where c is an arbitrary constant value and $c \in \mathbb{R}$.

8.3.2 Random Attack

The second type of cyber-attack is denoted as the random attack. It injects random false data into the local distributed controller of a compromised generation unit and makes its IC value change in an arbitrary manner. The adversary can lead all the generation units to operate at a nonoptimal, even unstable operating point by launching such an attack. Thus, the adversary input of the random attack is

$$u_{mi}[k] = c[k], \tag{8.20}$$

where $c[k]$ equals any arbitrary value that the adversary injects and $c[k] \in \mathbb{R}$.

8.4 Reputation-Based Resilient Distributed Energy Management Algorithm

The reputation-based resilient distributed energy management algorithm is developed to guarantee an accurate IC computation for EDP in the presence of misbehaving generation units. The algorithm embeds four extra steps (detection, mitigation, isolation, and adaption) of security mechanisms as a feedback loop into the conventional distributed energy management algorithms (i.e., the ICC algorithm in this chapter) and achieves attack resilience in a distributed fashion. At each step, every generation unit only uses local and two-hop neighbors' information to detect and identify the misbehaving units. As a result, all the well-behaving generation units will isolate the misbehaving units and reach the correct consensus state asymptotically.

A detailed description of each step is given below.

8.4.1 Detection Step

In the detection step, each generation unit will not only send its own IC information to the neighboring generation units, but also relay its neighbors' IC information to all the other neighbors. Therefore, all the generation units will receive
and utilize their two-hop neighbors' information to make verification.

Similar to the watchdog mechanism in the wireless sensor networks, each
generation unit executes a real-time anomaly detection mechanism locally for
all its neighboring generation units to monitor their shared IC information. For
example, the ith generation unit first calculates its neighbor j's kth time-step IC
value $IC_{ij}[k]$ redundantly using its two-hop neighbors' IC information based on
Equation 8.12 and then stores it in time-step $k-1$. In time-step k, it compares
$IC_{ij}[k]$ with the IC information $IC_j[k]$ sent by the neighboring generation unit j
and determines the correctness of the received information with the following
criteria:

1. If $|IC_{ij}[k] - IC_j[k]| \leq \zeta(k, \delta)$, the neighboring generation unit is behaving
   normally.

2. If $|IC_{ij}[k] - IC_j[k]| > \zeta(k, \delta)$, the neighboring generation unit is behaving
   anomalously.

$\zeta(k, \delta)$ is the detection threshold function. It depends on the time-step k and
the disturbance δ. Then, the following anomaly detection rule is executed:

$$G_{ij}[k] = \begin{cases} G_{ij}[k-1] + 1, |IC_{ij}[k] - IC_j[k]| \leq \zeta(k, \delta), \\ G_{ij}[k-1], |IC_{ij}[k] - IC_j[k]| > \zeta(k, \delta), \end{cases} \tag{8.21}$$

where G_{ij} is the number of times that the ith generation unit received verified
correct IC information from neighbor j up to the kth time-step.

8.4.2 Mitigation Step

In the mitigation step, a reputation metric is used to quantitatively measure the
credibility of the neighboring generation units. Reputation is widely used as an
index to indicate the credibility of a node in the *ad hoc* network [21]. Given a set
of verified correct and incorrect information from a neighbor, the probability of
receiving a particular combination of correct and incorrect information from it
follows a beta distribution [7]. Thus, the reputation metric can indicate how much
trust a generation unit should put in another unit's information. The Bayesian
reputation function is used here to calculate the reputation metric. It is given by
the ratio of the number of correct verified IC information to the number of total
received information:

$$rep_{ij}[k] = \frac{\eta G_{ij}[k] + 1}{\eta k + 2}, \tag{8.22}$$

where $rep_{ij}[k]$ is the reputation value of generation unit j in the ith generation unit's local distributed controller in time-step k, and η is the reputation coefficient that adjusts the reputation updating rate.

If the neighboring generation unit j always sends out the correct IC information, its reputation value in the ith generation unit is equal to 1.0. As it begins providing false IC information, its reputation value will start to drop, and then the ith generation unit can react accordingly.

8.4.3 Isolation Step

In the isolation step, the identified misbehaving generation unit is isolated from the rest of the system to avoid further system performance deterioration. When the generation unit j continues sending false IC information consecutively for a certain period of time, its rep_{ij} will fall below a certain threshold rep_{th} and it will be considered malicious; as a result, its neighbors will reject all the information it provides (i.e., rep_{ij} is set to 0). Meanwhile, if generation unit j is isolated by its neighbors, the distributed controllers of its neighboring generation units will trigger the circuit breakers to isolate it in the physical system layer as well.

Nevertheless, a timeout mechanism could be used to allow the misbehaving generation units to rejoin the system after it behaves normally again.

8.4.4 Adaption Step

The final step to close the feedback loop for the security mechanism is to adapt the consensus computation weights d_{ij} in the matrix D_n according to the reputation metric. The neighboring generation units that have higher reputation values should be weighted more. Therefore, the update rule for the weights of the ith generation unit is defined as

$$d_{ij}[k] = \frac{rep_{ij}[k]}{\sum_{h=1}^{|V_i|} rep_{ih}[k]}, \tag{8.23}$$

where V_i is the set of neighbors of the ith generation unit and $|V_i|$ is the cardinality of set V_i.

In Equation 8.23, the consensus computation weights are adaptively updated in each iteration. Accordingly, the update rule of the follower generation units in Equation 8.12 is then changed to

$$IC_i[k+1] = \sum_{i=1}^{n} d_{ij}[k]IC_j[k]. \tag{8.24}$$

Besides, in order to satisfy the system power balance constraint when excluding the power generation of the compromised generation units from the system

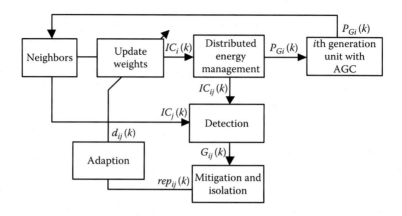

Figure 8.2: Block diagram of the reputation-based resilient distributed energy management algorithm for the *i*th generation unit.

after isolating them, the update rule for the leader generation unit in Equation 8.14 is also changed to

$$IC_i[k+1] = \sum_{i=1}^{n} d_{ij}[k]IC_j[k] + \epsilon(\Delta P[k] + P_{Gm}[k]),\qquad(8.25)$$

where P_{Gm} is the total power generation of the misbehaving generation units when they are being isolated.

In summary, by incorporating the four steps of security mechanisms discussed above, the reputation-based resilient distributed energy management algorithm is able to improve the resilience of the energy management system to adversaries by identifying and isolating the compromised generation units in the system in a distributed fashion. The block diagram of the algorithm is shown in Figure 8.2.

8.5 Simulation Case Studies

In this section, the IEEE 30-bus system from the Matlab MATPOWER library is used to demonstrate the effectiveness of the reputation-based resilient distributed energy management algorithm by solving the EDP under different adversary scenarios. Meanwhile, the EDP for the IEEE 30-bus system is also solved using the conventional centralized quadratic programming (QP) method as the reference to verify the correctness of IC solutions calculated by the reputation-based resilient distributed energy management algorithm. The sensitivities of the algorithm to several important algorithm coefficients and different communication topologies are also analyzed to illustrate the robustness of the algorithm.

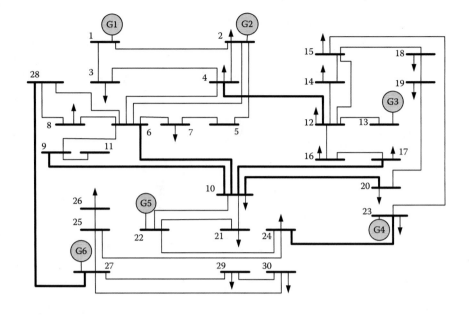

Figure 8.3: One-line diagram of the IEEE 30-bus system.

The one-line diagram of the 30-bus system is shown in Figure 8.3. It contains 6 generation units and 20 loads. The total demand of loads is 189.2 MW. The cost function parameters and generation limits of each generation unit are shown in Table 8.1. Figure 8.4 shows the communication topology among six generation units. Generation unit 1 is set as the leader of the system. All the simulation parameters are shown in Table 8.2.

Table 8.1   Cost Function Parameters of Generation Units in the IEEE 30-Bus System

Unit	α_i	β_i	γ_i	$P_{Gi,max}$ (MW)	$P_{Gi,min}$ (MW)
1	0.02	2	0	80	0
2	0.0175	1.75	0	80	0
3	0.0625	1	0	50	0
4	0.00834	3.25	0	55	0
5	0.025	3	0	30	0
6	0.025	3	0	40	0

Table 8.2   Simulation Parameters

Simulation Time (s)	Sampling Time (s)	ϵ	rep_{th}	η
30	0.1	0.001	0.4	5

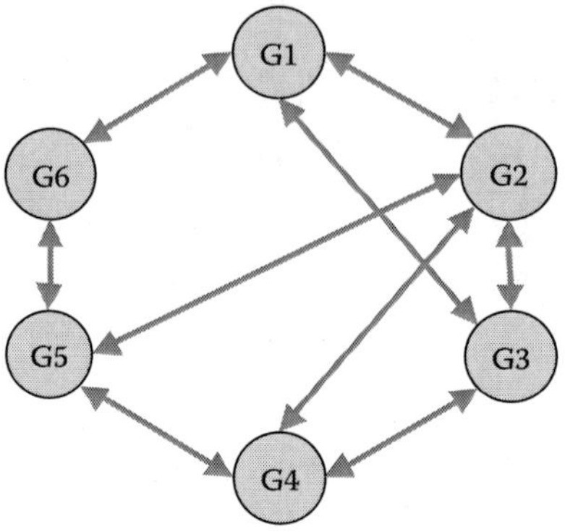

Figure 8.4: Communication topology of six generation units in the IEEE 30-bus system.

8.5.1   Single Compromised Generation Unit

In the first case study, we test the adversary scenario where one single generation unit is compromised in the system. Let us assume generation unit 5 is compromised by malicious cyber-attack when time $t \geq 2$ s ($k \geq 20$). For reference, we first use QP to solve this EDP when the total system power demand is 189.2 MW. The true optimal IC value of the rest of the system after the isolation of generation unit 5 is 3.901 $/MWh.

8.5.1.1   Fault Attack on a Single Generation Unit

A fault attack is injected on generation unit 5 when time $t \geq 2$ s. The simulation result in Figure 8.5 illustrates that the ICs calculated by the ICC algorithm do not converge when generation unit 5 is compromised. The ICs of all six generation units stay at different values, and thus the overall 30-bus system operates at a nonoptimal condition. In Figure 8.6, the reputation-based resilient distributed energy management algorithm is used in the same adversary scenario. The result shows that all the ICs proceed toward consensus initially, and then the compromised unit (5) stops updating its IC and diverges from the rest of the ICs. After

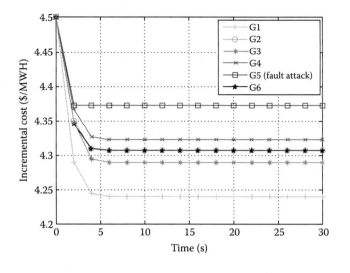

Figure 8.5: ICs of six generation units using the ICC algorithm when generation unit 5 is compromised by a fault attack.

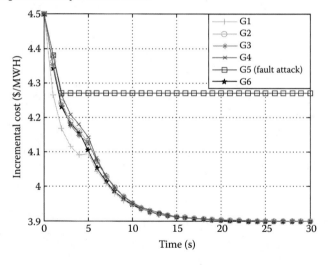

Figure 8.6: ICs of six generation units using the reputation-based resilient distributed energy management algorithm when generation unit 5 is compromised by a fault attack.

3 s of mitigation, the well-behaving generation units eventually determine that generation unit 5 is malicious, and therefore isolate it in the system. The remaining five generation units then proceed to achieve the convergence to the optimal IC (3.901 $/MWh), validated by the reference solution provided by QP.

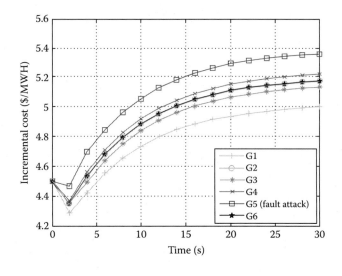

Figure 8.7: ICs of six generation units using the ICC algorithm when generation unit 5 is compromised by a random attack.

8.5.1.2   *Random Attack on a Single Generation Unit*

A random attack $c[k] = 0.2$ is injected on generation unit 5 when time $t \geq 2$ s in this case. Figures 8.7 and 8.8 show that the convergence of the ICC algorithm fails when generation unit 5 is compromised, while the reputation-based resilient distributed energy management algorithm successfully detects and isolates the misbehaving generation unit and allows the remaining five generation units to converge to the optimal IC value.

8.5.2   *Multiple Compromised Generation Units*

Let us test a different adversary scenario where multiple generation units are compromised in the second case study. Assume generation unit 4 is compromised by a malicious cyber-attack when time $t \geq 2$ s $(k \geq 20)$ and generation unit 5 is compromised by another malicious cyber-attack when time $t \geq 5$ s $(k \geq 50)$. For reference purpose, the QP gives the optimal IC of the remaining four generation units as 4.379 \$/MWh after isolating the compromised generation units.

8.5.2.1   *Fault Attack on Multiple Generation Units*

Two fault attacks are injected on generation units 4 and 5 when $t \geq 2$ s and $t \geq 5$ s, respectively. Figure 8.9 shows the simulation result of the reputation-

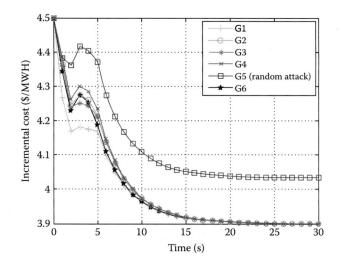

Figure 8.8: ICs of six generation units using the reputation-based resilient distributed energy management algorithm when generation unit 5 is compromised by a random attack.

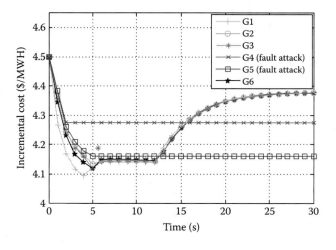

Figure 8.9: ICs of six generation units using the reputation-based resilient distributed energy management algorithm when generation units 4 and 5 are compromised by fault attacks.

based resilient distributed energy management algorithm in this scenario. With the embedded security mechanisms in place, the algorithm is able to isolate the two compromised generation units in sequence, and the rest of the system gradually converges to the optimal point as desired.

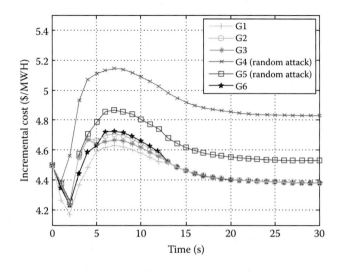

Figure 8.10: ICs of six generation units using the reputation-based resilient distributed energy management algorithm when generation units 4 and 5 are compromised by random attacks.

8.5.2.2 Random Attack on Multiple Generation Units

A random attack $c[k] = 0.3$ is injected on generation unit 4 when time $k \geq 20$, and another random attack $c[k] = 0.1$ is launched on generation unit 5 when $k \geq 50$. The convergence of the reputation-based resilient distributed energy management algorithm in the face of these two attacks is illustrated in Figure 8.10. As we can see, after isolating the comprised generation units (4 and 5), the remaining four generation units reach the optimal operating point ($IC = 4.379$ \$/MWh).

8.5.3 Sensitivity Analysis

The previous case studies demonstrate that the reputation-based resilient distributed energy management algorithm is effective in mitigating and isolating the compromised misbehaving generation units and guarantees that the well-behaving units will operate at the optimal condition. The following case studies focus on analyzing the sensitivity of the algorithm to two important coefficients and different communication topologies.

8.5.3.1 Effect of ICC Convergence Coefficient

In the ICC algorithm, the convergence rate can be controlled by adjusting the convergence coefficient ϵ. Thus, the convergence speed of the reputation-based resilient distributed energy management algorithm is also affected by the choice of ϵ.

Figures 8.11 and 8.12 show two different convergence rates using different values for ϵ under the single fault attack scenario. With the same sampling time, the system using $\epsilon = 0.005$ converges much faster than the system using $\epsilon = 0.0005$. Thus, a larger ϵ may lead to a faster convergence speed. However, Figure 8.11 also shows that the IC convergence has a larger oscillation when $\epsilon = 0.005$. Therefore, if ϵ is larger than a certain upper limit, it will cause the system to become unstable. The choice of ϵ value will depend on the generation unit dynamics, system and communication topology, and so forth.

8.5.3.2   Effect of Reputation Threshold

Figures 8.13 and 8.14 show the simulation results of the single fault attack scenario, but using two different reputation thresholds. As we can see, the system with a smaller reputation threshold $rep_{th} = 0.1$ reaches the IC consensus about 10 s slower than the system with $rep_{th} = 0.2$. This result suggests that the convergence rate of the reputation-based resilient distributed energy management algorithm is also affected by the value of the reputation threshold. A higher reputation threshold results in faster algorithm convergence; however, it may also lead to false accusation. The choice of reputation threshold depends on the trade-off between the convergence rate and the detection accuracy.

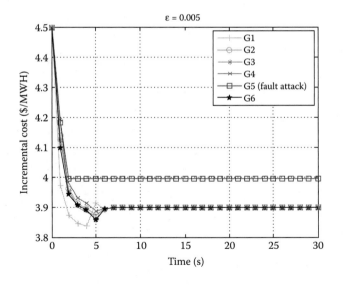

Figure 8.11: ICs of six generation units using the reputation-based resilient distributed energy management algorithm when $\epsilon = 0.005$.

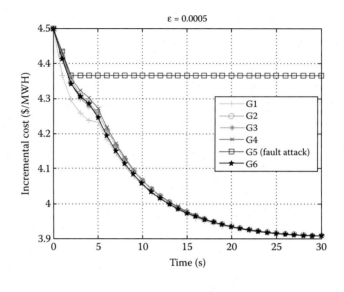

Figure 8.12: ICs of six generation units using the reputation-based resilient distributed energy management algorithm when $\epsilon = 0.0005$.

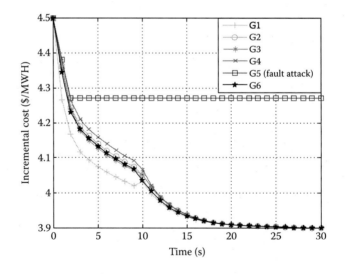

Figure 8.13: ICs of six generation units using the reputation-based resilient distributed energy management algorithm when $rep_{th} = 0.2$.

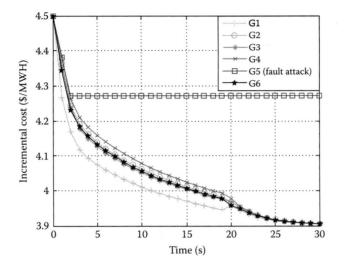

Figure 8.14: ICs of six generation units using the reputation-based resilient distributed energy management algorithm when $rep_{th} = 0.1$.

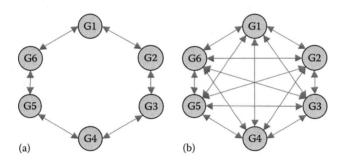

Figure 8.15: Six generation units with two different communications topologies: (a) ring topology and (b) fully connected topology.

8.5.3.3   Effect of Communication Topology

Different communication topologies also affect the convergence of the reputation-based resilient distributed energy management algorithm. In this case study, two communication topologies—ring topology and fully connected topology—are both tested for the six generation units in the system, as shown in Figure 8.15.

Figures 8.16 and 8.17 show the simulation results for the single random attack scenario while using the two different communication topologies in Figure 8.15.

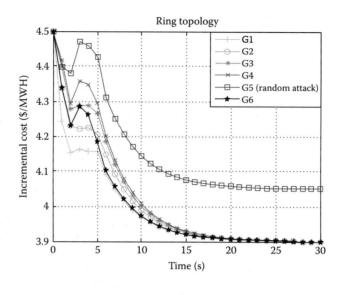

Figure 8.16: ICs of six generation units using the reputation-based resilient distributed energy management algorithm with the ring communication topology.

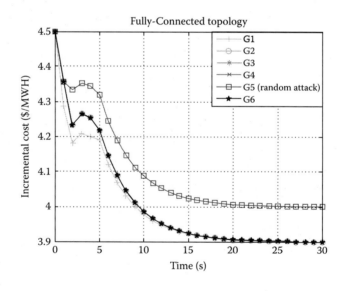

Figure 8.17: ICs of six generation units using the reputation-based resilient distributed energy management algorithm with the fully connected communication topology.

The results show that the fully connected topology has a faster convergence rate than the ring topology, and it also has fewer oscillations during the convergence process. Since the fully connected topology has a larger algebraic connectivity than the ring topology, the algebraic connectivity of the communications network may also be used to infer the speed, as well as the smoothness, of the algorithm convergence.

8.6 Conclusions

The vulnerability of distributed energy management algorithms in the smart grid to malicious cyber-attacks has been explored in this chapter. As an example, the ICC algorithm has been used to solve the classic EDP in a distributed manner using the consensus framework in the presence of adversaries. The malicious cyber-attacks have been modeled as two types of exogenous inputs to represent two types of adversary scenarios: (1) fault attack and (2) random attack. In both scenarios, the compromised generation units in the system affected other units and caused the ICC algorithms to fail.

In order to improve the resilience of distributed energy management systems, a reputation-based resilient distributed energy management algorithm has been introduced to ensure accurate IC convergence by identifying and isolating the misbehaving generation units during the IC computation process. Unlike the conventional supervisory monitoring schemes that are prone to a single point of failure, the reputation-based resilient distributed energy management algorithm realizes a fully distributed detection structure where each generation unit monitors its neighbors. Furthermore, it closes the control loop by embedding the anomaly detection mechanism as feedback inside the IC computation process, inherently enabling its resilience to misbehaving generation units. The effectiveness of the algorithm under the different adversary scenarios has been demonstrated in the IEEE 30-bus system. The sensitivity of the algorithm convergence to the ICC convergence coefficient, reputation threshold, and different communication topologies has also been discussed through simulation results.

References

1. A. Banerjee, K. Venkatasubramanian, T. Mukherjee, and S. Gupta. Ensuring safety, security, and sustainability of mission-critical cyber-physical systems. *Proceedings of the IEEE*, 100(1):283–299, 2012.

2. A. Boukerche, A. Zarrad, and R. Araujo. A cross-layer approach-based gnutella for collaborative virtual environments over mobile ad hoc networks. *IEEE Transactions on Parallel and Distributed Systems*, 21(7):911–924, 2010.

3. D. Choi and L. Xie. Ramp-induced data attacks on look-ahead dispatch in real-time power markets. *IEEE Transactions on Smart Grid*, 4(3):1235–1243, 2013.

4. G. Ericsson. Cyber security and power system communication: essential parts of a smart grid infrastructure. *IEEE Transactions on Power Delivery*, 25(3):1501–1507, 2010.

5. R. Gupta and M.-Y. Chow. Networked control system: Overview and research trends. *IEEE Transactions on Industrial Electronics*, 57(7):2527–2535, 2010.

6. M. Ilic, L. Xie, U. Khan, and J. Moura. Modeling of future cyber-physical energy systems for distributed sensing and control. *IEEE Transactions on Systems, Man and Cybernetics, Part A: Systems and Humans*, 40(4):825–838, 2010.

7. R. Ismail and A. Josang. The beta reputation system. In *Proceedings of Bled Conference on Electronic Commerce*, p. 41, 2002.

8. A. Jadbabaie, J. Lin, and A. Morse. Coordination of groups of mobile autonomous agents using nearest neighbor rules. *IEEE Transactions on Automatic Control*, 48(6):988–1001, 2003.

9. H. Khurana, M. Hadley, L. Ning, and D. Frincke. Smart-grid security issues. *IEEE Security and Privacy*, 8(1):81–85, 2010.

10. T. Kim and H. Poor. Strategic protection against data injection attacks on power grids. *IEEE Transactions on Smart Grid*, 2(2):326–333, 2011.

11. O. Kosut, L. Jia, R. Thomas, and L. Tong. Malicious data attacks on the smart grid. *IEEE Transactions on Smart Grid*, 2(4):645–658, 2011.

12. D. Kushner. The real story of Stuxnet. *IEEE Spectrum*, 50(3):48–53, 2013.

13. R. Langner. Stuxnet: Dissecting a cyberwarfare weapon. *IEEE Security and Privacy*, 9(3):49–51, 2011.

14. Y. Liu, P. Ning, and M. Reiter. False data injection attacks against state estimation in electric power grids. *ACM Transactions in Information and Systems Security*, 14(1), 2011.

15. R. Mudumbai, S. Dasgupta, and B. Cho. Distributed control for optimal economic dispatch of a network of heterogeneous power generators. *IEEE Transactions on Power Systems*, 27(4):1750–1760, 2012.

16. R. Olfati-Saber, J. Fax, and R. Murray. Consensus and cooperation in networked multi-agent systems. *Proceedings of the IEEE*, 95(1):215–233, 2007.

17. W. Ren and R. Beard. *Distributed Consensus in Multi-Vehicle Cooperative Control: Theory and Applications*. 1st ed. Springer, Berlin, 2010.

18. W. Ren, W. Beard, and M. Atkins. A survey of consensus problems in multi-agent coordination. In *Proceedings of American Control Conference*, Portland, OR, vol. 3, pp. 1859–1864, 2005.

19. S. Sridhar and M. Govindarasu. Model-based attack detection and mitigation for automatic generation control. *IEEE Transactions on Smart Grid*, 5(2):580–591, 2014.

20. S. Sridhar, A. Hahn, and M. Govindarasu. Cyber-physical system security for the electric power grid. *Proceedings of the IEEE*, 100(1):210–224, 2012.

21. S. Sundaram, J. Chang, K. Venkatasubramanian, C. Enyioha, I. Lee, and G. Pappas. Reputation-based networked control with data-corrupting channels. In *Proceedings of International Conference on Hybrid Systems: Computation and Control*, Chicago, IL, pp. 291–300, 2011.

22. L. Xie, Y. Mo, and B. Sinopoli. Integrity data attacks in power market operations. *IEEE Transactions on Smart Grid*, 2(4):659–666, 2011.

23. Y. Xu and W. Liu. Stable multi-agent-based load shedding algorithm for power systems. *IEEE Transactions on Power Systems*, 26(4):2006–2014, 2011.

24. Y. Yuan, Z. Li, and K. Ren. Modeling load redistribution attacks in power systems. *IEEE Transactions on Smart Grid*, 2(2):382–390, 2011.

25. W. Zeng and M.-Y. Chow. A reputation-based secure distributed control methodology in D-NCS. *IEEE Transactions on Industrial Electronics*, 61(11):6294–6303, 2014.

26. W. Zeng and M.-Y. Chow. Resilient distributed control in the presence of misbehaving agents in networked control systems. *IEEE Transactions on Cybernetics*, 44(11):2038–2049, 2014.

27. Z. Zhang and M.-Y. Chow. Convergence analysis of the incremental cost consensus algorithm under different communication network topologies in a smart grid. *IEEE Transactions on Power Systems*, 27(4):1761–1768, 2012.

28. W. Zeng, Y. Zhang, and M.-Y. Chow. Resilient distributed energy management subject to unexpected misbehaving generation units. *IEEE Transactions on Industrial Informatics*, 2015. DOI: 10.1109/TII.2015.2496228.

SECURITY ISSUES IN APPLICATION FIELDS

III

Chapter 9

Distributed Resilient Control of Operator–Vehicle Networks under Cyber Attacks

Minghui Zhu

Department of Electrical Engineering, Pennsylvania State University

Sonia Martínez

Department of Mechanical and Aerospace Engineering, University of California

CONTENTS

9.1   Introduction

The widespread deployment of smart communication, computing, and actuating devices has strengthened the coupling between the cyber and physical worlds, resulting in the generation of cyber-physical systems (CPS). Stronger cyber-physical integration produces new functions for control systems and new opportunities to improve their operating performance. However, cyber-physical integration also poses new security vulnerabilities. In particular, information and communications technology (ICT) systems are directly connected to control systems. Adversaries can exploit inherent vulnerabilities of ICT systems to compromise control systems and cause damages in the physical world. In addition, many CPS are large-scale systems where components are distributed over a geographically extended region. Wide-area deployment enlarges attack surfaces, and adversaries can potentially break down the whole network via individual node attacks. CPS play a role of vital importance in a number of engineering systems, ranging from the smart grid to autonomous vehicles. Hence, there is a pressing need to develop a holistic set of technologies to guarantee the security of CPS. From the point of view of decision making and control, attack detection, attack-resilient control, and security economics are three important aspects.

1. *Attack detection.* In ICT systems, intrusion detection systems are used to detect abnormal behavior in cyber space. However, current intrusion detection systems are not built for CPS and may not be able to detect misbehavior in the physical world. Hence, physical detectors are needed to complement existing intrusion detection systems. There has been recent work on attack detection of CPS [2, 10, 17, 19, 21, 22]. In particular, an

attack detection problem is formulated as unknown input estimation in control theory [19]. In [10], the problem is mapped into an ℓ_0 optimization problem, which is NP-hard. Thus, an ℓ_1 relaxation of the problem to a convex problem is considered for computational tractability. An extension of this approach to include disturbances with known bounds is given in [17]. In [22], a class of input-state filters is designed to detect switching attacks in stochastic dynamic systems.

2. *Attack-resilient control.* Current cyber-defense mechanisms are governed by slow and deliberative processes such as testing and patching. Apparently, this slow-time-scale paradigm mismatches the fast changes of physical systems. This demands new schemes to guarantee CPS resilience. That is, control systems are to be enabled to autonomously react to malicious attacks, protect physical systems from failure, and minimize the degradation of system performance before cyber vulnerabilities are fixed. Papers [1] and [13] formulate finite-horizon linear-quadratic-Gaussian (LQG) control problems as dynamic zero-sum games between a controller and jammer pair. Paper [30] utilizes game theory to develop a hierarchical framework and ensure cross-layer security. Event-triggered control under denial-of-service (DoS) attacks is studied in [12]. Our papers [27] and [28] discuss distributed attack-resilient formation control of multiple vehicles against DoS, replay, and deception attacks.

3. *Security economics.* Many CPS are large scale, and their components belong to heterogeneous stakeholders. The interests of stakeholders may not be aligned with systemwide security. One of the most basic principles of security is that no security solution is ultimately stronger than its weakest link. To prevent this, one solution is to provide stakeholders with certain incentives, for example, monetary rewards, to increase their contributions to CPS security. Some papers along this line include [3] and [29]. In [3], the authors investigate the security independency in infinite-horizon LQG against DoS attacks, and fully characterize the equilibrium of the induced game of security investment. In our paper [29], a class of competitive and cooperative resource allocation problems for resilient networked control systems is formulated.

9.1.1 Statement of Contributions

The current chapter contributes to the area of attack-resilient control. In particular, we investigate the distributed attack-resilient control of operator–vehicle networks, an important class of CPS. We present a novel formulation of the problem of distributed constrained formation control against replay attacks. In particular, each vehicle is remotely controlled by an operator and its actuation is limited. Vehicles aim to reach the desired formation within a given constraint set through real-time coordination with operators. Each operator–vehicle pair is attacked by an adversary, who is able to produce replay attacks by maliciously and consecutively repeating the control commands for a period of time. The information operators know about their opponents is limited and restricted to the maximum number, say τ_{max}, of consecutive attacks each adversary is able to launch. We focus on the design of distributed algorithms, which ensures mission completion in the presence of replay attacks. To achieve this goal, we come up with a novel distributed formation control algorithm that is based on receding-horizon control and leverages the idea of motion toward target points. We show that the input and state constraints are always enforced, and the desired formation can be asymptotically achieved provided that the union of communication graphs between operators satisfies certain connectivity assumptions. Under the same set of conditions, our proposed algorithm shows an analogous resilience to DoS attacks.

9.1.2 Organization

In Section 9.2, we formulate the problem investigated in the chapter. Section 9.3 develops a set of preliminary results. In Section 9.4, we present the main results of the chapter, including the formal description of our algorithm and the summary of its resilience properties. This is followed by the complete analysis of the algorithm in Section 9.5. In Section 9.6, we provide a set of remarks and numerical examples to illustrate the algorithm performance. Section 9.7 includes the conclusions and future work remarks.

9.2 Problem Formulation

In this section, we first present the architecture of the operator–vehicle network and the distributed constrained formation control problem of interest. After that, we introduce the model of replay attackers considered in the chapter. This is followed by a description of the prior knowledge operators possess about their rivals and the objective of this chapter.

9.2.1 Operator–Vehicle Network

Consider a group of vehicles in $\mathbb{R}^d$, for some $d \in \mathbb{Z}_{>0}$, labeled by $i \in V \triangleq \{1, \ldots, N\}$. The dynamics of each vehicle is governed by the following second-order, discrete-time dynamic system:

$$p_i(k+1) = p_i(k) + v_i(k), \quad v_i(k+1) = v_i(k) + u_i(k), \tag{9.1}$$

where $p_i(k) \in X \subseteq \mathbb{R}^d$ (respectively, $v_i(k) \in \mathbb{R}^d$) is the position (respectively, the velocity) of vehicle i, and $u_i(k) \in U \subseteq \mathbb{R}^d$ then stands for its input. Throughout the chapter, we assume the following on the constraint sets:

Assumption 9.1 (constraint sets).   The state constraint set X is convex and compact. The input constraint set U is a box; that is, $U \triangleq \{u \in \mathbb{R}^d \mid \|u\|_\infty \leq u_{\max}\}$* for some $u_{\max} > 0$.

Each vehicle i is remotely maneuvered by an operator i, and this assignment is one-to-one and fixed. Each vehicle is able to identify its location and velocity and send this information to its operator through a communication network. Within the vehicle team, vehicles cannot communicate with each other. Each operator can, on the one hand, exchange information with neighboring operators and, on the other hand, deliver control commands to her associated vehicle via the communication network. See Figure 9.1 for a sketch of the operator–vehicle network. The mission of the operator–vehicle network is to achieve a desired formation that is characterized by the formation digraph $\mathcal{G}^F \triangleq (V, \mathcal{E}^F)$. Each edge $(i,j) \in \mathcal{E}^F \subseteq V \times V \setminus \text{diag}(V)$, starting from vehicle j and pointing to vehicle i, is associated with a formation vector $v_{ij} \in \mathbb{R}^d$. Throughout this chapter, we impose the following on $\mathcal{G}^F$:

Assumption 9.2 (formation digraph).   The formation digraph $\mathcal{G}^F$ is strongly connected; that is, for any pair of $(j,i) \in \mathcal{E}^F$, there is a directed path starting from i and ending up with j.

Being a member of the team, each operator i is only aware of local formation vectors; that is, v_{ij} for $j \in \mathcal{N}_i$, where $\mathcal{N}_i \triangleq \{j \in V \setminus \{i\} \mid (j,i) \in \mathcal{E}^F\}$. The multivehicle constrained formation control mission can be formulated as a team optimization problem where each global optimum corresponds to the formation of interest. In particular, the constrained formation control problem is encoded as the following quadratic program:

$$\min_{p \in X^N} \sum_{(i,j) \in \mathcal{E}^F} \|p_i - p_j - v_{ij}\|^2,$$

whose solution set, denoted as $X^* \subseteq X^N$, satisfies the following:

Assumption 9.3 (feasibility).   The optimal solution set X^* is nonempty.

We assume that operators and vehicles are synchronized by using a single clock. The interconnection between operators at time $k \geq 0$ will be represented by a directed graph $\mathcal{G}(k) = (V, \mathcal{E}(k))$, where $\mathcal{E}(k) \subseteq \mathcal{E}^F$ is the set of edges. Here, $(i,j) \in \mathcal{E}(k)$ if and only if operator i is able to receive the message from operator j at time k. Denote by $\mathcal{N}_i(k) \triangleq \{j \in V \mid (i,j) \in \mathcal{E}(k)\}$ the set of (in-)neighboring operators of operator i at time k. In order to achieve network-wise objectives,

---

*In this chapter, the notation of $\|\cdot\|$ (respectively, $\|\cdot\|_\infty$) stands for the 2-norm (respectively, ∞-norm) of vectors.

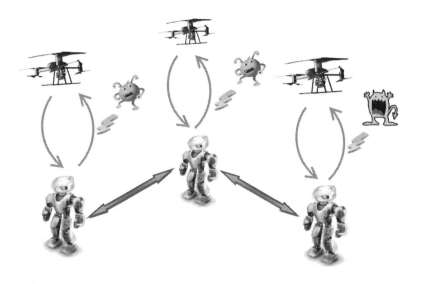

Figure 9.1: Architecture of the operator–vehicle adversarial network where the operator is represented by the humanoid robot.

interoperator topologies should be sufficiently connected such that decisions of any operator can eventually affect any other one. This is formally stated in the following assumption:

Assumption 9.4 (periodic communication).   There is a positive integer B such that, for any $k \geq 0$, $(i,j) \in \bigcup_{s=0}^{B-1} \mathcal{E}(k+s)$ for any $(i,j) \in \mathcal{E}^F$.

A direct result of Assumptions 9.2 and 9.4 is that $\bigcup_{s=0}^{B-1} \mathcal{E}(k+s)$ is a superset of $\mathcal{E}^F$, and thus $(V, \bigcup_{s=0}^{B-1} \mathcal{E}(k+s))$ is strongly connected.

9.2.2   Model of Adversaries

We now set out to describe the attacker model considered in the chapter. A group of N adversaries tries to abort the mission of achieving formation in X. An adversary is allocated to attack a specific operator–vehicle pair, and this assignment is fixed over time. Thus, we identify adversary i with the operator–vehicle pair i. In this chapter, we consider the class of *replay attacks* where the packages transmitted from operators to vehicles are maliciously repeated by adversaries. In particular, each adversary i is associated with memory storing past information, and its state is denoted by $M_i^a(k)$. If she launches a replay attack at time k, adversary i executes the following: (1) erases the data sent from operator i; (2) delivers the past control command stored in her memory, $M_i^a(k)$, to vehicle i; and (3) keeps the state of the memory, that is, $M_i^a(k+1) = M_i^a(k)$. In this case, $s_i^a(k) = 1$ indicates the occurrence of a replay attack, where the auxiliary variable $s_i^a(k) \in \{0,1\}$. If she does not produce any replay attack at time k, adversary i intercepts the data, say u_i, sent from operator i and stores it in her memory; $M_i^a(k+1) = u_i$. In this case, $s_i^a(k) = 0$ and u_i is successfully received by vehicle i. Without loss of any generality, we assume that $s_i^a(0) = 0$.

We define the variable $\tau_i^a(k)$ with initial state $\tau_i^a(0) = 0$ to indicate the consecutive number of attacks. The evolution of $\tau_i^a(k)$ is determined in the following way: if $s_i^a(k) = 1$, then $\tau_i^a(k) = \tau_i^a(k-1) + 1$; otherwise, $\tau_i^a(k) = 0$. It is noted that $\tau_i^a(k)$ is reset to zero when adversary i does not replay the data at time k. Hence, for any k with $s_i^a(k) = 0$, $\tau_i^a(k)$ represents the number of consecutive attacks produced by adversary i up to time k since the largest $0 \le k' < k$ with $s_i^a(k') = 0$.

Each adversary needs to spend a certain amount of energy to launch a replay attack. We assume that the energy of adversary i is limited, and adversary i is only able to launch at most $\tau_{\max} \ge 1$ consecutive attacks; see Assumption 9.5.

Assumption 9.5 (maximum number of consecutive attacks).   There is $\tau_{\max} \ge 1$ such that $\max_{i \in V} \sup_{k \ge 0} \tau_i^a(k) \le \tau_{\max}$.

Remark 9.1.   Replay attacks have been successfully used in the past and show a number of advantages to an adversary. Stuxnet was the latest cyber attack to control systems. In this accident, Stuxnet exploited replay attacks to compromise a nuclear facility; see [9, 16]. Replay attacks (and DoS attacks in Section 9.6) do not require any information of the operator–vehicle network and the algorithm exploited. This is in contrast to false data injection in [15, 18] and deception attacks in [2, 24, 25]. From the point of view of adversaries, replay attacks (and DoS attacks) are easier to launch, and thus more preferable when they lack the information of the target control systems. Note that replay attacks are less sophisticated than deception attacks and false data injection. However, the discussion in Section 9.2.3 demonstrates that replay attacks are still capable of making a mission fail if they are not explicitly taken into account in the algorithm design.

Finally, in comparison with DoS attacks, deception attacks, and false data injection, replay attacks demand more memory to store intercepted information.

Remark 9.2   For the ease of presentation, we assume that only the links from operators to vehicles are compromised. Our proposed algorithm can be readily applied to the scenario where the links from vehicles to operators are attacked.

9.2.3   *Motivating Scenario*

In this section, we use a simple scenario to illustrate the failure of the classic formation control algorithm under replay attacks. For the ease of presentation, we consider the following special case: (1) $\nu_{ij} = 0$; (2) the vehicle dynamics is first order; and (3) the input and state constraints are absent, that is, $X = U = \mathbb{R}$. The special case is the consensus or rendezvous problem that has been extensively studied.

The classic consensus algorithm, for example, in [7], is rephrased to fit in our setup as follows: at each time instant k, operator i receives $p_j(k)$ from neighboring operator $j \in \mathcal{N}_i(k)$ and sends the control command $u_i(k) = \sum_{j \in V} a_{ij}(k)p_j(k) - p_i(k)$ to vehicle i. If $s_i^a(k) = 1$, adversary i sends $M_i^a(k)$ to vehicle i and lets $M_i^a(k+1) = M_i^a(k)$. If $s_i^a(k) = 0$, adversary i then lets $M_i^a(k+1) = u_i(k)$. After receiving the data $u_i(k)$ (if $s_i^a(k) = 0$) or $M_i^a(k)$ (if $s_i^a(k) = 1$), vehicle i implements it and then sends the new location $p_i(k+1) = p_i(k) + u_i(k)$ (if $s_i^a(k) = 0$) or $p_i(k+1) = p_i(k) + M_i^a(k)$ (if $s_i^a(k) = 1$) to operator i.

In the above classic consensus algorithm, it is not difficult to verify that if the event of $s_i^a(k) = 1$ occurs infinitely often for any $i \in V$, then vehicles fail to reach any consensus. Even worse, the maximum deviation of $D(k) \triangleq \max_{i \in V} p_i(k) - \min_{i \in V} p_i(k)$ can be intentionally driven to infinity despite the limitation of $\tau_{\max}$. We further look into a simpler case to illustrate this point.

Consider two operator–vehicle pairs with $p_1(0) \neq p_2(0)$. Assume that the two operators communicate with each other all the time, and the update rule is $\frac{1}{2}(p_i(k) + p_j(k))$. Suppose $\tau_{\max} \geq 2$, and that adversaries adopt a periodic strategy: $s_1(k) = s_2(k) = 0$ if k is a multiple of $\tau_{\max} + 1$; otherwise, $s_1(k) = s_2(k) = 1$. It is not difficult to verify that $D(\kappa(\tau_{\max} + 1)) = \tau_{\max}^\kappa D(0)$ for integer $\kappa \geq 1$. Hence, $D(k)$ diverges to infinity at a geometric rate of $\tau_{\max}$.

The above discussion yields the following insights: First, the classic consensus algorithm can be easily prevented from reaching consensus by persistently launching replay attacks. Second, in the worst case, adversaries may be able to drive $D(k)$ to infinity if adversaries know the algorithm and are able to intelligently take advantage of this information. Further, if their energy restriction is smaller, that is, $\tau_{\max}$ is larger, adversaries can speed up the divergence of $D(k)$. These facts evidently motivate the design of new distributed resilient algorithms that explicitly take into account replay attacks.

The detection of replay attacks is not difficult when operators and vehicles are synchronized. A detection scheme consists of attaching a time index to each control command from the operator, and then the vehicle can detect replay attacks by simply comparing the current time instant and the time index of the received command. This simple detection scheme will be employed in our subsequent algorithm design.

9.2.4 Prior Information about Adversaries and Objective

In hostile environments, it would be reasonable to expect that operators have limited information about adversaries. In this chapter, we assume that the only

Table 9.1   Basic Notations

n	Computing Horizon
$\mathbf{u}_i(k \to k+n-1|k)$	Collection of $\{u_i(k+s|k)\}_{0 \le s \le n-1}$
$\bar{\mathbf{u}}_i(k \to k+n-1|k)$	Collection of $\{\bar{u}_i(k+s|k)\}_{0 \le s \le n-1}$
$\mathbf{K}_i(k \to k+n-1|k)$	Collection of $\{K_i(k+s|k)\}_{0 \le s \le n-1}$
$|\mathcal{N}_i(k)|$	Cardinality of $\mathcal{N}_i(k)$
p_{max} (respectively, v_{max})	Upper bound on the position (respectively, the velocity)
$\mathbb{P}_X$	Projection operator onto X

information operator i possesses is the quantity τ_{max} or any of its upper bounds. At each time, each operator i makes a decision before her opponent, adversary i. Hence, operator i cannot predict whether adversary i would produce an attack at this time. Our objective is to design a distributed algorithm that ensures formation control under the above informational restriction.

> Objective: Given the only information of τ_{max}, we aim to devise a distributed algorithm, including the distributed control law $u_i(k)$ for vehicle i, such that $p_i(k) \in X$ and $u_i(k) \in U$ for all $k \ge 0$ and $i \in V$, and $\lim_{k \to +\infty} \mathrm{dist}(p(k), X^*) = 0$ and $\lim_{k \to +\infty} \|v_i(k)\| = 0$.

To conclude this section, we summarize the main notations in Table 9.1 that will be used in Sections 9.3 and 9.4. In particular, $u_i(k+s|k)$ means the control command of time instant $k+s$ ($s \ge 0$), and this control command is generated at time instant k.

9.3   Preliminaries

In this section, we provide both notation and a set of preliminary results that will be used to state our algorithm and analyze its convergence properties in the sequel.

9.3.1   Coordinate Transformation

We pick any scalar $\beta > 1$ and define the change of coordinates $T : \mathbb{R}^{3d} \to \mathbb{R}^{3d}$ such that $T(p_i, v_i, u_i) = (p_i, q_i, \bar{u}_i)$, where $q_i = p_i + \beta v_i$ and $\bar{u}_i = v_i + \beta u_i = \frac{1}{\beta}(q_i - p_i) + \beta u_i$. Applying this coordinate transformation on dynamics (9.1), we obtain

$$p_i(k+1) = (1 - \frac{1}{\beta})p_i(k) + \frac{1}{\beta}q_i(k),$$

$$q_i(k+1) = (1 + \frac{1}{\beta})q_i(k) - \frac{1}{\beta}p_i(k) + \beta u_i(k) = q_i(k) + \bar{u}_i(k). \quad (9.2)$$

We refer to $\bar{u}_i(k)$ as the auxiliary control of vehicle i.

Remark 9.3. Since β is nonzero, the formation property of $\lim_{k \to +\infty} \text{dist}(p(k),$ $X^*) = 0$ and $\lim_{k \to +\infty} \|v_i(k)\| = 0$ is equivalent to

$$\lim_{k \to +\infty} \text{dist}(q(k), X^*) = 0, \quad \lim_{k \to +\infty} \|p_i(k) - q_i(k)\| = 0, \quad \forall i \in V.$$

This equivalence will be used for the algorithm design and analysis.

9.3.2 Constrained Multiparametric Program

In this part, we introduce a constrained multiparametric program (the n-OC problem, for short), which will be used in our distributed resilient formation control algorithms. Given any pair of $u_{\max} > 0$ and $\beta > 1$, we choose a pair of positive constants $v_{\max}$ and $\bar{u}_{\max}$ such that the following holds:

$$v_{\max} + \bar{u}_{\max} \le \beta u_{\max}, \quad \bar{u}_{\max} \le v_{\max}. \quad (9.3)$$

We then introduce the following notations:

$$\varrho \triangleq \min\{\frac{1}{2}, \frac{\bar{u}_{\max}}{2p_{\max} + \beta v_{\max}}\}, \quad (9.4)$$

$$W \triangleq \{v_i \in \mathbb{R}^d \mid \|v_i\|_\infty \le v_{\max}\}, \quad \bar{U} \triangleq \{\bar{u}_i \in \mathbb{R}^d \mid \|\bar{u}_i\|_\infty \le \bar{u}_{\max}\},$$

where W (respectively, $\bar{U}$) is the constraint set imposed on the velocity v_i (respectively, the auxiliary input $\bar{u}_i$) of vehicle i.

Choose $\hat{\varrho} \in (0, \varrho]$. One can see that a set of positive constants δ, α, and γ can be chosen so that

$$(1 + (1 - \hat{\varrho})^2)\alpha + \hat{\varrho}^2 \gamma < \min\{2\alpha, \alpha + \gamma\} - \delta. \quad (9.5)$$

The relation (9.5) will be used in the proof of Claim 3 of Proposition 9.1.

Remark 9.4. We now proceed to choose *one* set of parameters to satisfy (9.5). Choose $\gamma < \alpha$; then (9.5) is equivalent to $(1 + (1 - \hat{\varrho})^2)\alpha + \hat{\varrho}^2 \gamma < \alpha + \gamma - \delta$ and

$$\alpha < \frac{1 - \hat{\varrho}^2}{(1 - \hat{\varrho})^2}\gamma - \frac{1}{(1 - \hat{\varrho})^2}\delta.$$

Notice that

$$\frac{1 - \hat{\varrho}^2}{(1 - \hat{\varrho})^2} > 1.$$

Then one can always choose a set of δ, α, and γ, with $\gamma < \alpha$ and δ being sufficiently small to satisfy the above relation.

With the above notations in place, we then define the following n-horizon optimal control with the state and input constraints X and U (n-OC, for short) parameterized by the vector $(p_i, q_i, z_i, v_i) \in X^3 \times W$:

$$\min_{\bar{\mathbf{u}}_i \in \mathbb{R}^{d \times n}} \sum_{s=0}^{n-1} \left(\alpha \|z_i - q_i(s)\|^2 + \gamma \|\bar{u}_i(s)\|^2 \right) + \alpha \|z_i - q_i(n)\|^2,$$

$$\text{such that} \quad q_i(s+1) = q_i(s) + \bar{u}_i(s),$$

$$p_i(s+1) = (1 - \frac{1}{\beta}) p_i(s) + \frac{1}{\beta} q_i(s),$$

$$\bar{u}_i(s) = K_i(s)(z_i - q_i(s)),$$

$$v_i(s) = \frac{1}{\beta} (q_i(s) - p_i(s)),$$

$$q_i(s+1) \in X, \quad v_i(s) \in W,$$

$$\bar{u}_i(s) \in \bar{U}, \quad 0 \le s \le n-1. \tag{9.6}$$

The initial states are given by $p_i(0) = p_i$, $q_i(0) = q_i$, and $v_i(0) = v_i$. This problem is defined for sets satisfying Assumption 9.1 and constants α, γ satisfying the condition (9.5). In the n-OC problem (9.6), the state $z_i \in X$ will be some target point defined later. A detailed discussion on problem (9.6) will be given in Remark 9.5.

The following proposition characterizes the solutions to the n-OC and its proof will be given in Section 9.5:

Proposition 9.1 (characterization of the optimal solutions to the n-OC). Consider any of the optimal solutions to the n-OC parameterized by the vector $(p_i, q_i, z_i, v_i) \in X^3 \times W$; that is, $\bar{\mathbf{u}}_i = (\bar{u}_i(0), \dots, \bar{u}_i(n-1))^T \in \mathbb{R}^{d \times n}$, with $\bar{u}_i(s) = K_i(s)(z_i - q_i(s))$, for all $0 \le s \le n-1$. Then, there is a pair of $\vartheta_{\min}, \vartheta_{\max} \in (0, 1)$ independent of $(p_i, q_i, z_i, v_i) \in X^3 \times W$ such that $K_i(s) \in [\vartheta_{\min}, \vartheta_{\max}]$.

The n-OC problem for some $(p_i(k), q_i(k), z_i(k), v_i(k)) \in X^3 \times W$ will be used in our algorithms. When necessary, we will use the notation $\mathbf{u}_i(k \to k+n-1|k)$ and $\mathbf{K}_i(k \to k+n-1|k)$ to refer to the resulting control sequences.

9.4 Distributed Constrained Formation Control Algorithm against Replay Attacks

In this section, we propose a distributed constrained formation control algorithm. After this, we summarize the algorithm resilience to replay attacks.

9.4.1 Statement of Our Algorithm

In order to play against replay attackers, we exploit the receding-horizon control (RHC) methodology [8, 14] and the idea of motion toward target points [11] to synthesize a distributed algorithm. The usage of RHC in the proposed algorithm is motivated by two salient features of RHC: first, it can *explicitly* handle state and input constraints, which is a unique advantage of RHC; second, it is able to generate suboptimal control laws approximating an associated infinite-horizon optimal control problem. More importantly, RHC is able to produce a sequence of feasible control commands for the next few steps. These commands serve as backup and are used by vehicles in response to replay attacks. As mentioned before, operators cannot predict the occurrence of replay attacks and have to account for the worst case. That is, each operator assumes that her opponent could launch attacks at every time instant, and chooses $n \geq \tau_{\max} + 1$. The distributed algorithm is described as follows.

Algorithm statement. Each vehicle has a memory storing the backup control commands in response to replay attacks. The state of vehicle i's memory is denoted by $M_i^v(k) \in \mathbb{R}^{d \times n}$.

At each time k, operator i receives $p_j(k)$ from operator $j \in \mathcal{N}_i(k)$. Operator i assumes that the vehicles of her current neighbors do not move over a finite time horizon of length n, and then identifies $\phi_i(k)$, the target point that minimizes the local formation error of $\sum_{j \in \mathcal{N}_i(k)} \|q_i - p_j(k) - v_{ij}\|^2 + \|q_i - p_i(k)\|^2 + \|q_i - q_i(k)\|^2$:

$$\phi_i(k) \triangleq \frac{1}{2 + |\mathcal{N}_i(k)|} \left(q_i(k) + p_i(k) + \sum_{j \in \mathcal{N}_i(k)} (p_j(k) + v_{ij}) \right).$$

According to Remark 9.3, the local formation error of $\sum_{j \in \mathcal{N}_i(k)} \|q_i - p_j(k) - v_{ij}\|^2 + \|q_i - p_i(k)\|^2 + \|q_i - q_i(k)\|^2$ captures the sum of the distance of $q(k)$ to X^* and the disagreement between p_i and q_i.

If $v_{ij} = 0$, then $\phi_i(k)$ is a convex combination of the time-dependent states. If these time-dependent states are in X, so is $\phi_i(k)$. However, the formation vectors v_{ij} are nonzero, so $\phi_i(k)$ is potentially outside X. In order to enforce the state constraint X, operator i computes the target point $z_i(k)$ via projecting $\phi_i(k)$ onto X, that is, $z_i(k) \triangleq \mathbb{P}_X[\phi_i(k)]$, where $\mathbb{P}_X$ is the projection operator onto the set of X.

After obtaining the target point $z_i(k)$, operator i solves the n-OC parameterized by the vector of $(p_i(k), q_i(k), z_i(k), v_i(k))$ and obtains the auxiliary control sequence $\bar{\mathbf{u}}_i(k \to k+n-1|k)$.* Operator i then generates the real control sequence of $\mathbf{u}_i(k \to k+n-1|k)$ by simulating the dynamics of vehicle i over the time frame $[k, k+n]$ as follows:

$$p_i(k+s+1|k) = (1 - \frac{1}{\beta})p_i(k+s|k) + \frac{1}{\beta}q_i(k+s|k),$$

$$q_i(k+s+1|k) = q_i(s+k|k) + \bar{u}_i(k+s|k),$$

$$u_i(k+s|k) = \frac{1}{\beta}\bar{u}_i(k+s|k) - \frac{1}{\beta^2}(q_i(k+s|k)$$

$$- p_i(k+s|k)), \quad 0 \leq s \leq n-1, \tag{9.7}$$

where $q_i(k|k) = q_i(k)$ and $p_i(k|k) = p_i(k)$. After that, operator i sends the package including $\mathbf{u}_i(k \to k+n-1|k)$ to vehicle i, where each element $u_i(k+s|k)$ in the package is labeled by the time index $k+s$ for $0 \leq s \leq n-1$.

If $s_i^a(k) = 1$, adversary i launches a replay attack, sending the stored command $M_i^a(k)$ to vehicle i, and letting $M_i^a(k+1) = M_i^a(k)$. If $s_i^a(k) = 0$, adversary i then does not produce any attack; instead, it intercepts the package containing $\mathbf{u}_i(k \to k+n-1|k)$ and updates her memory as $M_i^a(k+1) = \mathbf{u}_i(k \to k+n-1|k)$.

After receiving the package, vehicle i checks the time index, which is $k - \tau_i^a(k)$. If the package is new (i.e., $\tau_i^a(k) = 0$), then vehicle i replaces it in her memory by the new arrival (i.e., $M_i^v(k+1) = \mathbf{u}_i(k \to k+n-1|k)$), implements $u_i(k) = u_i(k|k)$, and sends $p_i(k+1)$ and $v_i(k+1)$ to operator i. If the package is repeated (i.e., $\tau_i^a(k) \geq 1$), then vehicle i implements $u_i(k|k-\tau_i^a(k))$ in its memory, sets $M_i^v(k+1) = M_i^v(k)$, and sends $p_i(k+1)$ and $v_i(k+1)$ to operator i. At the next time $k+1$, every decision maker will repeat the above process.

Remark 9.5. In the n-OC parameterized by $(p_i(k), q_i(k), z_i(k), v_i(k))$, the solution $\bar{\mathbf{u}}_i(k \to k+n-1|k)$ is a suboptimal controller on steering the state $q_i(k)$ toward the target point $z_i(k)$ on saving the control effort $\bar{u}_i(k)$ in (9.2). The idea of moving toward target points for distributed RHC was first proposed and analyzed in [11]. Like any other RHC law, for example, in [8, 14], our proposed algorithm requires that each operator online solves an optimization problem, the n-OC, at each time instant. We will discuss what solving these optimization problems involves in Section 9.6.

We summarize the distributed replay-attack-resilient formation control algorithm in Algorithm 9.1.

---

*Here, we assume the feasibility of the n-OC parameterized by the vector of $(p_i(k), q_i(k), z_i(k), v_i(k))$. Later, we will verify this point in Lemma 9.1.

Algorithm 9.1 Replay-attack-resilient formation control

■ Initially, operators agree on $\beta > 1$ and a pair of positive constants $v_{\max}$ and $\bar{u}_{\max}$ such that (9.3) holds. In addition, operators agree on a set of positive constants δ, α, and γ such that (9.5) holds.

■ At each $k \geq 0$, adversary, operator, and vehicle i execute the following steps:

  1. Operator i receives the location $p_j(k)$ from her neighboring operator $j \in \mathcal{N}_i(k)$ and computes the target point $z_i(k)$. Operator i solves the n-OC parameterized by the vector of $(p_i(k), q_i(k), z_i(k), v_i(k))$ and obtains the solution of $\bar{\mathbf{u}}_i(k \to k+n-1|k)$. After that, operator i computes $\mathbf{u}_i(k \to k+n-1|k)$ via (9.7) and sends it to vehicle i.

  2. If $s_i^a(k) = 1$, adversary i sends $M_i^a(k)$ to vehicle i and lets $M_i^a(k+1) = M_i^a(k)$. If $s_i^a(k) = 0$, adversary i sets $M_i^a(k+1) = \mathbf{u}_i(k \to k+n-1|k)$.

  3. If $\tau_i^a(k) = 0$, then vehicle i sets $M_i^v(k+1) = \mathbf{u}_i(k \to k+n-1|k)$, implements $u_i(k|k)$, and sends $p_i(k+1)$ and $v_i(k+1)$ to operator i. If $\tau_i^a(k) \geq 1$, then vehicle i implements $u_i(k|k-\tau_i^a(k))$ in $M_i^v(k)$, sets $M_i^v(k+1) = M_i^v(k)$, and sends $p_i(k+1)$ and $v_i(k+1)$ to operator i.

  4. Repeat for $k = k+1$.

9.4.2 Resilience Properties of Our Algorithm

Theorem 9.1 provides the convergence properties of the distributed replay-attack-resilient formation control algorithm, and its analysis is given in Section 9.5.

Theorem 9.1

(convergence properties of the distributed replay-attack-resilient formation control algorithm). Suppose that Assumptions 9.1, on the constraint sets; 9.2, on the connected formation digraph; 9.3, on problem feasibility; 9.4, on periodic communication; and 9.5, on the maximum number of attacks, hold. Let vehicle i start from $(p_i(0), v_i(0))$, with $(p_i(0), p_i(0) + \beta v_i(0)) \in X^2$ and $v_i(0) \in W$ for $i \in V$. Then, the distributed replay-attack-resilient formation control algorithm with $n \geq \tau_{\max} + 1$ ensures the following properties:

Constraint satisfaction: $p_i(k) \in X$ and $u_i(k) \in U$ for all $i \in V$ and $k \geq 0$.

Achieving formation: $\lim_{k \to +\infty} \text{dist}(p(k), X^*) = 0$ and $\lim_{k \to +\infty} \|v_i(k)\| = 0$ for all $i \in V$.

9.5 Analysis

In this section, we provide complete analysis of Proposition 9.1 and Theorem 9.1. We start with the proof for Proposition 9.1.

Proof for Proposition 9.1.   To simplify notations in the proof, we assume that $k = 0$ and drop the conditional independency on the starting time k, for example, $q_i(s) = q_i(k + s|k)$.

By (9.2), one can verify that $v_i(s+1)$ is a convex combination of $v_i(s)$ and $\bar{u}_i(s)$ through the following relations:

$$v_i(s+1) = \frac{1}{\beta}(q_i(s+1) - p_i(s+1))$$

$$= \frac{1}{\beta}(q_i(s) + \bar{u}_i(s) - (1 - \frac{1}{\beta})p_i(s) - \frac{1}{\beta}q_i(s))$$

$$= (1 - \frac{1}{\beta})v_i(s) + \frac{1}{\beta}\bar{u}_i(s). \tag{9.8}$$

In order to simplify the notations of the n-OC, we define the coordinate transformation $y_i(s) = q_i(s) - z_i(0)$. Then the n-OC parameterized by $(p_i(0), q_i(0), z_i(0), v_i(0)) \in X^3 \times W$ becomes the following one:

$$\min_{\mathbf{K}_i(0 \to n-1) \in \mathbb{R}^n} \sum_{s=0}^{n-1} \left(\alpha\|y_i(s)\|^2 + \gamma\|\bar{u}_i(s)\|^2\right) + \alpha\|y_i(n)\|^2,$$

$$\text{s.t. } p_i(s+1) = (1 - \frac{1}{\beta})p_i(s) + \frac{1}{\beta}(y_i(s) + z_i(0)),$$

$$y_i(s+1) = y_i(s) + \bar{u}_i(s),$$

$$\bar{u}_i(s) = -K_i(s)y_i(s), \ 0 \leq s \leq n-1,$$

$$v_i(s+1) = (1 - \frac{1}{\beta})v_i(s) + \frac{1}{\beta}\bar{u}_i(s),$$

$$y_i(s+1) + z_i(0) \in X, \ v_i(s) \in W,$$

$$\bar{u}_i(s) \in \bar{U}, \ 0 \leq s \leq n-1, \tag{9.9}$$

where $y_i(0) \neq 0$, and we change the decision variable $\bar{u}_i(0 \to n-1)$ to $\mathbf{K}_i(0 \to n-1)$ for the ease of presentation. The remainder of the proof is divided into the following three claims to characterize the solutions to (9.9).

Claim 9.1.   Given $z_i(0) \in \mathbb{R}^d$, the set of $Y \triangleq \{y_i \in \mathbb{R}^d \mid y_i + z_i(0) \in X\}$ is convex.

Proof. Pick any $\bar{y}_i$ and $\tilde{y}_i$ from Y, and any $\mu \in [0,1]$. Since $\bar{y}_i$ and $\tilde{y}_i$ are in Y, $\bar{y}_i + z_i(0)$ and $\tilde{y}_i + z_i(0)$ are in X. Since X is convex, $\mu(\bar{y}_i + z_i(0)) + (1-\mu)(\tilde{y}_i + z_i(0)) = \mu\bar{y}_i + (1-\mu)\tilde{y}_i + z_i(0) \in X$. This implies that $\mu\bar{y}_i + (1-\mu)\tilde{y}_i \in Y$, and the convexity of Y follows.

With Claim 1, we are now ready to find a feasible solution to (9.9) that will produce an upper bound of the optimal value of (9.9).

Claim 9.2. Consider the scalar sequence of $\tilde{\mathbf{K}}_i(0 \to n-1) \triangleq \{\tilde{K}_i(s)\}_{0 \leq s \leq n-1}$. If $\tilde{K}_i(s) \in [0,\varrho]$ for $0 \leq s \leq n-1$ and ϱ satisfying (9.4), then $\tilde{\mathbf{K}}_i(0 \to n-1)$ is a feasible solution candidate to (9.9).

Proof. Consider (9.9) where $v_i(0) = \tilde{v}_i(0)$, $y_i(0) = \tilde{y}_i(0)$, and $\mathbf{K}_i(0 \to n-1) = \tilde{\mathbf{K}}_i(0 \to n-1)$. Let $\{\tilde{v}_i(s)\}_{0 \leq s \leq n}$ and $\{\tilde{y}_i(s)\}_{0 \leq s \leq n}$ be the generated states and $\{\bar{u}_i(s)\}_{0 \leq s \leq n-1}$ be the produced auxiliary inputs in (9.9).

In order to verify the feasibility of $\tilde{\mathbf{K}}_i(0 \to n-1)$ to (9.9), we will check by induction that the following property, say constraint verification (CV), holds for all $0 \leq \tau \leq n-1$: for $0 \leq s \leq \tau$, we have that $\tilde{y}_i(s+1) \in Y$, $\tilde{v}_i(s+1) \in W$, and $\bar{u}_i(s) \in U$.

Let us start from the case $\tau = 0$. Recall that $z_i(0) \in X$ and $\tilde{y}_i(0) + z_i(0) = q_i(0) \in X$. This implies that 0 and $\tilde{y}_i(0)$ are both in Y. Since Y is convex, shown in Claim 1, and $\tilde{K}_i(0) \in [0,1]$, $\tilde{K}_i(0) \times 0 + (1 - \tilde{K}_i(0)) \times \tilde{y}_i(0) = \tilde{y}_i(1) \in Y$. In addition, we notice the following estimates on $\|\tilde{y}_i(0)\|_\infty$:

$$\|\tilde{y}_i(0)\|_\infty \leq \|p_i(0) - z_i(0)\|_\infty + \beta\|\tilde{v}_i(0)\|_\infty \leq 2p_{max} + \beta v_{max},$$

where p_{max} (respectively, v_{max}) is the uniform bound on X (respectively, W). Since $\tilde{K}_i(0) \in [0,\varrho]$ and (9.4), we have

$$\|\bar{u}_i(0)\|_\infty \leq \tilde{K}_i(0)\|\tilde{y}_i(0)\|_\infty \leq \varrho(\beta v_{max} + 2p_{max}) \leq \bar{u}_{max},$$

that is, $\bar{u}_i(0) \in \bar{U}$. Note that $\tilde{v}_i(1)$ is a convex combination of $\tilde{v}_i(0) \in W$ and $\bar{u}_i(0) \in \bar{U} \subseteq W$. Hence, we have $\tilde{v}_i(1) \in W$ and CV holds for $\tau = 0$.

Assume that CV holds for some $0 \leq \tau \leq n-2$. One can follow the same arguments above by replacing the time instants 0 and 1 with τ and $\tau+1$, respectively, to show that CV holds for $\tau+1$. By induction, we conclude that $\tilde{\mathbf{K}}_i(0 \to n-1)$ consists of a feasible solution candidate to (9.9). This completes the proof for Claim 2.

It follows from Claim 2 that the n-OC is feasible; that is, one can build a candidate solution by taking $0 \leq \tilde{K}_i(s) \leq \varrho$, with $0 \leq s \leq n-1$. We now set out to further characterize its optimal solutions.

Claim 9.3. There is a pair of ϑ_{min} and ϑ_{max} in $(0,1)$ such that $K_i(s) \in [\vartheta_{min}, \vartheta_{max}]$ for any optimal solution $\mathbf{K}_i(0 \to n-1)$ to (9.9).

Proof. Let $\{y_i(s)\}_{0\leq s\leq n}$ be the states generated by the optimal solution $\mathbf{K}_i(0\to n-1)$ in (9.9). Pick any $1\leq\tau\leq n$ and assume that $y_i(n-\tau)\neq 0$. From Bellman's principle of optimality, we know that the last τ components, $\{K_i(s)\}_{n-\tau\leq s\leq n-1}$, of $\mathbf{K}_i(0\to n-1)$ define an optimal solution to the truncated version of n-OC. More precisely, $\{K_i(s)\}_{n-\tau\leq s\leq n-1}$ is an optimal solution to the $(n-\tau)$-OC parameterized by $(p_i(n-\tau),v_i(n-\tau),q_i(n-\tau),z_i(0))$, which is given by

$$\min_{\mathbf{K}_i(n-\tau\to n-1)\in\mathbb{R}^\tau}\sum_{s=n-\tau}^{n-1}\left(\alpha\|y_i(s)\|^2+\gamma\|\bar{u}_i(s)\|^2\right)+\alpha\|y_i(n)\|^2,$$

$$\text{s.t.}\quad p_i(s+1)=(1-\frac{1}{\beta})p_i(s)+\frac{1}{\beta}(y_i(s)+z_i(0)),$$

$$y_i(s+1)=y_i(s)+\bar{u}_i(s),$$

$$\bar{u}_i(s)=-K_i(s)y_i(s),\quad n-\tau\leq s\leq n-1,$$

$$v_i(s+1)=(1-\frac{1}{\beta})v_i(s)+\frac{1}{\beta}\bar{u}_i(s),$$

$$y_i(s+1)+z_i(0)\in X,\quad v_i(s)\in W,$$

$$\bar{u}_i(s)\in\bar{U},\quad n-\tau\leq s\leq n-1. \tag{9.10}$$

Denote by r_τ^* the optimal value of (9.10). It is easy to see that r_τ^* is lower bounded by the sum of the first two running states and the first input, that is,

$$r_\tau^*\geq\alpha\|y_i(n-\tau)\|^2+\gamma\|K_i(n-\tau)y_i(n-\tau)\|^2$$

$$+\alpha\|(1-K_i(n-\tau))y_i(n-\tau)\|^2$$

$$=h(K_i(n-\tau))\|y_i(n-\tau)\|^2, \tag{9.11}$$

where $(1-K_i(n-\tau))y_i(n-\tau)$ is the state by applying the auxiliary input $-K_i(n-\tau)y_i(n-\tau)$ to $y_i(n-\tau)$, and $h(v)\triangleq\alpha+\gamma v^2+\alpha(1-v)^2$.

Regarding the function $h(v)$, we notice that $h(v)$ is quadratic in v and reaches the minimum at $\frac{\alpha}{\alpha+\gamma}$. Then there is a pair of $\vartheta_{\min}$ and $\vartheta_{\max}$ in $(0,1)$ such that

$$h(v)\geq\min\{\alpha+\gamma,2\alpha\}-\delta,\quad v\notin[\vartheta_{\min},\vartheta_{\max}], \tag{9.12}$$

where $\alpha+\gamma=h(1)$, $2\alpha=h(0)$, and δ is given in (9.5).

We now set out to show that $K_i(n-\tau)\in[\vartheta_{\min},\vartheta_{\max}]$. To achieve this, we now construct a solution candidate $\{\tilde{K}_i(s)\}_{n-\tau\leq s\leq n-1}$ to (9.10) where $\tilde{K}_i(s)=\hat{\varrho}\in(0,\varrho]$ for $n-\tau\leq s\leq n-1$. It follows from Claim 2 that $\{\tilde{K}_i(s)\}_{n-\tau\leq s\leq n-1}$ is a feasible solution candidate to (9.9). Let $\tilde{r}_\tau$ be the value of (9.10) generated by

$\{\tilde{K}_i(s)\}_{n-\tau \leq s \leq n-1}$. We then have the following relations on $\tilde{r}_\tau$:

$$
\tilde{r}_\tau = \alpha \sum_{\kappa=0}^{\tau} (1 - \hat{\varrho})^{2\kappa} \|y_i(n-\tau)\|^2 + \gamma \sum_{\kappa=0}^{\tau-1} \hat{\varrho}^2 (1-\hat{\varrho})^{2\kappa} \|y_i(n-\tau)\|^2
$$

$$
\leq \frac{1}{1-(1-\hat{\varrho})^2} \left(\alpha(1-(1-\hat{\varrho})^4) + \gamma\hat{\varrho}^2(1-(1-\hat{\varrho})^2) \right) \times \|y_i(n-\tau)\|^2
$$

$$
= \left((1+(1-\hat{\varrho})^2)\alpha + \hat{\varrho}^2\gamma \right) \|y_i(n-\tau)\|^2. \tag{9.13}
$$

On the right-hand side of the first equality of (9.13), the first summation is the aggregation sum of running states and the second one is the accumulated control cost.

By (9.11), (9.12), (9.13), and (9.5), one can verify that if $K_i(n-\tau) \notin [\vartheta_{\min}, \vartheta_{\max}]$, then

$$
\tilde{r}_\tau = \left((1+(1-\hat{\varrho})^2)\alpha + \hat{\varrho}^2\gamma \right) \|y_i(n-\tau)\|^2
$$

$$
< (\min\{\alpha+\gamma, 2\alpha\} - \delta) \|y_i(n-\tau)\|^2 \leq r_\tau^*.
$$

That is, $\tilde{r}_\tau < r_\tau^*$, contradicting the optimality of $\{K_i(s)\}_{n-\tau \leq s \leq n-1}$ for (9.10). Hence, it must be the case that $K_i(n-\tau) \in [\vartheta_{\min}, \vartheta_{\max}] \subset (0,1)$. This holds for any $1 \leq \tau \leq n$, and thus this completes the proof for Claim 3.

The last claim establishes the result of Proposition 9.1.

The following lemma shows the property of constraint enforcement in Theorem 9.1:

Lemma 9.1

(constraint satisfaction and feasibility of the *n*-OC). The *n*-OC parameterized by $(p_i(k), q_i(k), z_i(k), v_i(k)) \in X^3 \times W$ is feasible for all $i \in V$ and $k \geq 1$. In addition, it holds that $p_i(k) \in X$, $v_i(k) \in W$, and $u_i(k) \in U$ for all $i \in V$ and $k \geq 0$.

Proof. It is trivial that $z_i(0) \in X$ due to the projection operator. Recall that X is convex, and $(p_i(0), q_i(0)) \in X^2$. Since $p_i(1)$ is a convex combination of $p_i(0)$ and $q_i(0)$ by (9.2), thus $p_i(1) \in X$. As a consequence of Claim 2 in the proof of Proposition 9.1, the *n*-OC parameterized by $(p_i(0), q_i(0), z_i(0), v_i(0)) \in X^3 \times W$ is feasible. This ensures that $q_i(1) \in X$, $v_i(1) \in W$, and $\bar{u}_i(0) \in W$. Note that $\|u_i(0)\|_\infty \leq \frac{1}{\beta}(\|v_i(0)\|_\infty + \|\bar{u}_i(0)\|_\infty) \leq \frac{1}{\beta}(\bar{u}_{\max} + v_{\max}) \leq u_{\max}$ by (9.3). Hence, it yields $u_i(0) \in U$.

The remainder of the proof can be derived by means of induction, and is omitted here.

With the above instrumental results, we are now ready to characterize the convergence properties of Algorithm 9.1 and complete the proof of Theorem 9.1.

Proof for Theorem 9.1. Consider $i \in V$ and time $k \geq 0$. Note that the control command $u_i(k) = u_i(k|k - \tau_i^a(k))$ is applied to (9.1), or equivalently, $\bar{u}_i(k)$ is applied to (9.2). Thus, the closed-loop dynamics of (9.2) is given by

$$p_i(k+1) = (1 - \frac{1}{\beta})p_i(k) + \frac{1}{\beta}q_i(k),$$

$$\begin{aligned}
q_i(k+1) &= q_i(k) + \bar{u}_i(k|k - \tau_i^a(k)) \\
&= q_i(k) + K_i(k|k - \tau_i^a(k))(z_i(k - \tau_i^a(k)) - q_i(k)) \\
&= q_i(k) + K_i(k|k - \tau_i^a(k))(\mathbb{P}_X[\phi_i(k - \tau_i^a(k))] - q_i(k)) \\
&= q_i(k) + K_i(k|k - \tau_i^a(k))(\phi_i(k - \tau_i^a(k)) - q_i(k)) + w_i(k).
\end{aligned} \tag{9.14}$$

The term $w_i(k)$ in (9.14) is the error induced by the projection operator $\mathbb{P}_X$ given by

$$w_i(k) \triangleq K_i(k|k - \tau_i^a(k))(\mathbb{P}_X[\phi_i(k - \tau_i^a(k))] - \phi_i(k - \tau_i^a(k))).$$

Substituting directly the definition of $\phi_i(k - \tau_i^a(k))$, as follows,

$$\begin{aligned}
\phi_i(k - \tau_i^a(k)) &= \frac{q_i(k - \tau_i^a(k)) + p_i(k - \tau_i^a(k))}{2 + |\mathcal{N}_i(k - \tau_i^a(k))|} \\
&+ \frac{\sum_{j \in \mathcal{N}_i(k - \tau_i^a(k))}(p_j(k - \tau_i^a(k)) + v_{ij})}{2 + |\mathcal{N}_i(k - \tau_i^a(k))|},
\end{aligned}$$

into (9.14) leads to the following:

$$\begin{aligned}
p_i(k+1) &= (1 - \frac{1}{\beta})p_i(k) + \frac{1}{\beta}q_i(k), \\
q_i(k+1) &= q_i(k) + K_i(k|k - \tau_i^a(k)) \\
&\times \Big\{ \frac{1}{2 + |\mathcal{N}_i(k - \tau_i^a(k))|}(q_i(k - \tau_i^a(k)) + p_i(k - \tau_i^a(k)) \\
&+ \sum_{j \in \mathcal{N}_i(k - \tau_i^a(k))}(p_j(k - \tau_i^a(k)) + v_{ij})) - q_i(k) \Big\} + w_i(k).
\end{aligned} \tag{9.15}$$

Pick any $p^* \in X^*$, and we define the errors $q_i^e(k) \triangleq q_i(k) - p_i^*$ and $p_i^e(k) \triangleq p_i(k) - p_i^*$. Subtract p_i^* on both sides of (9.15), and we rewrite (9.15) in terms of $p_i^e(k)$ and $q_i^e(k)$ as follows:

$$\begin{aligned}
p_i^e(k+1) &= (1 - \frac{1}{\beta})p_i^e(k) + \frac{1}{\beta}q_i^e(k), \\
q_i^e(k+1) &= (1 - K_i(k|k - \tau_i^a(k)))q_i^e(k) + \frac{K_i(k|k - \tau_i^a(k))}{2 + |\mathcal{N}_i(k - \tau_i^a(k))|}q_i^e(k - \tau_i^a(k)) \\
&+ \sum_{j \in \mathcal{N}_i(k - \tau_i^a(k)) \cup \{i\}} \frac{K_i(k|k - \tau_i^a(k))}{2 + |\mathcal{N}_i(k - \tau_i^a(k))|}p_j^e(k - \tau_i^a(k)) + w_i(k),
\end{aligned} \tag{9.16}$$

where we use $p_i^* - p_j^* = v_{ij}$ for $p^* \in X^*$ in (9.16). By Remark 9.3, we notice that the following consensus property for algorithm (9.16)

$$\lim_{k \to +\infty} \|p_i^e(k) - q_j^e(k)\| = 0, \quad i, j \in V, \tag{9.17}$$

is equivalent to achieving the formation control mission in (9.1). In particular, $\lim_{k \to +\infty} \|p_i^e(k) - q_i^e(k)\| = 0$ implies that $\lim_{k \to +\infty} \|p_i(k) - q_i(k)\| = 0$. And the property $\lim_{k \to +\infty} \|q_i^e(k) - q_j^e(k)\| = 0$ implies the following:

$$\lim_{k \to +\infty} \|(q_i(k) - p_i^*) - (q_j(k) - p_j^*)\| = \lim_{k \to +\infty} \|q_i(k) - q_j(k) - d_{ij}\| = 0.$$

We will show that $w_i(k)$ is diminishing in Claim 4. In this way, the dynamics of (9.16) are decoupled along different dimensions, and thus we will only consider the scalar case, that is, $d = 1$, for the ease of presentation in the remainder of the proof.

In order to show the consensus property (9.17), we transform the second-order algorithm (9.16) into an equivalent first-order one. To achieve this, we introduce a transformed system with two classes of agents: location agents labeled by $\{1, \ldots, N\}$ and velocity agents labeled by $\{N+1, \ldots, 2N\}$. With these, we define the state $x(k) \in \mathbb{R}^{2N}$ in such a way that $x_i(k) = p_i^e(k)$ for location agent $i \in \{1, \ldots, N\}$ and $x_i(k) = q_{i-N}^e(k)$ for velocity agent $i \in \{N+1, \ldots, 2N\}$. Let $V_T \triangleq \{1, \ldots, 2N\}$. Consequently, algorithm (9.16) can be transformed into the following first-order consensus algorithm subject to delays and errors $e_\ell(k)$:

$$x_\ell(k+1) = a_{\ell\ell}(k)x_\ell(k) + \bar{a}_{\ell\ell}(k)x_\ell(k - \tau_\ell^a(k))$$

$$+ \sum_{\ell' \in V_T \setminus \{\ell\}} a_{\ell\ell'}(k)x_{\ell'}(k - \tau_\ell^a(k)) + e_\ell(k), \tag{9.18}$$

where $e_\ell(k) = 0$, $\bar{a}_{\ell\ell}(k) = 0$ for $\ell \in \{1, \ldots, N\}$, and $e_\ell(k) = w_{\ell-N}(k)$ for $\ell \in \{N+1, \ldots, 2N\}$. Without loss of any generality, we assume that $x_\ell(k) = x_\ell(0)$ for $k = -1, \ldots, -\tau_{\max}$.

The weights in (9.18) induce the communication graph $\mathcal{G}_T(k) \triangleq \{V_T, \mathcal{E}_T(k)\}$ defined as $(\ell, \ell') \in \mathcal{E}_T(k)$ if and only if $a_{\ell\ell'}(k) \neq 0$ for $\ell \neq \ell'$. From (9.15), we can see that location agent i and velocity agent i can communicate to each other all the time. This observation in conjunction with Assumptions 9.2, on the connectedness of the formation digraph, and 9.4, on periodic communication, yields that the directed graph $(V_T, \bigcup_{k=0}^{B-1} \mathcal{E}_T(k_0 + k))$ is strongly connected for any $k_0 \geq 0$. Figure 9.2 shows an illustrative example with three operators where operators 1 and 2 communicate when k is odd, and operators 2 and 3 communicate when k is even.

Recall that $K_i(k|k - \tau_i^a(k)) \in [\vartheta_{\min}, \vartheta_{\max}] \subset (0, 1)$ by Proposition 9.1. For (9.18), one can verify that there is $\eta_{\min} \in (0, 1)$ such that

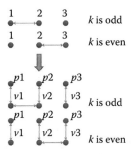

Figure 9.2: Three agents communicate as indicated by the first two graphs over even and odd times. This translates into the bottom communication graphs for the transformed system.

$$a_{\ell\ell}(k) \geq \eta_{\min}, \quad a_{\ell\ell'}(k) \neq 0 \Rightarrow a_{\ell\ell'}(k) \geq \eta_{\min}, \tag{9.19}$$

$$\bar{a}_{\ell\ell}(k) + \sum_{\ell' \in V_T} a_{\ell\ell'}(k) = 1, \tag{9.20}$$

where (9.19) is referred to as the nondegeneracy property and (9.20) is referred to as the stochasticity property.

In order to show the consensus property (9.17), we first show that the error term $e_\ell(k)$ is diminishing.

Claim 9.4: For any $\ell \in \{N+1, \ldots, 2N\}$, it holds that

$$\lim_{k \to +\infty} \|e_\ell(k)\| = 0.$$

Proof: Since X is convex and $p_i^* \in X$, it follows from the projection theorem (e.g., on p. 17 in [6]) that the following holds for $i \in V$:

$$\|z_i(k - \tau_i^a(k)) - p_i^*\|^2 = \|\mathbb{P}_X[\phi_i(k - \tau_i^a(k))] - p_i^*\|^2$$

$$\leq \|\phi_i(k - \tau_i^a(k)) - p_i^*\|^2 - \|w_i(k)\|^2. \tag{9.21}$$

For the term of $\|\phi_i(k - \tau_i^a(k)) - p_i^*\|^2$, the following relations hold:

$$\|\phi_i(k - \tau_i^a(k)) - p_i^*\|^2$$
$$= \left\| \frac{q_i^e(k - \tau_i^a(k)) + p_i^e(k - \tau_i^a(k))}{2 + |\mathcal{N}_i(k - \tau_i^a(k))|} + \frac{\sum_{j \in \mathcal{N}_i(k - \tau_i^a(k))} p_j^e(k - \tau_i^a(k))}{2 + |\mathcal{N}_i(k - \tau_i^a(k))|} \right\|^2$$
$$\leq \frac{\|q_i^e(k - \tau_i^a(k))\|^2 + \|p_i^e(k - \tau_i^a(k))\|^2}{2 + |\mathcal{N}_i(k - \tau_i^a(k))|} + \frac{\sum_{j \in \mathcal{N}_i(k - \tau_i^a(k))} \|p_j^e(k - \tau_i^a(k))\|^2}{2 + |\mathcal{N}_i(k - \tau_i^a(k))|}, \tag{9.22}$$

where in the equality we use $p_i^* - p_j^* = v_{ij}$, and in the inequality we use the fact that the function $\| \cdot \|^2$ is a convex function and the Jensen's inequality (e.g., inequality (1.7) on p. 19 in [6]).

Subtract p_i^* on both sides of the update rule for $q_i(k)$ in (9.14), and it renders the following:

$$q_i^e(k+1) = (1 - K_i(k|k - \tau_i^a(k)))q_i^e(k)$$
$$+ K_i(k|k - \tau_i^a(k))(z_i(k - \tau_i^a(k)) - p_i^*).$$

Since $\| \cdot \|^2$ is a convex function, then the following holds:

$$\|q_i^e(k+1)\|^2 \leq (1 - K_i(k|k - \tau_i^a(k)))\|q_i^e(k)\|^2$$
$$+ K_i(k|k - \tau_i^a(k))\|z_i(k - \tau_i^a(k)) - p_i^*\|^2. \tag{9.23}$$

Analogously, one can verify the following relation via the update rule for $p_i(k)$ in (9.15):

$$\|p_i^e(k+1)\|^2 \leq (1 - \frac{1}{\beta})\|p_i^e(k)\|^2 + \frac{1}{\beta}\|q_i^e(k)\|^2. \tag{9.24}$$

Recall that $K_i(k|k - \tau_i^a(k)) \in [\vartheta_{min}, \vartheta_{max}]$ by Proposition 9.1. Then the combination of (9.21), (9.22), (9.23), and (9.24) establishes that the following holds for all $\ell \in V_T$:

$$\|x_\ell(k+1)\|^2 \leq b_{\ell\ell}(k)\|x_\ell(k)\|^2 + \bar{b}_{\ell\ell}(k)\|x_\ell(k - \tau_\ell^a(k))\|^2$$
$$+ \sum_{\ell' \in V_T \setminus \{\ell\}} b_{\ell\ell'}(k)\|x_{\ell'}(k - \tau_\ell^a(k))\|^2 - \vartheta_{min}\|e_\ell(k)\|^2, \tag{9.25}$$

where the following properties hold for the weights:

$$b_{\ell\ell} \geq \bar{\eta}_{min}, \quad b_{\ell\ell'}(k) \neq 0 \Rightarrow b_{\ell\ell'}(k) \geq \bar{\eta}_{min}, \tag{9.26}$$

$$\bar{b}_{\ell\ell}(k) + \sum_{\ell' \in V_T} b_{\ell\ell'}(k) = 1, \tag{9.27}$$

for some $\bar{\eta}_{min} \in (0, 1)$.

The iterative relation (9.25) induces the communication graph $\bar{\mathcal{G}}_T(k) \triangleq \{V_T, \bar{\mathcal{E}}_T(k)\}$, where $(\ell, \ell') \in \bar{\mathcal{E}}_T(k)$ if and only if $b_{\ell\ell'}(k) \neq 0$, with $\ell \neq \ell'$. Recall that $K_i(k|k - \tau_i^a(k)) \in [\vartheta_{min}, \vartheta_{max}]$ by Proposition 9.1. Then $\mathcal{E}_T(k) = \bar{\mathcal{E}}_T(k)$, and thus the directed graph $(V_T, \bigcup_{k=0}^{B-1} \bar{\mathcal{E}}_T(k_0 + k))$ is strongly connected for any $k_0 \geq 0$.

We denote the maximum value of $\|x_\ell\|^2$ over the interval $[k - \tau_{\max}, k]$ asfollows:

$$\Pi(k) \triangleq \max_{0 \le s \le \tau_{\max}} \max_{\ell \in V_T} \|x_\ell(k - s)\|^2.$$

By (9.20) and $\tau_i^a(k) \le \tau_{\max}$, it then follows from (9.25) and (9.27) that the following holds for any $\ell \in V_T$:

$$\|x_\ell(k+1)\|^2 \le \Pi(k) - \vartheta_{\min}\|e_\ell(k)\|^2,$$

and thus, $\Pi(k+1) \le \Pi(k)$; that is, the sequence of $\{\Pi(k)\}$ is nonincreasing.

We now move to show by contradiction that $\|e_\ell(k)\|^2$ decreases to zero for all $\ell \in \{N+1, \ldots, 2N\}$ via studying the iterative relation (9.25). In particular, we assume that $\|e_\ell(k)\|^2$ is strictly away from zero infinitely often and derive that $\Pi(k)$ could be arbitrarily negative, contradicting $\Pi(k) \ge 0$.

Assume that there is some $\ell \in \{N+1, \ldots, 2N\}$ and $\varepsilon > 0$ such that the event $\vartheta_{\min}\|e_\ell(k)\|^2 \ge \varepsilon$ occurs infinitely often. Denote by the set $\{s_1, s_2, \ldots\}$ the collection of time instants when $\vartheta_{\min}\|e_{\bar\ell}(k)\|^2 \ge \varepsilon$ occurs. Without loss of any generality, we assume that $s_1 \ge 2NB + 1$, and $s_{\kappa+1} \ge s_\kappa + 2NB + 1$ for $\kappa \ge 1$.

We now consider the time instant s_1. Define the set $\mathcal{D}_0 = \{\bar\ell\}$. Since $(V_T, \bigcup_{k=0}^{B-1} \bar{\mathcal{E}}_T(s_1 + k))$ is strongly connected, there is a nonempty set $\mathcal{D}_1 \subset V_T \setminus \{\bar\ell\}$ of agents such that for all $\ell \in \mathcal{D}_1$, $b_{\ell\bar\ell}(k) \ne 0$ occurs at least once during the time frame $[s_1, s_1 + B - 1]$. By induction, a set $\mathcal{D}_{\kappa+1} \subset V_T \setminus (\mathcal{D}_0 \cup \cdots \cup \mathcal{D}_\kappa)$ can be defined by considering those agents $\ell \notin \mathcal{D}_0 \cup \cdots \cup \mathcal{D}_\kappa$ where there is some $\ell' \in \mathcal{D}_0 \cup \cdots \cup \mathcal{D}_\kappa$ such that $b_{\ell\ell'}(k) \ne 0$ occurs at least once during the time frame $[s_1 + \kappa B, s_1 + (\kappa + 1)B - 1]$. The graph $(V_T, \bigcup_{k=0}^{B-1} \bar{\mathcal{E}}_T(s_1 + \kappa B + k))$ is strongly connected; $\mathcal{D}_{\kappa+1} \ne \emptyset$ as long as $V_T \setminus (\mathcal{D}_0 \cup \cdots \cup \mathcal{D}_\kappa) \ne \emptyset$. Thus, there exists $\mathcal{L} \le 2N - 1$ such that the collection of $\mathcal{D}_0, \ldots, \mathcal{D}_\mathcal{L}$ is a partition of V_T.

For each time instant $k \ge s_1 + 1$, we define the set $\Omega(k) \subseteq V_T$ such that $\ell \in \Omega(k)$ if and only if $\|x_\ell(k+1)\|^2 \le \Pi(s_1) - \bar{\eta}_{\min}^{k-s_1-1}\varepsilon$. One then can verify that the following properties hold for the set $\Omega(k)$:

Property 1 (P1). If $\ell \in \Omega(K)$, then $\ell \in \Omega(k)$ for all $k \ge K + 1$.

Property 2 (P2). If $\ell' \in \Omega(k)$ and $b_{\ell\ell'}(k) \ne 0$, then $\ell \in \Omega(k+1)$.

In particular, P1 is a result of the following:

$$\|x_\ell(k+1)\|^2 \le b_{\ell\ell}(k)\|x_\ell(k)\|^2 - (1 - b_{\ell\ell}(k))\Pi(k)$$
$$\le b_{\ell\ell}(k)\|x_\ell(k)\|^2 - (1 - b_{\ell\ell}(k))\Pi(s_1),$$

where we use the monotonicity property of $\{\Pi(k)\}$ and the stochasticity property (9.27). Analogously, P2 is a result of the following:

$$\|x_\ell(k+1)\|^2 \le b_{\ell\ell'}(k)\|x_{\ell'}(k)\|^2 - (1 - b_{\ell\ell'}(k))\Pi(s_1).$$

One can see that $\mathcal{D}_0 = \{\bar{\ell}\} \subseteq \Omega(s_1 + 1)$. By (P1), $\mathcal{D}_0 = \{\bar{\ell}\} \subseteq \Omega(k)$ for all $k \geq s_1 + 1$. Assume that $\mathcal{D}_0 \cup \cdots \cup \mathcal{D}_\kappa \subseteq \Omega(s_1 + 1 + \kappa B)$ for some $0 \leq \kappa \leq \mathcal{L} - 1$. Pick any $\ell \in \mathcal{D}_{\kappa+1}$ for $\kappa \geq 0$. By construction of the set of $\{\mathcal{D}_0, \ldots, \mathcal{D}_\mathcal{L}\}$, there is some $\ell' \in \mathcal{D}_0 \cup \cdots \cup \mathcal{D}_\kappa$ such that $b_{\ell\ell'}(k') \neq 0$ at some time $k' \in [s_1 + \kappa B, s_1 + (\kappa+1)B - 1]$. Hence, $\{\ell\} \cup \mathcal{D}_0 \cup \cdots \cup \mathcal{D}_\kappa \subseteq \Omega(k' + 1)$, and thus $\mathcal{D}_0 \cup \cdots \cup \mathcal{D}_{\kappa+1} \subseteq \Omega(s_1 + 1 + (\kappa+1)B)$ by P1 and P2. By induction, we have $V_T = \Omega(s_1 + 1 + (\mathcal{L} + 1)B)$. By P1, we further have $V_T = \Omega(s_1 + 1 + 2NB)$ and thus

$$\Pi(s_1 + 1 + 2NB) \leq \Pi(s_1) - \overline{\eta}_{\min}^{2NB} \varepsilon. \tag{9.28}$$

Recall that $s_2 \geq s_1 + 2NB + 1$. By the monotonicity of $\{\Pi(k)\}$, we have

$$\Pi(s_2) \leq \Pi(s_1 + 1 + 2NB) \leq \Pi(s_1) - \overline{\eta}_{\min}^{2NB} \varepsilon. \tag{9.29}$$

Following analogous lines toward (9.29), one can verify the following by induction:

$$\Pi(s_{\kappa+1}) \leq \Pi(s_\kappa) - \overline{\eta}_{\min}^{2NB} \varepsilon, \quad \forall \kappa \geq 1.$$

This further gives

$$\Pi(s_{\kappa+1}) \leq \Pi(s_1) - \kappa \overline{\eta}_{\min}^{2NB} \varepsilon.$$

Since $\inf_{k \geq 0} \Pi(k) \geq 0$, we reach a contradiction by letting $\kappa \to +\infty$ in the above relation. Consequently, it establishes that $\{e_\ell(k)\}$ diminishes.

With Claim 4 at hand, we are now ready to show the consensus property (9.17).

Claim 5. The consensus property (9.17) holds.

Proof. We denote

$$M(k) \triangleq \max_{0 \leq s \leq \tau_{\max}} \max_{\ell \in V_T} x_\ell(k - s), \tag{9.30}$$

$$m(k) \triangleq \min_{0 \leq s \leq \tau_{\max}} \min_{\ell \in V_T} x_\ell(k - s), \tag{9.31}$$

$$D(k) \triangleq M(k) - m(k).$$

To summarize, algorithm (9.18) enjoys the nondegeneracy property (9.19), the stochasticity property (9.20), and the property that the directed graph $(V_T, \bigcup_{k=0}^{B-1} \mathcal{E}_T(k_0 + k))$ is strongly connected for any $k_0 \geq 0$. By using Claim 4 and following similar lines toward Corollary 3.1 in our paper [23], we show the maximum deviation of $D(k)$ is diminishing, and the desired result is established.

Here, we provide a sketch of the proof on $D(k)$ being diminishing. Let us fix $\ell \in V_T$ for every time instant k and define $\mathcal{D}_0 = \{\ell\}$. Recall that

$(V_T, \bigcup_{k=0}^{B-1} \mathcal{E}_T(k_0 + k))$ is strongly connected for any $k_0 \geq 0$. We replace B by $\max\{B, \tau_{\max}\}$ in the paragraph right before Lemma 3.1 in [23] and construct the collection of $\mathcal{D}_0, \ldots, \mathcal{D}_{\mathcal{L}}$ consisting of a partition of V_T with some $\mathcal{L} \leq 2N - 1$. Following the same lines in Lemma 3.1 and using the new definitions of $m(k)$ and $M(k)$ in (9.30) and (9.31), one can show that for every $\kappa \in \{1, \ldots, \mathcal{L}\}$, there exists a real number $\eta_\kappa > 0$ such that for every integer $s \in [\kappa \max\{B, \tau_{\max}\}, (\mathcal{L}\max\{B, \tau_{\max}\} + \max\{B, \tau_{\max}\} - 1)]$, and $\kappa' \in \mathcal{D}_\kappa$, it holds that for $t = k + s$,

$$x_{\kappa'}(t) \geq m(k) + \sum_{q=0}^{s-1} \min_{\ell' \in V_T} e_{\ell'}(k+q) + \eta_\kappa(x_\ell(s) - m(s)),$$

$$x_{\kappa'}(t) \leq M(k) + \sum_{q=0}^{s-1} \max_{\ell' \in V_T} e_{\ell'}(k+q) - \eta_\kappa(M(s) - x_\ell(s)).$$

The rest of the proofs can be finished by following the same lines in [23] and replacing B by $\max\{B, \tau_{\max}\}$.

By Remark 9.3, the consensus property (9.17) establishes the desired result. This completes the proof.

9.6 Discussion and Simulation

In this section, we discuss several aspects of the distributed replay-attack-resilient formation control algorithm and its possible variations. Numerical examples are provided to illustrate the algorithm performance.

9.6.1 Special Case of Consensus

In the constrained formation control problem, we cannot characterize the diminishing rate of projection errors, and this prevents us from finding an estimate of the convergence rate of Algorithm 9.1. When $v_{ij} = 0$, the formation control problem reduces to the consensus (or rendezvous) problem. Since X is convex and $\phi_i(k)$ is a convex combination of states in X, $z_i(k) = \phi_i(k)$ and the projection errors are absent. For this special case, we can guarantee that the algorithm converges at a geometric rate.

Corollary 9.1 Suppose that $v_{ij} = 0$ and Assumptions 9.1, 9.4, and 9.5 hold. Let vehicle i start from $(p_i(0), v_i(0))$, with $(p_i(0), p_i(0) + \beta v_i(0)) \in X^2$ and $v_i(0) \in W$ for $i \in V$. Algorithm 9.1 with $n \geq \tau_{\max}$ ensures that the vehicles converge to the consensus at a geometric rate of $(1 - \eta)^{\frac{1}{2NB-1}}$ for some $\eta \in (0, 1)$.

The readers are referred to our previous paper [26] for the complete analysis of Corollary 9.1.

9.6.2 Resilience to Denial-of-Service Attacks

Consider the class of denial-of-service (DoS) attacks, for example, in [3,4,13]. In particular, adversary i produces a DoS attack by erasing the control commands sent from operator i, and vehicle i receives nothing at this time. It is easy for vehicles to detect the occurrence of DoS attacks via verifying the receipt of control commands at each time instant. Algorithm 9.1 can be slightly modified to address the scenario where adversaries launch replay or DoS attacks on the data sent from vehicles to operators. If adversary i produces an attack at time k, then operator i does nothing at this time. In this way, the results of Theorem 9.1 apply as well provided that the computing horizon is larger than the maximum number of consecutive DoS attacks, that is, $n \geq \tau_{\max} + 1$.

9.6.3 Solving n-OC

As in Proposition 9.1, we will focus on the program (9.9) in order to simplify the notations. Now we convert the program (9.9) into a quadratic program through the following steps. By using the relation of $y_i(s) = y_i(0) + \sum_{\tau=0}^{s-1} \bar{u}_i(\tau)$, one can see that $\bar{u}_i(s) = -K_i(s) \prod_{\tau=0}^{s-1}(1 - K_i(\tau)) y_i(0)$ and $y_i(s) = \prod_{\tau=0}^{s-1}(1 - K_i(\tau)) y_i(0)$. We denote $J_i(s) \triangleq K_i(s) \prod_{\tau=0}^{s-1}(1 - K_i(\tau))$. By using $y_i(s) = y_i(0) + \sum_{\tau=0}^{s-1} \bar{u}_i(\tau)$ and $\bar{u}_i(s) = -J_i(s) y_i(0)$, one can simplify (9.9) to the following compact form after some algebraic manipulation:

$$\min_{\mathbf{J}_i(0 \to n-1) \in \mathbb{R}^n} \mathbf{J}_i(0 \to n-1)^T P_i \mathbf{J}_i(0 \to n-1)$$

$$+ y_i(0)^T Q_i \mathbf{J}_i(0 \to n-1)$$

$$\text{s.t.} \quad E_i \mathbf{J}_i(0 \to n-1) \leq F_i y_i(0) + G_i z_i(0) + H_i, \tag{9.32}$$

where a term independent of $\mathbf{J}_i(0 \to n-1)$ has been removed from the original objective function. In (9.32), the matrix P_i is symmetric and positive definite, and the matrices Q_i, E_i, F_i, G_i, and H_i have proper dimensions. One can see that the program (9.32) is a multiparametric quadratic program and a number of existing efficient algorithms can be used to solve it. Given the solution $\mathbf{J}_i(0 \to n-1)$ to (9.32), operator i then computes $\{\bar{u}_i(s)\}_{0 \leq s \leq n-1}$ by using $\bar{u}_i(s) = -J_i(s) y_i(0)$.

9.6.4 Pros and Cons of Distributed Resilient Formation Control Algorithms

By exploiting the RHC methodology, Algorithm 9.1 demonstrates its resilience to replay attacks and DoS attacks. Resilience is achieved under limited information about adversaries; that is, operators are only aware of $\tau_{\max}$, but do not need to know the attacking policy. In addition, employing the RHC method-ology explicitly guarantees that the state and input constraints are enforced all the time. These attractive advantages motivate future research effort on the applica-tion RHC methodologies to other cooperative control tasks in the presence of replay and DoS attacks.

On the other hand, notice that the resilience of our algorithms comes at the expense of higher computation, communication, and memory costs in compar-ison with the classic consensus algorithm. In particular, each operator needs to solve a multiparametric program at each time; a sequence of control commands has to be sent to each vehicle, and each vehicle is required to store a sequence of control commands as backup. If X is a polyhedron, the computational burden of solving the multiparametric quadratic program (9.32) can be traded with memory costs by means of explicit model predictive control initiated in [5].

9.6.5 Trade-Off between Computation, Memory, and Communication Costs

One can trade communication costs with computation costs by exploiting the idea of event or self-triggered control, for example, in [20]. In particular, each operator increases the computing horizon $n \geq \tau_{max} + 1$ and aperiodically com-putes and sends the control commands. Consider the time instant $k_{i,0} \geq 0$, and assume $s_i(k_{i,0}) = 0$. Then $\bar{\mathbf{u}}_i(k_{i,0} \to k_{i,0} + n - 1)$ is successfully delivered. After that, operator i does not compute and send any control command to vehicle i until the time instant $k_{i,0} + (n - \tau_{max} - 2)$. Since $k_{i,0} + (n - \tau_{max} - 2)$, operator i keeps executing Step 1 in Algorithm 9.1 at each time instant until $s_i(k_{i,1}) = 0$ for some $k_{i,1} \in [k_{i,0} + (n - \tau_{max} - 2), k_{i,0} + n]$. Operator i then repeats the above process after $k_{i,1}$.

Event or self-triggered control only requires operator i to perform local com-putation and communication to vehicle i at $\{k_{i,\ell}\}_{\ell \geq 0}$. However, on the other hand, self-triggered control increases the size of the n-OC and introduces larger delays into the system, potentially slowing down the convergence rate.

9.6.6 Numerical Examples

Consider a group of 10 vehicles restricted in the area $X \triangleq [-10, 10] \times [-10, 10]$. The input and velocity limits of each vehicle are $u_{max} = 5$ and $v_{max} = 2.5$, respec-tively. We study the following three cases via numerical simulations:

1. $n = 10$ and $\tau_{max} = 0$ (no attacks occur).

2. $n = 10$ and $\tau_{max} = 10$ (each adversary launches the attacks all the time except the time instants that are the multiples of 10).

3. $n = 30$ and $\tau_{max} = 30$ (each adversary launches the attacks all the time except the time instants that are the multiples of 30).

We now proceed to discuss the simulation results. Figure 9.3 is concerned with Case 3, and demonstrates that the vehicles start from four corners of the square X and eventually form the desired configuration at the center of X.

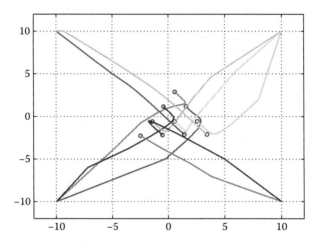

Figure 9.3: Vehicle trajectories for Case 3.

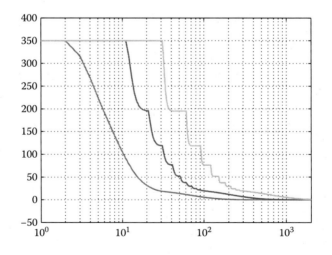

Figure 9.4: Comparison of error evolution where the lines, from left to right, are for Cases 1–3.

Figure 9.4 compares the error evolution of three cases. It is evident that a larger τ_{max} slows down the convergence rate and makes the error evolution less smooth. This coincides with the theoretic results in Theorem 9.1 and the intuition that a larger τ_{max} would produce greater damage to the operator–vehicle network.

9.7 Conclusions

We have formulated a distributed constrained formation control problem in operator–vehicle adversarial networks. To ensure the mission completion, we have proposed a novel distributed algorithm based on RHC and shown that the algorithm allows the vehicles to enforce the state and input constraints and asymptotically achieve the desired formation despite malicious replay attacks. The algorithm shows an analogous resilience to denial-of-service attacks.

In this chapter, we have focused on network attacks, including replay and DoS attacks. Future research directions include the investigation of attack-resilient control when vehicles are compromised by malware.

Acknowledgment

M.Z. is partially supported by Army Research Office grant W911NF-13-1-0421 (MURI) and National Science Foundation grant CNS-1505664. S.M. is partially supported by Air Force Office of Scientific Research grant 11RSL548.

References

1. S. Amin, A. Cardenas, and S. S. Sastry. Safe and secure networked control systems under denial-of-service attacks. In *12th ACM International Conference on Hybrid Systems: Computation and Control*, pp. 31–45, San Francisco, CA, 2009.

2. S. Amin, X. Litrico, S. Sastry, and A. Bayen. Stealthy deception attacks on water SCADA systems. In *13th ACM International Conference on Hybrid Systems: Computation and Control*, pp. 161–170, Stockholm, 2010.

3. S. Amin, G. A. Schwartz, and S. S. Sastry. Security of interdependent and identical networked control systems. *Automatica*, 49(1):186–192, 2013.

4. G. K. Befekadu, V. Gupta, and P. J. Antsaklis. Risk-sensitive control under a class of denial-of-service attack models. In *American Control Conference*, pp. 643–648, San Francisco, June 2011.

5. A. Bemporad, M. Morari, V. Dua, and E. N. Pistikopoulos. The explicit linear quadratic regulator for constrained systems. *Automatica*, 38(1):3–20, 2002.

6. D. P. Bertsekas. *Convex Optimization Theory*. Athena Scientific, Nashua, NH, 2009.

7. D. P. Bertsekas and J. N. Tsitsiklis. *Parallel and Distributed Computation: Numerical Methods*. Athena Scientific, Nashua, NH, 1997.

8. E. Camacho and C. Bordons. *Model Predictive Control*. London: Springer-Verlag, 2004.

9. N. Falliere, L. O. Murchu, and E. Chien. W32.stuxnet dossier. Symantec Corporation, Mountain View, CA, 2011.

10. H. Fawzi, P. Tabuada, and S. Diggavi. Secure estimation and control for cyber-physical systems under adversarial attacks. *IEEE Transactions on Automatic Control*, 59(6):1454–1467, 2014.

11. G. Ferrari-Trecate, L. Galbusera, M. P. E. Marciandi, and R. Scattolini. Model predictive control schemes for consensus in multi-agent systems with single- and double-integrator dynamics. *IEEE Transactions on Automatic Control*, 54(11):2560–2572, 2009.

12. H. Shisheh Foroush and S. Martínez. On multi-input controllable linear systems under unknown periodic DoS attacks. In *SIAM Conference on Control and Its Applications (CT)*, San Diego, CA, January 2013.

13. A. Gupta, C. Langbort, and T. Basar. Optimal control in the presence of an intelligent jammer with limited actions. In *IEEE International Conference on Decision and Control*, pp. 1096–1101, Atlanta, GA, December 2010.

14. D. Q. Mayne, J. B. Rawlings, C. V. Rao, and P. O. M. Scokaert. Constrained model predictive control: Stability and optimality. *Automatica*, 36(6):789–814, 2000.

15. Y. Mo, E. Garone, A. Casavola, and B. Sinopoli. False data injection attacks against state estimation in wireless sensor networks. In *IEEE International Conference on Decision and Control*, pp. 5967–5972, Atlanta, GA, December 2010.

16. Y. Mo, T. Kim, K. Brancik, D. Dickinson, L. Heejo, A. Perrig, and B. Sinopoli. Cyber-physical security of a smart grid infrastructure. *Proceedings of the IEEE*, 100(1):195–209, 2012.

17. M. Pajic, J. Weimer, N. Bezzo, P. Tabuada, O. Sokolsky, I. Lee, and G. Pappas. Robustness of attack-resilient state estimators. In *ACM/IEEE International Conference on Cyber-Physical Systems*, pp. 163–174, Berlin, April 2014.

18. F. Pasqualetti, R. Carli, and F. Bullo. A distributed method for state estimation and false data detection in power networks. In *IEEE Int. International Conference on Smart Grid Communications*, pp. 469–474, Brussels, October 2011.

19. F. Pasqualetti, F. Dorfler, and F. Bullo. Attack detection and identification in cyber-physical systems. *IEEE Transactions on Automatic Control*, 58(11): 2715–2729, 2013.

20. P. Tabuada. Event-triggered real-time scheduling of stabilizing control tasks. *IEEE Transactions on Automatic Control*, 52(9):1680–1685, 2007.

21. A. Teixeira, S. Amin, H. Sandberg, K. Johansson, and S. Sastry. Cyber security analysis of state estimators in electric power systems. In *49th IEEE Conference on Decision and Control*, Atlanta, GA, pp. 5991–5998, 2010.

22. S. Yong, M. Zhu, and E. Frazzoli. Resilient state estimation against switching attacks on stochastic cyber-physical systems. In *IEEE Conference on Decision and Control*, Osaka, Japan, 2015.

23. M. Zhu and S. Martínez. Discrete-time dynamic average consensus. *Automatica*, 46(2):322–329, 2010.

24. M. Zhu and S. Martínez. Attack-resilient distributed formation control via online adaptation. In *IEEE International Conference on Decision and Control*, pp. 6624–6629, Orlando, FL, December 2011.

25. M. Zhu and S. Martínez. Stackelberg game analysis of correlated attacks in cyber-physical system. In *American Control Conference*, pp. 4063–4068, San Francisco, CA, June 2011.

26. M. Zhu and S. Martínez. On distributed resilient consensus against replay attacks in adversarial networks. In *American Control Conference*, pp. 3553–3558, Montreal, Canada, June 2012.

27. M. Zhu and S. Martínez. On distributed constrained formation control in operator–vehicle adversarial networks. *Automatica*, 49(12):3571–3582, 2013.

28. M. Zhu and S. Martínez. On attack-resilient distributed formation control in operator–vehicle networks. *SIAM Journal on Control and Optimization*, 52(5):3176–3202, 2014.

29. M. Zhu and S. Martínez. On the performance of resilient networked control systems under replay attacks. *IEEE Transactions on Automatic Control*, 59(3):804–808, 2014.

30. Q. Zhu, C. Rieger, and T. Basar. A hierarchical security architecture for cyber-physical systems. In *International Symposium on Resilient Control Systems*, pp. 15–20, Boise, ID, 2011.

Chapter 10

Privacy-Preserving Data Access Control in the Smart Grid

Kan Yang

Department of Electrical and Computer Engineering, University of Waterloo

Xiaohua Jia

Department of Computer Science, City University of Hong Kong

Xuemin (Sherman) Shen

Department of Electrical and Computer Engineering, University of Waterloo

CONTENTS

The smart grid, as the next generation of power grids, has attracted tremendous attention from both academia and the industrial community. Due to the two-way flows of energy and information, the smart grid enables utilities to sense and monitor the status in the power grid and respond to changes in power demand, supply, and cost in a close-loop way. The two-way communication, however, also raises new security and privacy challenges to power systems. In this chapter, we first describe a conceptual framework of the smart grid and propose some critical data security and privacy issues in the smart grid. Then, we review several popular data access control models and discuss their suitability to be adopted for ensuring data security and privacy in the smart grid. After that, we propose a privacy-preserving energy consumption data access control (ECD-AC) scheme as a cryptographic implementation of attribute-based access control in the smart grid. Finally, we summarize this chapter and point out some open problems in the design of data access control schemes for the smart grid.

10.1 Introduction

The smart grid is regarded as the revolutionary and evolutionary generation of the power grid, which integrates power system engineering with information and communication technologies. Due to bidirectional flows of energy and two-way communication of information, the smart grid enables utilities to actively sense and monitor each interconnected element in the power grid, ranging from power generation to distribution and consumption, and respond to changes in power demand, supply, and costs in a close-loop way [13, 18]. On the other hand, the smart grid also enables consumers to manage their energy use efficiently and conveniently [14].

The smart grid not only can increase the energy efficiency and reliability of the power grid significantly [6], but also can reduce greenhouse gas emissions by using renewable energy resources, such as solar, wind, and biomass. According to Institute of Electrical and Electronics Engineers (IEEE) smart grid [9], in the United States, the cost of nationwide smart grid ranges from $20 billion to $25 billion per year (over a 20-year period), while the benefits it can bring are $70 billion per year.

However, the two-way functionality and digital intelligence of the smart grid may also increase the risk of cyber attacks and vulnerabilities [2, 12, 15, 20]. The more endpoints and interconnected networks that are involved in the smart grid, the more ways for security problems to get in. For example, smart meters were shown to be the most vulnerable element of the smart grid during the Black Hat Security Conference in 2009. Therefore, in order to make the smart grid truly "smart," it is critical to build an intrinsic security strategy to safeguard smart grid infrastructures, including smart meters, switches, gateways, supervisory control and data acquisition (SCADA) control centers, databases, and energy participants.

Electrical consumption data collected from remote terminal units (RTUs) can be used to monitor the status of the power system, distribute electricity, predict future conditions, and calculate costs. However, these data may not only contain sensitive information (e.g., home address, name, and account information). but also reveal users' personal behaviors. For example, the real-time energy and appliance usage data may reveal personal activities at home such as showering, cooking, sleeping, and surfing the Internet, or even determine if anyone is home. Energy usage patterns over time can also determine the number of people in the household, work schedule, vocation, and other lifestyle habits. Sometimes, customers may employ third parties to assist them in better managing energy consumption by analyzing their energy usage data, which also provides the opportunity for dishonest third parties to abuse or misuse smart grid data, for example, theft of physical property and surveillance of residences or business. On the other hand, customer home area networks (HANs) and building energy management (BEM) systems may also want to interact with the electric utilities as well as third-party energy service providers to access their own energy profiles, usage, pricing, and so forth. Thus, data access control becomes one of the most critical and challenging issues in the smart grid.

In this chapter, we investigate the data access control problem in the smart grid. Specifically, we first describe the system architecture of the smart grid and abstract several critical data security and privacy issues in the smart grid. Then, we review some access control models in the literature and discuss their suitability to be adopted into the smart grid. We also take the energy consumption data as an example and propose a cryptographic implementation for attribute-based access control in the smart grid. Finally, we summarize this chapter and provide some open problems in the design of data access control schemes in the smart grid.

10.2   Architecture of the Smart Grid

Extensive studies have been conducted to design the architecture of the smart grid from academia [25], industry [7], and government [17, 18]. The conceptual reference model proposed by the U.S. National Institute of Standards and Technology (NIST) [18] is considered the most acceptable framework for the smart grid. As shown in Figure 10.1, NIST identifies seven domains in the smart grid: customers, markets, service providers, operations, bulk generation, transmission, and distribution.

- **Customers:** The end users of electricity, who may also generate, store, and manage the use of energy. Traditionally, there are three types of customers: residential, commercial, and industrial.

- **Markets:** The operators and participants in electricity markets.

- **Service providers:** The organizations providing services to electrical customers and utilities.

- **Operations:** The managers of the movement of electricity.

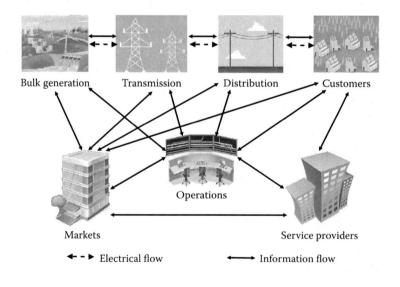

Figure 10.1: A conceptual reference model of the smart grid framework from NIST.

- **Bulk generation:** The generators of electricity in bulk quantities, such as by fossil, nuclear, and hydropower plants and wind farms. They may also store energy for later distribution.

- **Transmission:** The carriers of bulk electricity over long distances, who may also store and generate electricity.

- **Distribution:** The distributors of electricity to and from customers, who may also store and generate electricity.

10.3 Data Security and Privacy Issues in the Smart Grid

In the smart grid, a benefit from the advanced communication and information technologies is that a wide range of data from all levels of the power grid can be accessed, analyzed, and responded to. In this section, we mainly consider two major types of data in the smart grid: grid status data and energy consumption data.

- *Grid status data*: Represents the data related to electrical behavior and status of the power grid, which includes voltage, current phasors, real and reactive power flows, demand response capacity, distributed energy capacity, power flows, and price changes.

- *Energy consumption data:* Represents the data related to the energy consumption collected from customers, which include reactive power, peak power, voltage, appliance, time of usage, and energy consumption.

Data security and privacy are essential to provide stable and sustainable development of the smart grid. From the security perspective, the data in the smart grid may suffer from the following security attacks:

- *Loss of confidentiality:* Data may be disclosed to unauthorized users.

- *Loss of integrity:* Data may be modified or destroyed.

- *Loss of availability:* Data may not be timely and reliably accessible.

For grid status data, availability and integrity are the most important security concerns in the smart grid from the perspective of system reliability.

With the increasing availability of customer information online, data confidentiality and privacy are becoming more and more significant when the data are related to individual consumers or dwellings. For example, the energy consumption data may contain personal information such as house number, street address, name of homeowner or resident, and date of birth. It may also reflect the timing and energy usage for each appliance, which could provide a detailed timeline of activities occurring inside the home, such as showering, cooking, sleeping, and surfing the Internet, or even determine if anyone is home. Energy usage patterns over time can also determine the number of people in the household, work schedule, vocation, and other lifestyle habits. Specifically, the energy consumption data can be applied to

■ **Determine personal behavior patterns:** Smart meters may track the use of specific appliances, including the specific time and location, as well as the amount of electricity consumption. These data can also indicate the timeline of activities inside the home. By analyzing the energy consumption data, it is easy to determine personal behavior patterns and user living habits.

■ **Perform real-time remote surveillance:** The real-time energy usage data can reveal whether there are any people at home and what they are doing,

■ **Provide commercial use:** The energy consumption data can reveal lifestyle information that may be valuable for many entities. For example, the data may be purchased by vendors of a wide range of products and services for targeting sale and marketing campaigns.

Due to the large amount of data generated in the smart grid and the flexibility and scalability of cloud computing, it is a trend to employ cloud computing for data storage and processing. However, special attention should be paid when using cloud computing in the smart grid, as cloud service providers may not be fully trusted. When storing data in the cloud, users lose physical control of their data, so they may worry about their data being corrupted or deleted in the cloud [21] or given to unauthorized users for more profit gains [24]. Some customers may also employ third parties to analyze their energy consumption profiles and provide better energy use strategies. These third parties may also be dishonest and disclose users' personal information or abuse users' data. Therefore, effective data access control is desired to guarantee confidentiality and privacy of data in the smart grid, especially for the energy consumption data collected from customers.

10.4 Data Access Control in the Smart Grid

Data access control is an effective approach to guarantee data confidentiality and privacy. As shown in Figure 10.2, we review several popular access control models in the literature, including access control lists, role-based access control, and attribute-based access control.

10.4.1 Access Control Lists

Access control lists (ACLs) are one of the most basic models of access control, where data are associated with a list of mappings between the set of entities

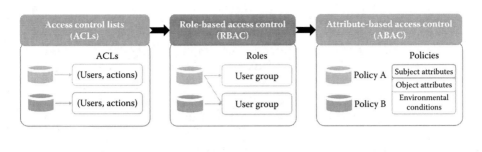

Figure 10.2: Access control models.

and the set of actions that each entity can take on the data. For example, the home address can be accessed by operators (read only), markets (read only), and utility service providers (read, modify, and delete). Some grid statuses can only be accessed by power control centers. Whenever a user tries to perform any action (e.g., modify) on the data (e.g., home address), the system or server checks the ACL of this data and determines whether to allow this action for that user. The ACLs can also be applied for a group of data (e.g., account information: name, home address, and contact information), which may have the same mappings between actions and entities.

The major limitation of the ACL model is that every user is treated as a distinct entity with distinct sets of privileges for each piece of data. That means ACLs have to be defined separately for each piece of data (or group of data), which would be a cumbersome process when many users have different levels of access permissions to a large amount of data. It is time-consuming and error-prone to selectively add, delete, and change ACLs on individual data, or even groups of data.

10.4.2 Role-Based Access Control

Role-based access control (RBAC) determines whether access will be granted or denied based on a user's role or function. Since many people may have the same role (e.g., market researcher), RBAC may group them into one category with a particular role, which means that the access control permission on a particular piece of data can be set only once for all members in the group with the same role. Users can also be members of multiple groups, which may have different access permissions, where more restrictive permissions override general permissions in RBAC.

Although RBAC has many advantages compared with the ACL model, it still suffers from some limitations. One of the most important is that it is difficult to achieve fine-grained access control for each user when grouping them into categories based on roles. So, how to differentiate individual members of a group is a challenging problem in RBAC.

10.4.3 Attribute-Based Access Control

In order to define specific access permissions for individual users, attribute-based access control (ABAC) is proposed to control data access based on the attributes associated with the user. According to the definition of NIST [10], in ABAC, the access policy is defined based on both subject attributes and object attributes under some environmental conditions. ABAC is more flexible in defining access policies than RBAC.

A key advantage of the ABAC model is that no specific users need to be known in advance when defining access policies. The access can be granted as long as the attributes of users can satisfy access policies. Thus, ABAC is useful for applications where organizations or data owners allow unanticipated users to access the data if their attributes can satisfy some policies. This feature makes ABAC well suited for large enterprises. An ABAC system can implement existing role-based access control policies and support a migration from role-based to a more granular access policy-based control of many different attributes of individual users. Gartner predicts that "by 2020, 70% of all businesses will use ABAC as the dominant mechanism to protect critical assets, up from 5% today [in 2013]" [26].

10.4.4 Encryption-Based Implementation

The ACLs, RBAC, and ABAC all require the system or server to evaluate access rules or policies, and make access decisions. Imagining that a server is not fully trusted by data owners, for example, cloud servers, how to protect data confidentiality and privacy becomes a challenging problem. In order to cope with this challenge, cryptographic techniques are applied to implement the above-mentioned three access control models. Data owners encrypt files by using the symmetric encryption approach with content keys and then use every user's public key to encrypt the content keys. However, traditional public key encryption methods may produce many copies of ciphertexts for multiple users, which is not efficient in large-scale systems, for example, the smart grid. Moreover, the key management is also a complex issue in public key infrastructures (PKIs).

To eliminate the need for distributing public keys (or maintaining a certificate directory), identity-based encryption (IBE) [5] is a new cryptographic primitive that allows data owners to encrypt their data with identities of users instead of their public keys. To make access policies more flexible and expressive, attribute-based encryption (ABE) [4, 8] is proposed for access control of encrypted data. There are two complementary forms of ABE: key-policy ABE (KP-ABE) [8] and ciphertext-policy ABE (CP-ABE) [4]. In KP-ABE systems, keys are associated with access policies and ciphertext is associated with a set of attributes; in CP-ABE systems, keys are associated with a set of attributes and ciphertext is associated with access policies.

10.5   Privacy-Preserving Access Control Scheme for Energy Consumption Data

In this section, we propose a cryptographic implementation of attribute-based access control (ABAC) for the privacy-preserving access control of energy consumption data based on a multiauthority ciphertext-policy attribute-based encryption method [11].

10.5.1   Cloud-Based Energy Consumption Data Control Model

We consider the cloud-based energy consumption data control model, as shown in Figure 10.3, which usually consists of the following types of entities: customers (data owners), utility company, cloud servers, data users, and attribute authorities (AAs).

> **Customers:** Customers are the owners of the energy consumption data. As customers may not trust the utility company (or the cloud server) to control the access of their data, before sending the data to the cloud server, the customer (data owner) defines an access policy over some attributes and encrypts the data under this access policy.

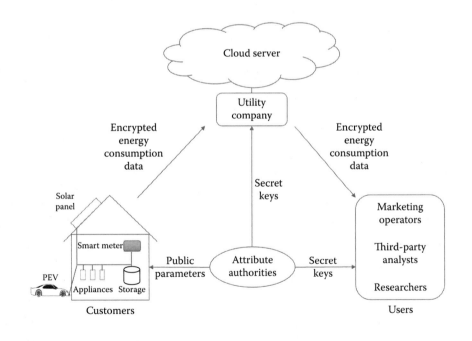

Figure 10.3: System model of attribute-based access control for energy consumption data. PEV, plug-in electric vehicle.

Utility company: The utility company provides utility services for customers. Due to the large volume of energy consumption data and the flexibility and scalability of cloud computing, a cloud server is a natural fit for storing and processing energy consumption data in the smart gird.

Cloud server: The cloud server stores the energy consumption data of customers and provides data access service to data users. However, the cloud server may be curious about the data content and disclose the data to unauthorized users.

Data users: Data users may be marketing operators, third-party analysts, researchers, and so forth. Each user may hold attributes from multiple attribute authorities. For each attribute, the user may receive a secret key associated with this attribute from the corresponding attribute authority. Users are dishonest in the sense that they may collude to try to access unauthorized data, but they cannot collude with the cloud server or other adversaries.

Attribute authorities (AAs). Each AA is responsible for managing attributes in its domain in the system. It assigns attributes to users according to the role or identity in its domain, and releases public parameters that are used for data encryption. Each AA is fully trusted in the system, and the channels between each AA and users are secure.

10.5.2 Preliminary Definitions

We first give some preliminary definitions of bilinear pairing and the linear secret-sharing schemes (LSSS) structure [3].

10.5.2.1 Bilinear Pairing

Let G_1, G_2, and G_T be three multiplicative groups with the same prime order p. A bilinear map is a map $e : G_1 \times G_2 \to G_T$ with the following properties:

1. Bilinearity: $e(u^a, v^b) = e(u, v)^{ab}$ for all $u \in G_1$, $v \in G_2$, and $a, b \in \mathbb{Z}_p$.

2. Nondegeneracy: There exist $u \in G_1$, $v \in G_2$ such that $e(u, v) \neq I$, where I is the identity element of G_T.

3. Computability: e can be computed in an efficient way.

Such a bilinear map is called a bilinear pairing.

If g_1 and g_2 are the generators of G_1 and G_2, respectively, $e(g_1, g_2)$ is the generator of G_T. The bilinear pairing applied in our proposed scheme is symmetric, where $G_1 = G_2 = G$.

10.5.2.2 LSSS Structure

Definition 10.1 LSSS A secret-sharing scheme Π over a set of parties $\mathcal{P}$ is called linear (over $\mathbb{Z}_p$) if

1. The shares for each party form a vector over $\mathbb{Z}_p$.

2. There exists a matrix M called the share-generating matrix for Π. The matrix M has l rows and n columns. For all $i = 1, \ldots, l$, the ith row of M is labeled by a party $\rho(i)$ (ρ is a function from $\{1, \ldots, l\}$ to $\mathcal{P}$). When we consider the column vector $v = (s, r_2, \ldots, r_n)$, where $s \in \mathbb{Z}_p$ is the secret to be shared and $r_2, \ldots, r_n \in \mathbb{Z}_p$ are randomly chosen, Mv is the vector of l shares of the secret s according to Π. The share $(Mv)_i$ belongs to party $\rho(i)$.

Every linear secret-sharing scheme according to the above definition also enjoys the *linear reconstruction* property: Suppose that Π is an LSSS for the access structure $\mathbb{A}$. Let $S \in \mathbb{A}$ be any authorized set, and let $I \subset \{1, 2, \ldots, l\}$ be defined as $I = \{i : \rho(i) \in S\}$. Then, there exist constants $\{w \in \mathbb{Z}_p\}_{i \in I}$ such that for any valid shares $\{\lambda_i\}$ of a secret s according to Π, we have $\sum_{i \in I} w_i \lambda_i = s$. These constants $\{w_i\}$ can be found in time polynomial in the size of the share-generating matrix M. We note that for unauthorized sets, no such constants $\{w_i\}$ exist.

10.5.3 Definition of Framework

We define the framework of the energy consumption data access control (ECD-AC) scheme as

Definition 10.2 ECD $-$ AC. The ECD-AC scheme consists of the following algorithms:

- **GlobalSetup**$(\lambda) \rightarrow$ PP. The global setup algorithm takes no input other than the implicit security parameter λ. It outputs the public parameters PP for the system.

- **AASetup**$(\text{PP}, aid) \rightarrow (\text{SAK}_{aid}, \text{PAK}_{aid})$. The AA setup algorithm is run by each AA with its identity aid. It takes the public parameters PP and the authority identity aid as inputs, and outputs a pair of secret–public authority keys $(\text{SAK}_{aid}, \text{PAK}_{aid})$ for this AA.

- **SKeyGen**$(\text{SAK}_{aid}, \text{PP}, S_{aid,uid}) \rightarrow \text{SK}_{aid,uid}$. The secret key generation algorithm takes as inputs the secret authority key SAK_{aid}, the public parameters PP, and a set of attributes $S_{aid,uid}$. It outputs a secret key $\text{SK}_{aid,uid}$ for the user uid.

- **Encrypt**$(m, \text{PP}, \{\text{PAK}_{aid}\}, \mathbb{A}) \rightarrow \text{CT}$. The encryption algorithm takes as inputs the message m, the public parameters PP, a set of relevant public authority keys $\{\text{PAK}_{aid}\}$, and an access policy $\mathbb{A}$. It outputs a ciphertext CT.

- **Decrypt**$(\text{CT}, \mathbb{A}, \{\text{SK}_{aid,uid}\}) \rightarrow m$. The decryption algorithm takes as inputs the ciphertext CT, the access policy $\mathbb{A}$, and a set of secret keys $\text{SK}_{aid,uid}$

corresponding to the access policy. It outputs the data m if the user's attributes satisfy the access policy. Otherwise, the decryption fails and outputs $\perp$.

10.5.4 Definition of Security Model

The security model for the ECD-AC scheme is defined against *indistinguishable chosen plaintext attacks* (IND-CPAs) by a security game between a challenger and an adversary. In this game, the adversaries can corrupt attribute authorities only statically, but key queries can be made adaptively:

- **Setup:** The system is initialized by running the global setup algorithm. The adversary specifies a set $S'_A \subset S_A$ of corrupted attribute authorities. The challenger generates the pairs of public authority key and secret authority key by running the AA setup algorithm. For uncorrupted attribute authorities in $S_A - S'_A$, the challenger only sends public authority keys to the adversary. For corrupted authorities in S'_A, the challenger sends both public authority keys and secret authority keys to the adversary.

- **Phase 1:** The adversary makes secret key queries by submitting pairs $(uid, S_{aid,uid})$ to the challenger, where uid is a user identity and $S_{aid,uid}$ is a set of attributes belonging to an uncorrupted authority aid. The challenger retrieves the corresponding secret keys $\text{SK}_{aid,uid}$ from the $SKeyGen(*)$ oracle and sends them to the adversary.

- **Challenge:** The adversary submits two equal-length messages m_0 and m_1. In addition, the adversary gives a set of challenge access structures $\{(M_1^*, \rho_1^*), \ldots, (M_q^*, \rho_q^*)\}$ that must satisfy the constraint that the adversary cannot ask for a set of keys that allow decryption, in combination with any keys that can obtained from corrupted authorities. The challenger then flips a random coin b, and encrypts m_b under all access structures $\{(M_1^*, \rho_1^*), \ldots, (M_q^*, \rho_q^*)\}$. Then, the ciphertexts $\{CT_1^*, \ldots, CT_q^*\}$ are given to the adversary.

- **Phase 2:** The adversary may query more secret keys, as long as they do not violate the constraints on the challenge access structures.

- **Guess:** The adversary outputs a guess b' of b.

The advantage of an adversary $\mathcal{A}$ in the above security game is defined as

$$\mathbf{Adv}_{ECD-AC}^{IND-CPA} = Pr[b' = b] - \frac{1}{2}.$$

Definition 10.3 The ECD-AC scheme is IND-CPA secure against static corruption of authorities if all polynomial time adversaries have at most a negligible advantage in the security game.

10.5.5 Construction of ECD-AC Scheme

10.5.5.1 System Initialization

The system initialization consists of both global setup phase and an AA setup phase:

- *Phase 1: Global setup.* The global setup algorithm takes no input other than the implicit security parameter λ. It chooses two multiplicative groups $\mathbb{G}$ and $\mathbb{G}_T$ with the same prime order p and the bilinear map $e : \mathbb{G} \times \mathbb{G} \to \mathbb{G}_T$ between them. Let g be a generator of $\mathbb{G}$. A random oracle H maps global user identity *uid* to an element in $\mathbb{G}$. The public parameters PP are published as

$$\text{PP} = (p, \mathbb{G}, \mathbb{G}_T, g, H).$$

- *Phase 2: AA setup.* Each AA with *aid* runs the AA setup algorithm to generate its secret–public key pair. Let S_{aid} denote the whole set of attributes managed by AA with *aid*. For each attribute $x \in S_{aid}$, the AA with *aid* chooses two random exponents $\alpha_{aid,x}, \beta_{aid,x} \in \mathbb{Z}_p$ and publishes its public key as

$$\text{PAK}_{aid} = \{e(g,g)^{\alpha_{aid,x}}, g^{\beta_{aid,x}}\}_{\forall x \in S_{aid}}.$$

  It keeps

$$\text{SAK}_{aid} = \{\alpha_{aid,x}, \beta_{aid,x}\}_{\forall x \in S_{aid}}.$$

10.5.5.2 Key Generation by AAs

For each user (including the utility company) with *uid*, each *AA* with *aid* first evaluates the role or identity in its domain and assigns a set of attributes $S_{aid,uid}$ to this user. Then, the AA with *aid* generates a secret key $\text{SK}_{aid,uid}$ for this user as

$$\text{SK}_{aid,uid} = \{(K_{x,aid,uid} = g^{\alpha_{aid,x}} H(uid)^{\beta_{aid,x}}, L_{x,aid,uid} = g^{\beta_{aid,x}})\}_{x \in S_{aid,uid}}.$$

10.5.5.3 Data Encryption by Customers

The customer (data owner) encrypts the energy consumption data before sending them to the cloud server. In real application, the data are first encrypted with a content key by using symmetric encryption methods. Then, the content key is further encrypted by running the encryption algorithm Encrypt. For simplification, here we directly use the data in the following description. To encrypt data m, the data owner first defines an access policy over attributes from multiple authorities. The access policy is described by an LSSS structure (M, ρ), where M is an $n \times l$ access matrix and ρ maps the rows of M to attributes. The data owner then runs the following encryption algorithm to encrypt the data m.

Encrypt$(m, \text{PP}, \{\text{PAK}_{aid}\}, (M, \rho)) \to \text{CT}$. The algorithm takes as inputs the data m, the public parameters PP, the public authority keys $\{\text{PAK}_{aid}\}$ of relevant AAs, and the access policy (M, ρ). It first chooses a random encryption secret $s \in \mathbb{Z}_p^*$ and shares it by a random vector $\vec{v} = (s, y_2, \ldots, y_l)$. Moreover, it also chooses another random vector $\vec{w}$ with 0 as its first entry. For $i = 1$ to n, it computes $\lambda_i = M_i \cdot \vec{v}$ and $w_i = M_i \cdot \vec{w}$, where M_i is the vector corresponding to the ith row of M. It outputs the ciphertext CT as

$$\text{CT} = (\ C = m \cdot e(g,g)^s,$$

$$\text{for } i = 1 \text{ to } n:$$

$$C_{1,i} = e(g,g)^{\lambda_i} \cdot \left(e(g,g)^{\alpha_{\phi(\rho(i)),\rho(i)}}\right)^{r_i},$$

$$C_{2,i} = g^{r_i},\ C_{3,i} = g^{\beta_{\phi(\rho(i)),\rho(i)} r_i} \cdot g^{w_i}\ ),$$

where $r_i \in \mathbb{Z}_p^*$ is randomly selected and ϕ maps the attribute to the identity of its corresponding AAs, that is, $\phi(x) = aid$ if attribute x belongs to AA with aid.

Then, the encrypted energy consumption data CT are sent to the utility company, which may be stored in cloud servers.

10.5.5.4 Data Decryption by Users

If the attributes of the user with uid satisfy the access policy, that is, the secret keys contain sufficient $\{K_{\rho(i),\phi(\rho(i)),uid}\}$ for a subset of rows i of M such that $(1, 0, \ldots, 0)$ is in the span of these rows, the user can decrypt the ciphertext as follows.

For each such i, the user computes

$$\frac{C_{1,i} \cdot e(H(uid), C_{3,i})}{e(K_{\rho(i),\phi(\rho(i)),uid}, C_{2,i})} = \frac{e(g,g)^{\lambda_i} e(g,g)^{\alpha_{\phi(\rho(i)),\rho(i)} r_i} e(H(uid), g^{\beta_{\phi(\rho(i)),\rho(i)} r_i} \cdot g^{w_i})}{e(g^{\alpha_{\phi(\rho(i)),\rho(i)}} H(uid)^{\beta_{\phi(\rho(i)),\rho(i)}}, g^{r_i})}$$

$$= e(g,g)^{\lambda_i} \cdot e(H(uid), g)^{w_i}.$$

Since the attributes satisfy the access policy, the user can find a set of constants $\{c_i\}$, such that

$$\sum_i c_i \cdot M_i = (1, 0, \ldots, 0).$$

Recall that

$$\lambda_i = M_i \cdot \vec{v}$$

and

$$w_i = M_i \cdot \vec{w};$$

we have

$$\sum_i c_i \lambda_i = s$$

and

$$\sum_i c_i w_i = 0.$$

The user then decrypts the data as

$$\frac{C}{\prod_i \left(e(g,g)^{\lambda_i} \cdot e(H(uid),g)^{w_i} \right)^{c_i}} = \frac{m \cdot e(g,g)^s}{e(g,g)^{\sum_i c_i \lambda_i} \cdot e(H(uid),g)^{\sum_i c_i w_i}}$$

$$= \frac{m \cdot e(g,g)^s}{e(g,g)^s}$$

$$= m.$$

We can see that access policies are enforced by the cryptography, which means that no entity is required to evaluate access policies and make access decisions. Moreover, only one ciphertext is produced for fine-grained access control with multiple users.

10.5.6 Security Analysis

Theorem 10.1
The ECD-CA scheme is indistinguishable secure against static corruption of authorities and chosen plaintext attacks under the generic bilinear group model [11] and random oracle model.

Proof 10.1 The ECD-CA scheme is constructed based on the Multi-Authority Ciphertext-Policy Attribute-Based Encryption (MA-CP-ABE) method with primer group order (MA-CP-ABE-Primer) in [11], which is proved to be secure under the generic bilinear group model and random oracle model. At an intuitive level, if there are any vulnerabilities in the scheme, these vulnerabilities must exploit specific mathematical properties of elliptic curve groups or cryptographic hash functions used when instantiating the scheme. If $\mathcal{A}$ is an adversary who can break our scheme with nonnegligible advantage, it can also break the MA-CP-ABE-Primer scheme in [11] with nonnegligible advantage. □

10.5.7 Performance Evaluation

We first summarize the size of each component in the ECD-AC scheme. As shown in Table 10.1, $|p|$ is the element size of $\mathbb{G}$ and $\mathbb{G}_T$, $N_{att,aid}$ is the total number of attributes managed by attribute authority *aid*, $n_{uid,aid}$ represents the number of attributes assigned to a user with *uid* from AA with *aid*, and $n_{att,m}$ is the number of attributes associated with data *m*.

Table 10.1   Size of Each Component in ECD-AC

Component	PK	PAK_{aid}	SAK_{aid}	$SK_{uid,aid}$	CT
Size ($\|p\|$)	5	$2N_{att,aid}$	$N_{att,aid}$	$2n_{uid,aid}$	$3n_{att,m}+1$

10.5.7.1   Storage Overhead

Customers only need to store the public parameters PP and the public authority key PAK of each authority. Users only hold the secret keys issued from each attribute authority, which is linear to the number of attributes they possess. Each authority stores the secret authority key SAK and the public parameters PP. The ciphertexts are stored in the cloud server.

10.5.7.2   Communication Cost

Toward the communication cost, we also describe it by using the size of each component. The communication cost between the authority and the customer is only the public parameters. The secret keys contribute the communication cost between the authority and the users. The communication cost between the cloud server and customers and users comes from the ciphertexts.

10.5.7.3   Computation Complexity

To evaluate the computation complexity of the ECD-AC scheme, we conduct the simulation on a Unix system with an Intel Core i5 CPU at 2.4 GHz and 8.00 GB RAM. The code uses the pairing-based cryptography (PBC) library version 0.5.12 and a symmetric elliptic curve α-curve, where the base field size is 512 bits and the embedding degree is 2. All the simulation results are the mean of 20 trials.

Figure 10.4 shows the computation overhead of the data encryption at the customer's end. As we can see in the figure, the time for data encryption is linear with the number of attributes involved in the ciphertexts. Figure 10.5 shows that the decryption time at the user's end is also linear with the number of attributes that satisfy the access policy.

10.6   Summary and Outlook

To summarize this chapter, we have investigated the system architecture of the smart grid and abstracted some critical data security and privacy issues in the smart grid. We have also reviewed some access control schemes in the literature and discussed their suitability to be adopted into the smart grid. Taking the energy consumption data as an example, we have further proposed an energy consumption data access control (ECD-AC) scheme for the smart grid and provided the security analysis and performance evaluation.

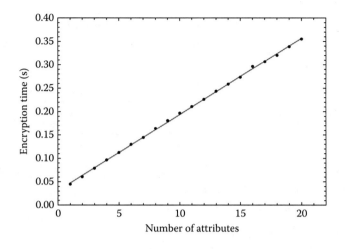

Figure 10.4: Computation time of data encryption.

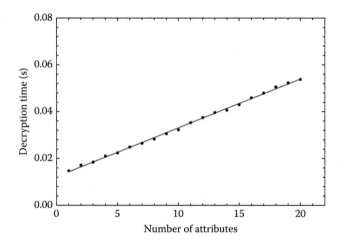

Figure 10.5: Computation time of data decryption.

There are still many research problems in the design of attribute-based access control schemes. For example, how do we efficiently revoke the attribute when a user's attributes have been changed? How do we update the access control policies for those published data? Although many studies have been done in data access control for cloud storage systems [1, 19, 21–25], how to apply these technologies for the smart grid is still an open problem. For example, in a cloud storage system, people care more about the efficiency of data decryption, while in the smart grid, the encryption efficiency is more interesting and significant, as smart meters or gateways in home area networks (HANs) are not so powerful [16].

References

1. H. Bao, R. Lu, B. Li, and Deng R. Blithe: Behavior rule based insider threat detection for smart grid. *IEEE Internet of Things Journal*, [Epub ahead of print].

2. A. Bari, J. Jiang, W. Saad, and A. Jaekel. Challenges in the smart grid applications: An overview. *International Journal of Distributed Sensor Networks*, pp. 1–11, 2014.

3. A. Beimel. Secure schemes for secret sharing and key distribution. PhD thesis, Israel Institute of Technology, Haifa, 1996.

4. J. Bethencourt, A. Sahai, and B. Waters. Ciphertext-policy attribute-based encryption. In *Proceedings of S&P*, pp. 321–334, Berkeley, CA, 2007.

5. D. Boneh and M. Franklin. Identity-based encryption from the weil pairing. In *Proceedings of CRYPTO*, pp. 213–229, Santa Barbara, CA, 2001.

6. B. Chai, J. Chen, Z. Yang, and Y. Zhang. Demand response management with multiple utility companies: A two-level game approach. *IEEE Transactions on Smart Grid*, 5(2):722–731, 2014.

7. Cisco. Internet protocol architecture for the smart grid. Cisco, San Jose, CA, 2009.

8. V. Goyal, O. Pandey, A. Sahai, and B. Waters. Attribute-based encryption for fine-grained access control of encrypted data. In *Proceedings of CCS*, pp. 89–98, Alexandria, VA, 2006.

9. IEEE (Institute of Electrical and Electronics Engineers). Smart grid consumer benefits. http://smartgrid.ieee.org/questions-and-answers/964-smart-grid-consumer-benefits.

10. V. Hu, D. Ferraiolo, R. Kuhn, A. Schnitzer, K. Sandlin, R. Miller, and K. Scarfone. Guide to attribute based access control (ABAC) definition and considerations. NIST Special Publication 800–162. National Institute of Standards and Technology, Gaithersburg, MD, 2014.

11. A. Lewko and B. Waters. Decentralizing attribute-based encryption. In *Proceedings of EUROCRYPT*, pp. 568–588, Tallinn, Estonia, 2011.

12. X. Li, X. Liang, R. Lu, X. Shen, X. Lin, and H. Zhu. Securing smart grid: Cyber attacks, countermeasures, and challenges. *IEEE Communications Magazine*, 50(8):38–45, 2012.

13. H. Liang, A. Tamang, W. Zhuang, and X. Shen. Stochastic information management in smart grid. *IEEE Communications Surveys and Tutorials—Special Issue on Energy and Smart Grid*, 16(3):1746–1770, 2014.

14. X. Liang, X. Li, R. Lu, X. Lin, and X. Shen. UDP Usage-based dynamic pricing with privacy preservation for smart grid. *IEEE Transactions on Smart Grid*, 4(1):141–150, 2013.

15. R. Lu, X. Liang, X. Li, X. Lin, and X. Shen. EPPA: An efficient and privacy-preserving aggregation scheme for secure smart grid communications. *IEEE Transactions on Parallel and Distributed Systems*, 23(9):1621–1631, 2012.

16. U.S. Department of Energy. The smart grid: An introduction. U.S. Department of Energy, Washington, DC, 2009.

17. U.S. Department of Energy. 2014 smart grid system report. http://energy.gov/sites/prod/files/2014/08/f18/SmartGrid-SystemReport2014.pdf.

18. National Institute of Standards and Technology. NIST framework and roadmap for smart grid interoperability standards. Release 1.0. National Institute of Standards and Technology, Gaithersburg, MD, 2010, p. 33. http://www.nist.gov/public_affairs/releases/upload/smartgrid_interoperability_final.pdf.

19. X. Shen. Empowering the smart grid with wireless technologies. *IEEE Network*, 3:2–3, 2012.

20. W. Wang and Z. Lu. Cyber security in the smart grid: Survey and challenges. *Computer Networks*, 57(5):1344–1371, 2013.

21. K. Yang and X. Jia. Data storage auditing service in cloud computing: Challenges, methods and opportunities. *World Wide Web*, 15:409–428, 2012.

22. K. Yang and X. Jia. Expressive, efficient, and revocable data access control for multi-authority cloud storage. *IEEE Transactions on Parallel and Distributed Systems*, 25(7):1735–1744, 2014.

23. K. Yang, X. Jia, and K. Ren. Secure and verifiable policy update outsourcing for big data access control in the cloud. *IEEE Transactions on Parallel and Distributed Systems*, 26(12):3461–3470, 2015.

24. K. Yang, X. Jia, K. Ren, B. Zhang, and R. Xie. DAC-MACS: Effective data access control for multiauthority cloud storage systems. *IEEE Transactions on Information Forensics and Security*, 8(11):1790–1801, 2013.

25. A. Zaballos, A. Vallejo, and J. Selga. Heterogeneous communication architecture for the smart grid. *IEEE Network*, 25(5):30–37, 2011.

26. Gartner. Gartner IAM Summit, Los Angeles, CA, November 2013.

Index

AAs, *see* Attribute authorities (AAs)
AA setup algorithm, 295, 296, 297
ABAC, *see* Attribute-based access control (ABAC)
ABE, *see* Attribute-based encryption (ABE)
Abstract model, of industrial control system, 60
Access control lists (ACLs), 290–291
Access policy, 292, 293, 297, 298, 299
ACLs, *see* Access control lists (ACLs)
Ad hoc network, 234
Adjacency matrix, 229
Adversaries, 296
 model of, 258–259
 and objective, 260–261
 scenarios, 231–233
  fault attack, 233
  random attack, 233
AGC, *see* Automatic generation control (AGC)
Anomaly-based detection methods, 147
A priori system model knowledge, 60
Attack detection, 254–255
Attack model, 101–102, 129–130
Attack-resilient control, 255
Attack schedule
 different, 108, 109
 optimal, 105–107, 109, 110, 114

system control performance under, 102–105
Attribute authorities (AAs), 294, 297, 300
Attribute-based access control (ABAC), 292, 293
Attribute-based encryption (ABE), 292
Automatic generation control (AGC), 227

β-dominance attack, 85–87
Bayes error, 83
Bayesian network, 169
Bayesian reputation function, 234
BC, *see* Bhattacharyya coefficient (BC)
BD, *see* Bhattacharyya distance (BD)
BD-detector, 88–89, 90
Behavior rules, 130–131
Beta distribution, 134, 137
Bhattacharyya coefficient (BC), 82
Bhattacharyya distance (BD), 82–83, 84
Bilinear pairing, 294
Binomial random graphs
 1-intersection, 203–206
 s-intersection, 187, 210–212
Black Hat Security Conference, 287

Caltech multivehicle wireless testbed (MVWT-II), 11